Green Supply Chain Management for Sustainable Business Practice

Mehmood Khan
Abu Dhabi University, UAE

Matloub Hussain
Abu Dhabi University, UAE

Mian M. Ajmal
Abu Dhabi University, UAE

A volume in the Advances in Logistics, Operations, and Management Science (ALOMS) Book Series

IGI GLOBAL
DISSEMINATOR OF KNOWLEDGE
www.igi-global.com

Published in the United States of America by
IGI Global
Business Science Reference (an imprint of IGI Global)
701 E. Chocolate Avenue
Hershey PA 17033
Tel: 717-533-8845
Fax: 717-533-8661
E-mail: cust@igi-global.com
Web site: http://www.igi-global.com

Library of Congress Cataloging-in-Publication Data

Names: Khan, Mehmood, 1973- editor. | Hussain, Matloub, 1973- editor. | Ajmal, Mian M., editor.
Title: Green supply chain management for sustainable business practice / Mehmood Khan, Matloub Hussain, and Mian M. Ajmal, editors.
Description: Hershey, PA : Business Science Reference, [2017] | Includes bibliographical references and index.
Identifiers: LCCN 2016017489| ISBN 9781522506355 (hardcover) | ISBN 9781522506362 (ebook)
Subjects: LCSH: Materials management--Environmental aspects. | Industrial ecology. | Business logistics--Environmental aspects.
Classification: LCC TS161 .G753 2017 | DDC 658.7--dc23 LC record available at https://lccn.loc.gov/2016017489

This book is published in the IGI Global book series Advances in Logistics, Operations, and Management Science (ALOMS) (ISSN: 2327-350X; eISSN: 2327-3518)

British Cataloguing in Publication Data
A Cataloguing in Publication record for this book is available from the British Library.

Advances in Logistics, Operations, and Management Science (ALOMS) Book Series

ISSN: 2327-350X
EISSN: 2327-3518

Mission

Operations research and management science continue to influence business processes, administration, and management information systems, particularly in covering the application methods for decision-making processes. New case studies and applications on management science, operations management, social sciences, and other behavioral sciences have been incorporated into business and organizations real-world objectives.

The **Advances in Logistics, Operations, and Management Science** (ALOMS) Book Series provides a collection of reference publications on the current trends, applications, theories, and practices in the management science field. Providing relevant and current research, this series and its individual publications would be useful for academics, researchers, scholars, and practitioners interested in improving decision making models and business functions.

Coverage

- Risk Management
- Services management
- Operations Management
- Networks
- Finance
- Computing and information technologies
- Organizational Behavior
- Production management
- Political Science
- Information Management

IGI Global is currently accepting manuscripts for publication within this series. To submit a proposal for a volume in this series, please contact our Acquisition Editors at Acquisitions@igi-global.com or visit: http://www.igi-global.com/publish/.

The Advances in Logistics, Operations, and Management Science (ALOMS) Book Series (ISSN 2327-350X) is published by IGI Global, 701 E. Chocolate Avenue, Hershey, PA 17033-1240, USA, www.igi-global.com. This series is composed of titles available for purchase individually; each title is edited to be contextually exclusive from any other title within the series. For pricing and ordering information please visit http://www.igi-global.com/book-series/advances-logistics-operations-management-science/37170. Postmaster: Send all address changes to above address.

Titles in this Series

For a list of additional titles in this series, please visit: www.igi-global.com

Multinational Enterprise Management Strategies in Developing Countries
Mohammad Ayub Khan (Tecnológico de Monterrey, Mexico)
Business Science Reference • copyright 2016 • 490pp • H/C (ISBN: 9781522502760) • US $200.00 (our price)

Strategic Management of Sustainable Manufacturing Operations
Rameshwar Dubey (Symbiosis International University, India) and Angappa Gunasekaran (University of Massachuesetts, Dartmouth, USA)
Business Science Reference • copyright 2016 • 419pp • H/C (ISBN: 9781522503507) • US $215.00 (our price)

Handbook of Research on Global Fashion Management and Merchandising
Alessandra Vecchi (London College of Fashion, U.K.) and Chitra Buckley (London College of Fashion, U.K.)
Business Science Reference • copyright 2016 • 862pp • H/C (ISBN: 9781522501107) • US $335.00 (our price)

Handbook of Research on Managerial Strategies for Achieving Optimal Performance in Industrial Processes
Giner Alor-Hernández (Instituto Tecnológico de Orizaba, Mexico) Cuauhtémoc Sánchez-Ramírez (Instituto Tecnológico de Orizaba, Mexico) and Jorge Luis García-Alcaraz (Universidad Autónoma de Ciudad Juárez, Mexico)
Business Science Reference • copyright 2016 • 674pp • H/C (ISBN: 9781522501305) • US $365.00 (our price)

Supply Chain Strategies and the Engineer-to-Order Approach
Richard Addo-Tenkorang (University of Vaasa, Finland) Jussi Kantola (University of Vaasa, Finland) Petri Helo (University of Vaasa, Finland) and Ahm Shamsuzzoha (Sultan Qaboos University, Oman)
Business Science Reference • copyright 2016 • 283pp • H/C (ISBN: 9781522500216) • US $200.00 (our price)

Optimal Inventory Control and Management Techniques
Mandeep Mittal (Amity School of Engineering and Technology, India) and Nita H. Shah (Gujarat University, India)
Business Science Reference • copyright 2016 • 406pp • H/C (ISBN: 9781466698888) • US $170.00 (our price)

Table of Contents

Detailed Table of Contents

This chapter discusses sustainability in supply chains. It starts with a review of the concepts of sustainability and supply chain management; provides a discussion on the globalization of supply chains, its influence on sustainability, benefits of sustainable supply chains followed by the factors affecting the sustainability of supply chains, challenges in integrating sustainability into supply chains and managerial implications. It discusses the influence of supply chains on the triple bottom line. Examples of major factors affecting environmental and social sustainability include travel distances, modes of transportation, vessels used, frequency of trips, packaging materials, efficiencies of manufacturing processes, efficiencies in power generation/distribution, treatments given to polluted air/water generated during various processing from industries, expected product lifetime, manufactured quality of items, management of items failed during warranty or transportation, and the management of disposal of items, among others. The chapter concludes with a future research agenda.

The chapter will explain the concept of sustainability in supply chain management. Sustainability in business operations is fast becoming a vital element to successful global acceptance. It is also one of the major marketing strategies for Points-of-

Difference (POD) product positioning. It is necessary to have a basic knowledge of how sustainability is brought about in supply chains in order to truly comprehension the opportunities from it. An understanding of the basic elements to sustainability in supply chain management and global models adopted for implementation of sustainability by corporations is essential. The chapter further focuses on introducing the concept of reverse logistics as sustainability supply chain process and the strategic advantage for industries from adoption of reverse logistics across the supply chain. An insight into United Nations Global Compact program for sustainability across businesses and sustainability governance has been provided.

Section 2
Measuring the Performance of a Sustainable Supply Chain

Inventory management and supply chain management are integral issues and prime areas of concern in production and operations management, and industrial engineering. There is a drive for businesses and organizations to be accountable for their environmental impacts. Guaranteeing environmentally conscious supply chain operations is strongly linked to an organization's sustainability and overall success. The responsible management of forward and return product flows in production and inventory environments is a rapidly increasing requirement. This can be attributed to economic, environmental and/ or regulatory motivations. Mathematical modeling of such systems assists decision-making processes and provided a better understanding of the behavior of such production and inventory environments.

The United Arab Emirates (UAE) construction industry has been witnessing an unprecedented growth in the last two decades. As a result, the sector is facing a major challenge of reducing the carbon footprint and thus creating major concern for the governments and the environmental agencies in the UAE. The chapter discusses the role of Green Supply Chain Management (GSCM) in achieving sustainability in the UAE construction industry. The various stages of GSCM applicable to the construction industry are critically assessed along with green performance measures in achieving environmental, economic and operational performance. Finally, the chapter provides a list of recommendations that could be used by practitioners and policy makers in implementing and measuring sustainability practices in the construction industry.

The continuous improvement using waste elimination has been emphasized as the most important task of modern organizations. Lean manufacturing system has proved its capability to eliminate waste and produce environmental gains. The eight types of waste identified under lean have positive relations to green waste. In this chapter, a waste elimination process is suggested as an approach for lean sustainable outcomes. The process structure contains three consecutive phases: waste documentation, waste analysis, and waste removal. The techniques and tools of each phase are inclusively discussed.

The purpose of this chapter is to identify the dimensions of green supply chain and their impact on manufacturing practices. In this study, the authors used two extended strategies. First thorough review of literature was done considering articles from reputed journals. Second the factors identified from literature review was further refined through experts by forming a problem solving group consisting of seven experts from the manufacturing sector. These factors were used to develop the green supply chain management model using Interpretive structural modeling. Further MICMAC analysis was used to identify the driving and dependence power of the factors. The results of the analysis are very encouraging. Finally, the authors have presented the relationship management strategy for sustainable manufacturing practices.

Preface

The amount and scope of research in sustainability has so far been quite scholastic. The practitioners and academics have explored this fertile area through their own discourse and thus have paved path for a divide between the scopes of their definition of sustainability. One of the most important challenges faced by business managers today is the integration of sustainability into their core functions. The contemporary enterprise is forced to leap forward from the mere adoption of green practices towards re-thinking, re-designing, and re-developing of business practices in a more sustainable way. An organization is thus required to change their business models/operations linked to products, services, resources, processes, reporting, and behavior. The next logical move is to bring along these changes in the supply chain of this organization. This book addresses this big challenge of designing/implementing business sustainability in supply chains. In a span of nine chapters, this book presents views and findings of authors from all over the world. While some authors have adopted a practical perspective, others approach the topic more academically. They may have different ideas, consider different aspects, or reach different conclusions. All of them have shared their insights on the design and implementation of sustainability. The book is designed to help operations and supply chain managers understand how sustainability practices can be incorporated in their core functions. The book will also provide a guide to researchers, graduate students and practitioners on incorporating sustainability criteria into the functions such as procurement, manufacturing/production, logistics/transportation and disposal.

The first group of professionals that this book addresses is supply chain professionals. It is this group that will be faced with an increasing demand for demonstrating sustainability in their work. The responsibility for sustainability in operations, however, also rests with all the stakeholders in the context of their service and or product. The book therefore also aims to address (senior) management, executives, and sustainability professionals. It is this group that is faced with the implementation of sustainability in the core business practices of their organizations. The third and final group of professionals this book targets are educators and students in the field of business and supply chain management. They represent the future of their professions and will need to respond with the sustainability challenges that we come across.

This book is structured in three sections, with the following themes:

1. Introduction to Sustainable Supply Chain Management.
2. Measuring the Performance of a Sustainable Supply Chain.
3. Future of Sustainable Supply Chains.

OVERVIEW

As the book is devoted to a very diverse range of topics contributed mostly by academics, it would be appropriate to offer a bird's eye view of the contents of the chapters.

Section 1: Introduction to Sustainable Supply Chain Management

Chapter 1: "Sustainable Supply Chains"

This chapter highlights the significance of integrating sustainability into supply chains. Since supply chains generally constitute the largest cost (and environmental and social impact) for organizations, the integration of sustainability into supply chains has the highest potential for improving sustainable performance of organizations. It is a new and evolving field and there is a great need for establishing framework on all the dimensions of sustainability in supply chains. This chapter demonstrates the need for standardizing a comprehensive approach to measure the overall sustainability and defining sustainability indicators/index for supply chains considering all the three dimensions of sustainability.

Chapter 2: "An Introduction to Sustainable Supply Chain Management and Business Implications"

This chapter presents business operations is a challenging issue that corporates have been facing due to growing awareness of environmental issues and cause factors by public and governments. The increasing contribution of business operations such as sourcing, manufacturing and logistics to environmental issues, created a need for optimization of business operations with reduced negative effects. The chapter aims at (1) summarizing sustainability and its ranking system, (2) understanding sustainability in supply chain management and its parameters, (3) determining the importance of performance measurement in business sustainability, and (4) exploring the UN Global Compact Management for sustainable supply chain management.

Section 2: Measuring the Performance of a Sustainable Supply Chain

Chapter 3: "The Development and Analysis of Environmentally Responsible Supply Chain Models"

This chapter addresses the importance of environmental awareness in firms and organizations, and the integration of environmental consciousness into supply chains as a necessity for sustainability and continuous improvement. The chapter focuses on the mathematical modeling of supply chains highlighting research gaps and providing insights to future work to achieve environmental responsibility. The major recommendations presented in the chapter were to address the need for appropriate environmental performance measures in a supply chain context. These measures could be used to expand on existing reverse logistic inventory models and closed-loop supply chain models covering both economic and environmental aspects. Some environmental issues of immediate concern are the amounts of energy used, material used (and disposed of), and air emissions released to the environment.

Chapter 4: "Assessing the Possibilities of Green Supply Chain Management for the United Arab Emirates Construction Industry"

This chapter emphasized on the fact that global supply chain management (GSCM) in construction industry is still in infancy. The chapter identified (*i*) the potential drivers and challenges faced by the construction industry in the UAE in managing a global supply chain, (*ii*) the various stages in the supply chain for reducing the environmental impacts; for example design, purchasing, transportation and onsite and offsite construction, (*iii*) the importance of having an internal environmental management (IEM) for implementing GSCM, (*iv*) performance measures namely environmental, economic and operational to assess the performance improvement post GSCM implementation.

Chapter 5: "A Waste Elimination Process: An Approach for Lean and Sustainable Manufacturing Systems"

This chapter introduces a waste removal process as a technique for sustaining lean manufacturing to result in gaining green environment. The chapter outlined the background of lean manufacturing and sustainability, waste identification tools, waste elimination process and future directions towards green and sustainable manufacturing systems. The objectives of this chapter were to (1) identify the commonalities between lean and green manufacturing, (2) review the tools used for locating waste or non-value added activities inside a manufacturing supply chain, and (3) propose a waste elimination process as a lean sustainable approach.

Chapter 6: "Green Supply Chain Management Model for Sustainable Manufacturing Practices"

This chapter presented a systematic approach to develop a global supply chain management model for manufacturing sector. Their Interpretive Structural Model (ISM) identified a hierarchy of factors for decision making and building a clean system with minimum environmental impact. Their model arranged the levels for factors as market share and profitability (level 1), reduction in carbon emission (level 2), green technology adoption (level 3), total quality environmental management (level 4), supplier relationship management and Customer relationship management (level 5), top management commitment (level 6), and regulatory pressures and market pressures at bottom level (level 7). They identified customer relationship management, top management commitment, regulatory pressures and total quality environmental management as driving factors or independent variables. The rest of the factors were weak drivers and dependent ones.

Section 3: Future of Sustainable Supply Chains

Chapter 7: "Recent Developments in Green Supply Chain Management: Sourcing and Logistics"

This chapter reviews recent research on three classes of decision models that are used in the logistics and sourcing functions of sustainable supply chains. Specifically, they examine and summarize recent developments in sustainable decision models for supplier selection, inventory lot-sizing, and last-mile delivery. This review of the green and sustainable features found in the three models results in a single-source consolidation of models used in sustainable sourcing and logistics. This review may prove useful to researchers who focus on sustainable supply chain management and to practitioners who seek to expand their toolkit of models for the management and control of their supply chain.

Chapter 8: "Integrated Sustainable Supply Chain Management: Current Practices and Future Direction"

This chapter explores the role of integrated sustainable supply chain and a firm's non-financial performance. The chapter aims at providing an effective means to reduce costs and improve customer service. Organizations that aim to become part of an extended, integrated supply network can also expect that this would require an infrastructure enabling effective information flows and streamlined manufacturing and logistics activities. A key component of this infrastructure would be based on robust and durable collaborative arrangements with trading partners. The most effective of these networks would be those that are able to get the mix of information requirements, physical logistics and collaboration right, providing shared benefits to a majority of partner organizations. The chapter demonstrates that a successful integrated green supply chain organization would be able to develop a competitive business environment leading to firm sustainability.

Chapter 9: "Mastering Sustainable Supply Chain Management in the Business World"

This chapter aims to master Sustainable Supply Chain Management (SSCM) in the business world, thus describing the theoretical and practical concepts of sustainability, Supply Chain Management (SCM), and SSCM; the overview of Triple Bottom Line (TBL); and the applications of SSCM in the business world. The chapter performs a large-scale literature review and uses conceptual theory building to introduce sustainability beyond the TBL and to consider key supporting facets to implementing SSCM practices to the field of SCM and to demonstrating the relationships among environmental, social, and economic performances within the SSCM context. The chapter argues that mastering SSCM will significantly enhance organizational performance and achieve business goals in the global supply chain environments.

CONCLUSION

This book is a treatise on those various disaggregated aspects of supply chains that are connected with sustainability issues, and lend a holistic dimension to the notion by relating it with business practices. However, the book takes an informal style to connect to the academics as well as giving a theoretical discourse on certain instances to establish a connection to the managers and decision makers. The chapters enhance the theoretical paradigm of sustainability science by addressing how these theories need to be applied to model the coordination between stakeholders to build a greener supply chain. The book touches upon the design parameters of such a supply chain as well as outlines the future research avenues in this field.

Section 1
Introduction to Sustainable Supply Chain Management

Chapter 1
Sustainable Supply Chains

Amulya Gurtu
University of Wisconsin – Green Bay, USA

Cory Searcy
Ryerson University, Canada

Mohamad Y. Jaber
Ryerson University, Canada

ABSTRACT

This chapter discusses sustainability in supply chains. It starts with a review of the concepts of sustainability and supply chain management; provides a discussion on the globalization of supply chains, its influence on sustainability, benefits of sustainable supply chains followed by the factors affecting the sustainability of supply chains, challenges in integrating sustainability into supply chains and managerial implications. It discusses the influence of supply chains on the triple bottom line. Examples of major factors affecting environmental and social sustainability include travel distances, modes of transportation, vessels used, frequency of trips, packaging materials, efficiencies of manufacturing processes, efficiencies in power generation/distribution, treatments given to polluted air/water generated during various processing from industries, expected product lifetime, manufactured quality of items, management of items failed during warranty or transportation, and the management of disposal of items, among others. The chapter concludes with a future research agenda.

DOI: 10.4018/978-1-5225-0635-5.ch001

INTRODUCTION

The activities associated with modern supply chains have been carried out for centuries, though the use of the term supply chain is a relatively recent phenomenon. Spice trade from the Far East to Europe via Africa existed for many centuries prior to the industrial revolution and flourished even in the absence of modern means of transportation and communication (information technology). Despite this long history, the focus of research in the area of supply chain management (SCM) is relatively new as compared to other areas of engineering and management. The study of supply chain management has generated a lot of interest among academicians and practitioners in the last three decades. Perhaps for this reason, Feller, Shunk, and Callarman (2006) wrote that the term supply chain was first used in 1982. However, the earliest documented use of the term supply chains could be traced back to the middle of the 20th century in the context of petroleum products for a Shell refinery in Essex (Petroleum Times, 1951). Later, this term was used in the context of electricity supply (Banbury, 1975) from generation to distribution to final consumers. As a further illustration, Burns and Sivazlian (1978) performed an analysis for multi-echelon supply chains for optimization of inventory well before 1982.

The globalization of manufacturing and trading activities resulted in an increase in the complexities of supply chains and helped in popularizing the term "supply chain management". These increased complexities required a better understanding of many variables involved in global supply chains. The challenges of optimizing supply chain performance have made supply chain management a key competitive advantage in a number of industries over the last two decades. The scope of supply chain management, as defined in most of the earlier literature, is from the acquisition of the raw materials and parts through delivery of the finished products to the end customers. One of the earliest definitions of supply chains was "the supply chain is the connected series of activities which is concerned with planning, coordinating and controlling material, parts and finished goods from supplier to customer" (Stevens, 1989; pp 3). Until a decade ago, there was an active debate about the boundaries of supply chain management and its role with respect to the other functions of an organization.

In an effort to address this gap, Mentzer, DeWitt, Keebler, Min, Nix, Smith, and Zacharia (2001) defined a supply chain "as a set of three or more entities (organizations or individuals) directly involved in the upstream and downstream flows of products, services, finances, and/or information from a source to a customer." As a second example, the Canadian Supply Chain Sector Council (CSCSC, 2014) explained that a supply chain encompasses "supply of materials to a manufacturer", "the manufacturing process", and "the distribution of finished goods through a network of distributors and retailers to a final customer". The various organizations and customers in these processes are connected through a supply chain. The purpose of a supply chain is to integrate these processes and to ensure seamless flow of information across connected organizations. Building on these definitions of a supply chain, Mentzer *et al.* (2001, pp. 18) further defined supply chain management as "the systemic, strategic coordination of the traditional business functions and the tactics across these business functions within a particular company and across businesses within the supply chain, for the purposes of improving the long-term performance of the individual companies and the supply chain as a whole". This definition is widely cited in many papers on supply chain management (Svensson, 2002; Burgess, Singh, & Koroglu, 2006; Gundlach, Bolumole, Eltantawy, & Frankel, 2006; Gaiardelli, Saccani, & Songini, 2007; Ahi & Searcy, 2013a) and aligns with many other published definitions. For example, the Council of Supply Chain Management Professionals (Mentzer, Stank, & Esper, 2008, pp. 32), defined SCM as "the planning and management of all activities involved in sourcing and procurement, conversion, and all logistics management activities. Importantly, it also includes coordination and collaboration with channel partners, which can be suppliers, intermediaries, third-party service providers, and customers. In essence, supply chain management integrates supply and demand management within and across companies." Although different authors may emphasize different aspects of supply chain management, the definitions provided by Mentzer *et al.* (2001, 2008) and the CSCMP are fairly representative examples of how supply chain management is defined in the literature.

Having reached a general agreement on the scope of supply chains and its importance in organizations, the focus in the literature has shifted to optimizing various elements and addressing challenges in supply chains. For example, reducing the operating costs and improving the profitability of an organization is one of the primary functions of a business. There are only two ways of improving profitability, namely by either increasing revenues or reducing costs (McFadden, 1978). Revenue increases dependent on research and development, marketing, sales and service functions. Cost reductions depend on minimizing expenses of the activities within supply chains and other functions. It has also been recognized that supply chains, for industrial organizations, particularly in manufacturing, constitute the single largest cost for the organization, which makes it the single most important function for reducing costs. According to Bryceson (2006; pp. 110) "the supply chain makes up the single largest cost component of a corporation's revenues—typically 55-85% of the sales dollars". Therefore, organizations have overwhelmingly focused on supply chains as a tool for reducing operating cost and improving profitability. However, the research has highlighted that there are a number of other challenges associated with supply chain management that merit attention.

One area that is the subject of a rapidly growing number of articles is the sustainability of supply chains. Sustainable supply chains focus on improving the environmental and social performances, in addition to economic performance, of supply chains. Research in this area explicitly recognizes that supply chains influence our environment and society in many ways, both positively and negatively. Many authors have acknowledged the impacts of supply chains on the environment and society (Kolk & Pinkse, 2005; Bonney & Jaber, 2011; Gunady, Biswas, Solah, & James, 2012; Schieberl & Nickles, 2014). Supply chains have many issues. However, not all can be addressed at the same time due to constraints of resources. Therefore, organizations need to prioritize the areas to be addressed. Building on the principle of Pareto, the sustainability of supply chains is the most important aspect of sustainability of an organization because as stated earlier supply chains account for the biggest cost in organizations and, therefore, have the highest potential for cost reductions. The need for sustainability in supply chains is the central theme of this chapter. The benefits and challenges in the implementation of sustainable supply chains are also discussed later in the chapter. The next section explains sustainability and its relationship to supply chains.

BACKGROUND OF SUSTAINABILITY IN SUPPLY CHAINS

In the mid-1980s, a number of authors identified the need for integrating sustainability into various aspects of businesses, particularly with respect to ensuring broad coverage of large geographic areas (DeLaet, 1985; Jacobs, 1985; Lowry & Carpenter, 1985; Khosla, Prakash, & Revi, 1986; Westing, 1986). However, publication of the "Report of the World Commission on Environment and Development: Our Common Future" (WCED, 1987), famously known as Brundtland Report, marked a historic moment in popularizing the concept of sustainability and underlined the urgent need to integrate sustainability into businesses. The literature on sustainability began a rapid expansion following the publication of this report. The Brundtland report defined sustainable development as "the development that meets the needs of the present without compromising the ability of future generations to meet their own needs" (WCED, 1987; pp. 16). It highlighted the need for development that supports the environment by minimizing pollution in all countries. In other words, sustainability was defined as the use of natural resources in such a way that future generations are not deprived of these natural resources. This is possible when the rate of consumption of natural resources is the same as or less than the rate of renewal for those resources. Another corollary to this definition is that the rate of generating pollution (to air, water, and land) must be kept the same as or less than the time required to absorb and completely disintegrate that pollution in the environment (Daly, 1990; Costanza & Daly, 1992). Further, sustainability has been discussed based on the degree to which it focuses on "natural capital". Sustainability where natural capital and man-made capital are treated in a similar manner, i.e. there is no special consideration for natural capital, is referred to as "weak sustainability". On the other hand, the concept of "strong sustainability" suggests that natural and man-made capital need to be kept separate and treated separately because it takes a very long time to generate natural capital (Gutés, 1996; Ahi & Searcy, 2013b). These concepts still form the basis for different approaches to sustainability management.

By the late 1990s, sustainability made its way into the literature on supply chains. However, it was not until the last decade that both academicians and practitioners began widely embracing the idea of sustainability of supply chains (Taticchi, Tonelli, & Pasqualino, 2013). There were many prefixes, such as "eco", "green", and "sustainable" added to distinguish supply chain activities that considered the environmental impacts (and, in some cases, social) of these activities. However, a great variety of definitions was available and there was little clarity on the definition of each term (Gurtu, Searcy, & Jaber, 2015). One term that featured prominently in the early literature in this area was "green supply chains", which explicitly focused on building environmental considerations into supply chain management. One of the most prominent definitions of green supply chain management (GSCM) adopted the approach of reducing the burden on the shared natural resources throughout the lifecycle of a product. According to this definition, "green supply chain management ranges from reactive monitoring of the general environment management programmes to more proactive practices implemented through various Rs (Reduce, Re-use, Rework, Refurbish, Reclaim, Recycle, Remanufacture, Reverse logistics, etc.)" (Srivastava, 2007; pp. 54). It is clear from the above definition that its emphasis is on reducing environmental impacts, which may or may not include its influences on either the economic or the social aspects that also feature prominently in the concept of sustainability. To address this need, a new term "sustainable supply chains" began to appear in the literature (Seuring & Müller, 2008). Other definitions of GSCM also started to emerge and included terms such as 'reverse logistics' (Hervani, Helms, & Sarkis, 2005), 'lifecycle management' (H'Mida & Lakhal, 2007) and 'closed-loop supply chains' (Lau, 2011), which gave rise to literature on reverse logistics, reverse supply chains and closed-loop supply chains as an extension of environmentally friendly supply chains. For example, one of the definitions of reverse supply chains is "the series of activities required to retrieve a used product from a customer and either dispose of it or reuse it" (Guide & Wassenhove, 2002; pp. 52) and "a closed loop supply chain includes the returns processes and the manufacturer has the intent of capturing additional value and further integrating all supply chain activities"

(Guide, Harrison, & Wassenhove, 2003; pp. 3). However, these definitions did not adequately address the economic, environmental and social impacts of supply chains (Seuring & Müller, 2008; Ahi & Searcy, 2013a). Definitions of sustainable supply chains have been developed to address these points.

In an organizational context, sustainability is often closely linked to the concept of corporate social responsibility (CSR) (Hutchins & Sutherland, 2008). Both of these concepts tend to emphasize triple bottom line (TBL) performance as essential aspects of business success. The TBL explicitly requires emphasis on the organization's performance in the economic, environmental, and social dimensions of sustainability. Sustainability in a supply chain context can, therefore, no longer be viewed as being limited to economic or environmental sustainability alone. Sustainable supply chains are expected to address the TBL of sustainable performance, i.e. economic, environmental, and social sustainability. Despite a rapid growth in the literature, the concept of sustainable supply chains remains in its embryonic state and is still the subject of much debate regarding its definition, scope, and practical application. Seuring and Müller (2008; pp 1700) addressed these issues and provided a definition of sustainable supply chain management (SSCM) as "the management of material, information and capital flows as well as cooperation among companies along the supply chain while taking goals from all three dimensions of sustainable development, i.e., economic, environmental and social, into account which are derived from customer and stakeholder requirements." A more recent review of the literature on the definitions and gaps in green and sustainable supply chains was carried out by Ahi and Searcy (2013a). The authors also provided a new definition of sustainable supply chain management as "the creation of coordinated supply chains through the voluntary integration of economic, environmental, and social considerations with key inter-organizational business systems designed to efficiently and effectively manage the material, information, and capital flows associated with the procurement, production, and distribution of products or services in order to meet stakeholder requirements and improve the profitability, competitiveness, and resilience of the organization over the short and long-term" (Ahi & Searcy, 2013a; pp. 339). Many authors view sustainable supply chains as an extension of green supply chains (Ahi & Searcy, 2013a).

The development of global supply chains due to offshore production and outsourcing has gained a lot of momentum since the beginning of the 21st century, though it is important to stress that the concepts of outsourcing and global supply chains are not entirely new. However, ongoing reductions in communication costs and transportation time, along with the lack of stringent environmental laws and the availability of cheap labor in economically less developed places, have greatly facilitated the move to globalized supply chains. To date, the move to globalize supply chains has been driven by economic advantages. Historically, little attention has been paid to the adverse impacts of the globalization of supply chains on the environment and society. However, as demonstrated above, there is a growing awareness of the need to accommodate environmental and social considerations into supply chain-related decision-making. Cuthbertson (2011) has discussed the need for sustainable supply chains. Once environmental and social aspects of supply chains are considered along with economic factors in decision-making, the resulting supply chains will be truly sustainable supply chains.

CURRENT PRACTICE AND BENEFITS OF SUSTAINABLE SUPPLY CHAINS

Sustainable supply chains provide a holistic approach to optimizing the consumption of natural resources while minimizing emissions, providing sustainable economic value across supply chains rather than limited optimization of local economic issues, and considering the broad social impacts along the entire chain rather than centering solely on the focal firm. One of the biggest advantages of this perspective is an overall reduction in the burden on the shared regional and global commons, such as the atmosphere (Hardin, 2009). Integrating sustainability considerations into supply chain management may help in minimizing the gap in manufacturing technology used in different parts of the world, thus potentially resulting in lower emissions and higher product quality (Chao, 1998; Kaya & Özer, 2009). Tapiero (2007) stated that problems related to quality control in the context of outsourcing and supply chains have been given less importance. Further, according to Ghadge, Dani, and Kalawsky (2012; pp. 323), "researchers identify outsourcing activity as being responsible for product quality risk". Competitive advantages in the majority of existing supply chains are heavily weighted towards economic considerations.

One of the other key advantages of bringing supply chains under the umbrella of sustainability is that it provides new opportunities for competitive differentiation. This issue has been discussed by a number of authors. For instance, Harris (2006) argued that integrating sustainability in supply chains could serve as a strategy to increase the value of an organization. Bag (2013) explained that including sustainability in supply chain management is a strategy to enhance competitiveness in the global marketplace. Some of the advantages of this strategy cited by the author include a reduction in health risks, a cleaner working environment, and an increase in profitability. As a third example, Rao & Holt (2005) analyzed the linkages between green supply chain management and economic performance of organizations in South East Asian countries. The authors surveyed a small number of organizations on 64 parameters using a linear structural equation modeling approach and developed relationships among "greening the inbound function, greening production, greening the outbound function, competitiveness and economic performance" (Rao & Holt, 2005; pp. 900). They concluded that green supply chains are a source of competitive advantage to organizations, which results in better economic performance. Markley and Davis (2007) also evaluated the possibility of achieving competitive advantages through sustainable supply chains and found that sustainable supply chains have a positive correlation between environmental and social/ethical ratings, and profitability of organizations. The authors also found that sustainable supply chains in the future would not only provide competitive advantages, but also be a differentiating factor. Xu and Wang (2008) also reached a similar conclusion that sustainability provides competitive advantages. They further added that integrating sustainability is a necessity for survival rather than a luxury or a financial drain on corporate resources. Finally, as an example of value perceived by customers on sustainable supply chains, Faust (2013) presented a case study on the use of cashmere fiber by surveying 200 American young consumers. According to the author's finding, consumers demonstrated little awareness of the supply chain processes. However, sharing information about the entire process of supply chains from producing raw material to final products for consumers provided consumers a new perspective and, in turn, new advantages to retailers of sustainable products. Faust also concluded that consumers needed to be told the story of sustainable supply chains for improving the economic performance.

There is a growing emphasis on including environmental and social performance in assessing the overall performance of a business (Yakovleva, Dani, & Kalawsky, 2012). As one tangible illustration of this point, Yakovleva *et al.* presented a framework for evaluating the sustainability of supply chains in the food industry. Their study of the food industry included everything from agricultural production to food consumption and food-catering services. The reason for choosing the food industry was that it serves one of the basic physiological needs for humans and is important to society (Barrett, Ilbery, Brown, & Binns, 1999). Yakovleva *et al.* presented a case for sustainable indicators as part of the development of sustainability benchmarking exercise of the food industry in the UK. The process started with identifying sustainability indicators, on all three dimensions of sustainability for the food industry, such as energy consumption, water consumption, waste, wages among others and then conducting an analysis of those indicators. They observed that industry experts focused on indicators from the financial perspective and suggested involving stakeholders from other areas, such as society and public policy organizations, to avoid biases. The authors argued in favor of either involving stakeholders concerned with all three areas of sustainability or assigning different weights to the three areas of sustainability to arrive at a more reliable state of sustainability in supply chains.

Many organizations do not pay sufficient attention to the need for investments in sustainable technology because the cost of air, water and land pollution is not clearly defined (Sheng, 2012). Nidumolu, Prahalad, and Rangaswami (2009) considered investments in sustainable technology as an opportunity for doing business in the future. However, it should be acknowledged that there is a cynical viewpoint that addressing the TBL in organizational sustainability will adversely affect economic returns due to investments in sustainable technology and social programs. Carter and Rogers (2008) addressed this by presenting a framework for sustainable supply chain management (Figure 1). One of their research questions was (pp. 361) "is there a relationship between the integration of the concepts of sustainability and supply chain management, and long-term economic success?" Their analysis showed (pp. 371), "firms that strategically undertake SSCM will achieve higher economic performance than firms that pursue only one or two of the three components of the triple bottom line".

There are many approaches suggested for adopting sustainability in supply chains. For example, Nidumolu *et al.* (2009) have argued that, "There's no alternative to sustainable development" (pp. 57). They presented a 5-stage model for embracing all round sustainable development namely (pp. 60) (1) "viewing compliance as op-

Figure 1. Triple bottom line of sustainability (Carter & Rogers, 2008)

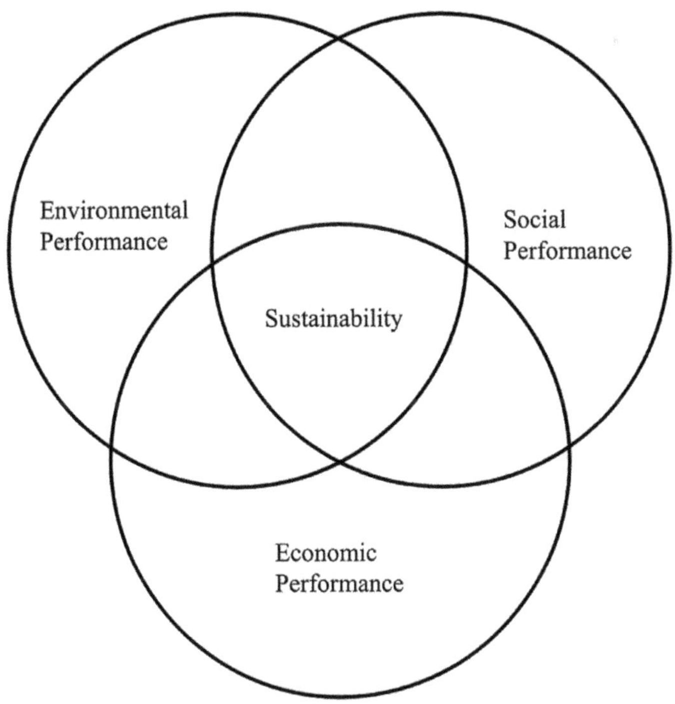

portunity", (2) "making value chains sustainable"; (3) "designing sustainable products and services"; (4) "developing new business models"; and (5) "creating next-practice platforms". The authors also highlighted the need for a proactive and collaborative global approach for success. However, it is acknowledged that making supply chains environmentally and socially sustainable, while simultaneously improving economic sustainability is not easy, and there is yet no guaranteed formula for success. It is clear, however, that it requires a vision, top management's commitment and support, and perseverance in implementation. Cetinkaya (2011) found similar causes for failure in developing sustainable supply chains. The author specified that the lack of (pp. 25) (1) "specific sustainability goals in the corporate vision and strategy", (2) "specific sustainability goals in the supply chain strategy", (3) "end-to-end responsibility of responsible logistics and supply chain managers", (4) "implementation experience and shared knowledge" and (5) "top management commitment" are the reasons of failures in integrating sustainability into supply chains. Although the majority of practitioners now understand the concept of the TBL of sustainable performance, the road to achieving it is still not clear (Gobble, 2012). The author argued that innovation would play a major role in achieving this equilibrium. Designing sustainable supply chains is not limited to an incremental improvement over the existing supply chains, though it is better than ignoring completely and not doing anything about it. However, for achieving better results of integrating sustainability in supply chains, it requires innovation and a fundamentally different perspective. Fiksel (2003) reached a similar conclusion and suggested an approach in following steps (pp. 5330, 5337): (1) "identifying system function and boundaries", (2) "establishing requirements", (3) "selecting appropriate technologies", (4) "developing a system design", (5) "evaluating the anticipated performance", and (vi) "devising a practical means for system deployment".

FACTORS AFFECTING SUSTAINABILITY OF SUPPLY CHAINS

There are many factors that affect the sustainability of supply chains. Economic sustainability is clearly fundamental to a business and has traditionally been the prime focus of businesses. Without economic sustainability, an organization will not be able to remain in business over the long-term. Therefore, it goes without saying that focus on environmental and social sustainability does not mean neglecting or excluding economic sustainability. Here the focus is mainly on environmental and social sustainability because these are generally ignored in the overriding emphasis on economic sustainability. Some examples of major factors affecting environmental and social sustainability include travel distances, modes of transportation, types of vessels used, frequency of trips (which depends on lot size), packaging materials, packing sizes, efficiencies of manufacturing processes, efficiencies in power generation/transmission/distribution, treatments given to polluted air/water generated during various processing from industries, expected product lifetime, manufactured quality of items, management of items failed during warranty or transportation, and the management of disposal of items (after use, due to failure, or in manufacturing), among others. These factors clearly affect the economic sustainability also and it is recognized that their significance may vary between supply chains. However, it is also important to stress that including the impact of environmental and social sustainability in supply chain decisions may alter the economic sustainability and may lead to different choices in supply chain management. In any case, it is clear that the globalization of supply chains has definitely increased transport distances, increased material handling/logistics activities, and increased the need for robust packaging, all of which adversely affect environmental sustainability.

Today's business environment is more global than ever before. It is increasingly difficult to find a product that is truly made in a single country, i.e. all inputs for a finished product are from a single country. Therefore, national and regional policies alone are no longer sufficient to address the environmental and social requirements of today's global supply chains. Variations in policies across national and regional boundaries affect sustainability in important ways. International organizations, such as the World Trade Organization (WTO), designed policies that have encouraged the globalization of trade and supply chains. However, these policies are generally overwhelmingly oriented towards the economic aspects of business success. There remains a lack of global policies to guide the incorporation of environmental and social considerations into supply chain management. Unsurprisingly, the current policy situation has lead businesses to focus overwhelmingly on economic measures for a business success. The available data clearly indicates that the globalization of supply chains has led to growing environmental impacts, such as increases in greenhouse gas (GHG) emissions (IEA, 2014). There are few coordinated policies to address such challenges.

Supply chain professionals and corporate leaders are overwhelmingly focused on reducing the cost of supply chains. They generally tend to ignore the bigger picture of their supply chains and how these affect the global environment and societies in different countries. This is complicated by the fact that corporate leaders' visions to integrate environmental and social sustainability of supply chains into CSR are not unanimous (Lepoutre & Heene, 2006). Corporate initiatives are generally influenced by either likely economic outcomes or government regulations. The lack of common laws and policies on sustainability-related issues, such as emissions, in different parts of the world and exclusion of international transport emissions from national emissions inventories (UNFCCC, 2013) encourages the overlooking of the environmental and social impacts of supply chains. The growth in CSR reporting by major corporations around the world has helped shed light and prompted thinking on these issues, but the parameters of CSR reporting (particularly on supply chain issues) remain heavily debated (Searcy, 2012). There is a need for further development in CSR reporting standards with respect to supply chain management. The lack of a standard to measure and report CSR makes it almost impossible for organizations to compare environmental and social performances of different organizations in the same way economic performances are compared using quarterly and annual financial reports.

Last, but not least in terms of importance, is awareness among consumers on the environmental and social aspects of supply chains. Consumers are at the extreme ends of supply chains. A negligible number of consumers care to know the details of supply chains for the products they buy or intend to buy. The prices of environmentally friendly products are generally higher and do not result in higher sales realization (Hassini, Surti, & Searcy, 2012). Consumers' decision parameters are limited to cash flow, aesthetics, operating costs, if any, and the useful life of the products they intend to buy. The price of a product plays a more important role than quality considerations in the buying decisions of consumers (Varki and Colgate, 2001; Papista & Krystallis, 2013). Marketing and advertising functions increasingly focused on transitioning society towards a culture of using a product for a one time or for a very short duration and then throwing it away to buy a new one. The culture of use and discard has created many challenges in environment management (Joshi & Dave, 2012; Overcash, 2012; Romer, 2012; Accorsi, Cascini, Cholette, Manzini, & Mora, 2014; Woods & Bakshi, 2014). This concept is driven primarily by economic aspirations and has almost no consideration to its impact on the environment or society. There are probably only a handful of examples, such as medical syringes, that may justify the approach of a single use on the three pillars of sustainability (economic, environment and society). The decision to replace a product after a short useful life can be justified as long as the environmental burden of manufacturing and usage of a new product is less than the environmental burden of recurring usage. While a consumer may prefer using new items every couple of days to every couple of years and as a result shareholders enjoy bigger dividends and government realize the collection of taxes on repeated sales, the environment is negatively impacted from reliance on products with short lifespans. Generating awareness about the environmental impact of products to consumers is crucial to bring a positive change in organizational approaches to supply chain sustainability (Galbreth & Gosh, 2013). Awareness of environmental issues is increasing among consumers but awareness still remains low. Yates (2008) found low awareness among UK consumers on sustainable consumption and the effects of their consumption behavior on the environment. About 3 decades ago, there was almost no information available to consumers about the ingredients, sodium/fat contents and calories of a processed food item. However, food labels evolved to fulfil this need and are still evolving. The U.S. Food and Drug Administration (FDA) has recently proposed changes to the nutrition label (FDA, 2014). It has been suggested that the same thing has to happen on product labels for environmental impact too (Hanss & Böhm, 2012). One of the well-known product sustainability labels is "ENERGY STAR" by the U.S. Environmental Protection Agency for promoting high-efficiency products for consumers and businesses. The "ENERGY STAR" program was started by the Environment Protection Agency (EPA) in 1992 for reducing energy consumption and thus reducing the impact of GHG on the environment (EPA, 2014). The "ENERGY STAR" rating facilitates consumers in purchasing energy efficient products. Consumers may rely on the "ENERGY STAR" rating when making a purchasing decision without the need of knowing any details of sustainability. Non-Government Organizations (NGOs) and media have a very important role in bringing this awareness among consumers and investors.

CHALLENGES IN SUSTAINABLE SUPPLY CHAINS

The challenges of integrating sustainability into supply chains are wide ranging. This ranges from the individual consumer level on one end to the systemic level at the other end. Therefore, processes for integrating sustainability into supply chains are complex, which can make the progress on these initiatives slower than other corporate initiatives. While it is difficult to change individual perceptions easily, it is very difficult and time-consuming to bring about systemic changes. Challenges on the consumer level range from generating awareness about sustainability to understanding the impacts of their actions to bringing changes in their consumption preferences. Similarly, changes required at systemic levels range from developing standard definitions for terms used in sustainable supply chains to accounting standards for emissions to policies for emissions based on consumption rather than production at international levels to including emissions from international transportation in national emissions inventories to adding total emissions on product labels (similar to showing calories and ingredients) for generating awareness among consumers and for easy comparison of products on sustainability parameters to developing sustainability indicators for various product categories, similar to the Energy Star rating for energy efficiency for some products categories. Abbasi and Nilsson (2012) reviewed the literature on identifying challenges in making supply chains environmentally sustainable and identified five major areas of challenges, i.e., costs, complexity, operationalization, mindset and cultural changes, and uncertainties. Cost is the first, and perhaps the biggest, barrier in integrating sustainability in supply chains (Seuring & Müller, 2008; Govindan, Kaliyan, Kannan, & Haq, 2014).

There is always a perception that improving environmental performance will lead to financial burden. This is partly true in the short term due to the need to make investments in better technology. Supply chains and sustainability are complex areas and combining them further increases complexity. As discussed earlier, there is a wide range of issues for integrating sustainability in supply chains, which require a holistic approach and support of top management (Giunipero, Hooker, & Denslow, 2012). Despite a broad agreement on reducing environmental impacts, making supply chains environmentally sustainable in practice is not easy due to a large number of partners and a lack of focus on sustainability (Carter & Rogers, 2008). Altering perceptions and bringing cultural changes are the hardest parts (Ellinger & Ellinger, 2014; Silvestre, 2014). Sustainability in supply chains is an emerging area and uncertainties are natural, though not desirable, at any stage. Uncertainties come from the fluid nature of pollution prevention policies, frequently changing policies, and lack of consistent policies across countries in the world. Finding a solution for technical problems is, in many cases, relatively easy when compared to the challenges of overcoming psychological barriers to finding those solutions. This, however, is not to say that technical solutions to sustainability challenges in supply chains are always straightforward. For example, collection and analysis of data for global supply chains is very difficult and time-consuming, if not impossible. Not all the players in a supply chain have environmental data and not everyone may be willing to share it for various reasons. When it comes to sharing demand forecast information between players of a supply chain, which help in proper planning and reducing supply chain costs, there is a lack of trust in providing this information accurately (Hopp & Ebrahim-Khanjari, 2012). Therefore, sharing of energy and environmental data between various players in supply chains presently appears a little farfetched. However, Ramanathan and Gunasekaran (2014) stated that supply chain collaboration improves environmental performance in manufacturing, which may serve as a driver of such initiatives.

Measuring the social impacts of global supply chains is a very complex process and the cycle time to measure these impacts takes a long period. In the short term, undoubtedly, globalization of supply chains has created many positive social impacts, such as new jobs in countries that are in need of employment and investments. However, it has also displaced many people from their jobs. There is little literature available on the effects of globalization of supply chains on the social aspects of people from outsourcing countries, who either lost their jobs or are working in less skilled or lower paid jobs or are unable to find a suitable job because their skills are no longer required in those countries. Schieberl and Nickles (2014) presented a detailed study on the impact of outsourcing on employment opportunities for citizens as well as foreign students in the USA. The authors also attributed the growth in outsourcing as a cause for the relatively slow economy in the USA in recent years. Hiring and retraining the experienced workforce is an expensive and time-consuming process. The skilled and experienced workforce has an additional and a huge challenge of overcoming a psychological barrier to accepting the situation and to demonstrate a willingness to accept low-skilled and less paid work. It is difficult for an experienced, skilled workforce to accept jobs that pay less. In fact, many skilled workers choose to remain unemployed rather than accepting a less skilled job. Linn, Sandifer, and Stein (1985), Mossakowski (2009) and, more recently, Bilgiç, and Yilmaz (2013) examined the effects of unemployment on health and concluded that the psychological pressure of not being able to find a job manifests in various physical ailments, which leads to higher health care expenses. Even in cases where skilled workers are willing to move to a different type of work, it takes time for them to reach the desired level of productivity because of the unlearning of their prior skills, which employers generally do not appreciate. They will be likely less productive in the immediate term, though this may depend on the extent to which their new job draws on similar skills to their old position.

Measuring performance of sustainability of a supply chain is another important area, which is developing fast (Taticchi, 2013; Bai, Sarkis, Wei, & Koh, 2012). In the absence of a proper system of measuring sustainability on the same parameters, tracking internal performance or comparing the performance of different organizations is difficult. It is also prudent to include the parameters of all the three bottom lines in the performance measurement systems. Some of the earlier methods include both quantitative and qualitative measures. The method of measuring performances using the balanced scorecard accounts for both qualitative and quantitative parameters in measuring organizational performance (Kaplan & Norton, 1992). Hassini *et al.* (2012) acknowledged the lack of common sustainability metrics for supply chains. The authors discussed various hurdles, such as selecting the right kind of metrics, focus on production and dynamic nature of supply chains among other, in finalizing a few common sustainability metrics. They also presented a framework for developing composite sustainability indicators (CIs) using TBL of sustainability and all the players of supply chains.

MANAGERIAL IMPLICATIONS

The role of supply chain management (SCM) professionals, in light of the challenges discussed above, has become very strategic in organizations. As a consequence, the day to day decisions on choosing a vendor or supplier have become more complex than ever. It is no longer the price (for an acceptable quality) that is the sole deciding factor in a sourcing decision. Managers have to think about the long-term impact of their sourcing decisions on the environment and/or society besides the economic viability of their organizations. It is important for the top management and SCM leaders in an organization to be aligned with the organizational philosophy regarding competing needs of economic viability, and environmental and social sustainability. The requirements for roles in SCM teams have also changed. The new rules require redefining the expectations from existing as well as new employees. A key challenge is in finding employees with skills in simultaneously addressing a wide range of priorities and training the existing team members to face the new and altered realities. However, the challenges are not limited to alignment within an organization and skills development for internal resources. SCM managers will have to motivate suppliers and vendors to embrace the new realities of the environment and social sustainability and educate them on improving their environmental and social performances as well. The SCM managers will also be required to collaborate with educational institutions and supply chain professional organizations for integration of sustainability into the core education curricula. Moreover, the managers will be expected to manage this transition in their department, organization and with vendors, while managing all other routine work such as budgeting, hiring, training, performance appraisals, and managing their own work-life balance, which creates a lot of pressure for managing their time effectively.

CONCLUSION

The discussion above touched on various elements of supply chains and sustainability. It highlighted the significance of integrating sustainability into supply chains. Since supply chains generally constitute the largest cost (and environmental and social impact) for organizations, the integration of sustainability into supply chains has the highest potential for improving sustainable performance of organizations. It is a new and evolving field and there is a great need for establishing a framework on all the dimensions of sustainability in supply chains. There is also a need for standardizing a comprehensive approach to measure the overall sustainability and defining a sustainability indicators/index for supply chains considering all the three dimensions of sustainability.

Sustainability has different meanings to people in academics and in business. Similarly, sustainability also means various things to different consumers (Hanss & Böhm, 2012). However, according to Simpson and Radford (2012), consumers perceive sustainability of products as protection of environment only due to the way media presents sustainability to consumers. Sustainability of products is rarely presented as sustainability of three pillars – economic, environmental and social sustainability. This brings us back to the discussion earlier about the significance of the role of NGOs and media in educating consumers. NGOs, media and government agencies have a major role in increasing awareness among consumers about sustainability, the meaning of sustainable products, how to differentiate sustainable products from non-sustainable products and how to compare sustainability of products. On the other extreme, a lot of work is required at the international level, which includes developing an international framework for integrating sustainability into supply chains, such as policies linking trade and environmental impacts to minimize the environmental and social impacts while maximizing economic output. Due to the globalization of businesses, the role of international organizations, such as the United Nations (UN), the World Trade Organization (WTO), the World Bank (WB), the International Monetary Fund (IMF) and the International Organization for Standardization (ISO), has become more important than ever. Each of these organizations has to play a vital role in reviewing their existing policies to include sustainability and enforce it across countries so that development in countries is not traded off with negative impacts on the environment and societies in any parts of the world. Between these two extremes could be initiatives focused on developing national programs to help organizations transition from conventional technology to invest in sustainable technology. Institutions which provide education, skills and training in the area of supply chains, be it focused on engineering or management, also need to include sustainability in their curricula.

There are a number of areas related to sustainable supply chains in need of further research. Some of these areas are the impacts of warranty failures due to additional transport on the cost of products and the environment; effects of disposal mechanisms of products, which are broken in transit or within the warranty period, on the environment and its related costs; cost of losing certain skills (such as engineering, manufacturing, management, accounting) due to offshore outsourcing; cost of training different skills to skilled workforce because jobs requiring their skills have been outsourced to different countries; impacts of outsourcing on unemployment, underemployment and on healthcare costs. Globalization and offshore outsourcing have resulted in increased movement of ships. This creates air pollution and water pollution. Air pollution has been covered in the literature but there is a need to study its effects on marine ecology. There is a need for standardization of CSR parameters for all types of organizations in any part of the world. Further, there is a need to develop performance indices on sustainability and developing product labels showing sustainability index to generate awareness among consumers.

REFERENCES

Abbasi, M., & Nilsson, F. (2012). Themes and challenges in making supply chains environmentally sustainable. *Supply Chain Management*, *17*(5), 517–530. doi:10.1108/13598541211258582

Accorsi, R., Cascini, A., Cholette, S., Manzini, R., & Mora, C. (2014). Economic and environmental assessment of reusable plastic containers: A food catering supply chain case study. *International Journal of Production Economics*, *152*, 88–101. doi:10.1016/j.ijpe.2013.12.014

Ahi, P., & Searcy, C. (2013a). A comparative literature analysis of definitions for green and sustainable supply chain management. *Journal of Cleaner Production*, *52*(1), 329–341. doi:10.1016/j.jclepro.2013.02.018

Ahi, P., & Searcy, C. (2013b). A stochastic approach for sustainability analysis under the green economics paradigm. *Stochastic Environmental Research and Risk Assessment*, 1–11.

Bag, S. (2013). Designing the green supply chain strategy for Indian manufacturing firm. *Journal of Supply Chain Management Systems*, *2*(1), 8–18.

Bai, C., Sarkis, J., Wei, X., & Koh, L. (2012). Evaluating ecological sustainable performance measures for supply chain management. *Supply Chain Management: An International Journal*, *17*(1), 78–92. doi:10.1108/13598541211212221

Banbury, J. G. (1975). Distribution—the final link in the electricity-supply chain. *Electronics and Power*, *21*(13), 773. doi:10.1049/ep.1975.0796

Barrett, H. R., Ilbery, B. W., Brown, A. W., & Binns, T. (1999). Globalization and the changing networks of food supply: The importation of fresh horticultural produce from Kenya into the UK. *Transactions of the Institute of British Geographers*, *24*(2), 159–174. doi:10.1111/j.0020-2754.1999.00159.x

Bilgiç, R., & Yilmaz, N. (2013). The correlates of psychological health among the turkish unemployed: Psychological burden of financial help during unemployment. *International Journal of Psychology*, *48*(5), 1000–1008. doi:10.1080/00207594.2012.717702 PMID:23151222

Bonney, M., & Jaber, M. Y. (2011). Environmentally responsible inventory models: Non-classical models for a non-classical era. *International Journal of Production Economics*, *133*(1), 43–53. doi:10.1016/j.ijpe.2009.10.033

Bryceson, K. P. (2006).*E'Issues for Agribusiness: The 'what', 'why', 'how*. CABI.

Burgess, K., Singh, P. J., & Koroglu, R. (2006). Supply chain management: A structured literature review and implications for future research. *International Journal of Operations & Production Management*, *26*(7), 703–729. doi:10.1108/01443570610672202

Burns, J. F., & Sivazlian, B. D. (1978). Dynamic analysis of multi-echelon supply systems. *Computers & Industrial Engineering*, *2*(4), 181–193. doi:10.1016/0360-8352(78)90010-4

Carter, C. R., & Rogers, D. S. (2008). A framework of sustainable supply chain management: Moving toward new theory. *International Journal of Physical Distribution & Logistics Management*, *38*(5), 360–387. doi:10.1108/09600030810882816

Cetinkaya, B. (2011). Developing a Sustainable Supply Chain Strategy. In *Sustainable supply chain management: Practical ideas for moving towards best practice* (pp. 17–55). Worcester, MA: Springer.

Chao, P. (1998). Impact of country-of-origin dimensions on product quality and design quality perceptions. *Journal of Business Research*, *42*(1), 1–6. doi:10.1016/S0148-2963(97)00129-X

Costanza, R., & Daly, H. E. (1992). Natural capital and sustainable development. *Conservation Biology*, *6*(1), 37–46. doi:10.1046/j.1523-1739.1992.610037.x

CSCMP – Supply Chain Management. (n.d.). *Council of Supply Chain Management Professionals*. Retrieved from http://cscmp.org/about-us/supply-chain-management-definitions

CSCSC – Supply Chain Definitions. (n.d.). *Canadian Supply Chain Sector Council*. Retrieved from http://www.supplychaincanada.org/en/supply-chain

Daly, H. E. (1990). Toward some operational principles of sustainable development. *Ecological Economics*, *2*(1), 1–6. doi:10.1016/0921-8009(90)90010-R

DeLaet, C. (1985). Developing networks and institutions in support of sustainable development. *Landscape Planning*, *12*(3), 301–309. doi:10.1016/0304-3924(85)90008-5

Ellinger, A. E., & Ellinger, A. D. (2014). Leveraging human resource development expertise to improve supply chain managers' skills and competencies. *European Journal of Training and Development*, *38*(1/2), 118–135.

EPA. (n.d.). *About ENERGY STAR*. Retrieved on 24 Mar. 2014, from https://www.energystar.gov/about/

Faust, M. (2013). Cashmere: A lux-story supply chain told by retailers to build a competitive sustainable advantage. *International Journal of Retail & Distribution Management*, *41*(11), 973–985.

FDA, U.S. Food and Drug Administration. (n.d.). *Proposed Changes to the Nutrition Facts Label*. Retrieved on 26 Mar. 2014 from http://www.fda.gov/Food/GuidanceRegulation/GuidanceDocumentsRegulatoryInformation/LabelingNutrition/ucm385663.htm

Feller, A., Shunk, D., & Callarman, T. (2006, March). Value chains versus supply chains. *BPTrends*.

Fiksel, G. (2003). Designing resilient, sustainable systems. *Environmental Science & Technology*, *37*(23), 5330–5339. doi:10.1021/es0344819 PMID:14700317

Gaiardelli, P., Saccani, N., & Songini, L. (2007). Performance measurement of the after-sales service network—Evidence from the automotive industry. *Computers in Industry*, *58*(7), 698–708. doi:10.1016/j.compind.2007.05.008

Galbreth, M. R., & Ghosh, B. (2013). Competition and sustainability: The impact of consumer awareness. *Decision Sciences*, *44*(1), 127–159. doi:10.1111/j.1540-5915.2012.00395.x

Ghadge, A., Dani, S., & Kalawsky, R. (2012). Supply chain risk management: Present and future scope. *International Journal of Logistics Management*, *23*(3), 313–339. doi:10.1108/09574091211289200

Giunipero, L. C., Hooker, R. E., & Denslow, D. (2012). Purchasing and supply management sustainability: Drivers and barriers. *Journal of Purchasing and Supply Management*, *18*(4), 258–269. doi:10.1016/j.pursup.2012.06.003

Gobble, M. M. (2012). Innovation and sustainability. *Research Technology Management*, *55*(5), 64–67. doi:10.5437/08956308X5505005

Govindan, K., Kaliyan, M., Kannan, D., & Haq, A. N. (2014). Barriers analysis for green supply chain management implementation in Indian industries using analytic hierarchy process. *International Journal of Production Economics*, *147*, 555–568. doi:10.1016/j.ijpe.2013.08.018

Guide, V. D. R., Harrison, T. P., & Wassenhove, L. N. V. (2003). The Challenge of Closed-Loop Supply Chains. *Interfaces*, *33*(6), 3–6. doi:10.1287/inte.33.6.3.25182

Guide, V. D. R., & Wassenhove, L. N. V. (2002). The reverse supply chain. *Harvard Business Review*, *80*(2), 25–26. PMID:12964466

Gunady, M. G. A., Biswas, W., Solah, V. A., & James, A. P. (2012). Evaluating the global warming potential of the fresh produce supply chain for strawberries, romaine/cos lettuces (lactuca sativa), and button mushrooms (agaricus bisporus) in Western Australia using life cycle assessment (LCA). *Journal of Cleaner Production, 28*, 81–87. doi:10.1016/j.jclepro.2011.12.031

Gundlach, G. T., Bolumole, Y. A., Eltantawy, R. A., & Frankel, R. (2006). The changing landscape of supply chain management, marketing channels of distribution, logistics and purchasing. *Journal of Business and Industrial Marketing, 21*(7), 428–438. doi:10.1108/08858620610708911

Gurtu, A., Searcy, C., & Jaber, M. Y. (2015). An analysis of keywords used in the literature on green supply chains. *Management Research Review, 38*(2), 166–194. doi:10.1108/MRR-06-2013-0157

Gutés, C. M. (1996). The concept of weak sustainability. *Ecological Economics, 17*(3), 147–156. doi:10.1016/S0921-8009(96)80003-6

H'Mida, S., & Lakhal, S. Y. (2007). A model for assessing the greenness effort in a product supply chain. *International Journal of Global Environmental Issues, 7*(1), 4–24. doi:10.1504/IJGENVI.2007.012074

Hanss, D., & Böhm, G. (2012). Sustainability seen from the perspective of consumers. *International Journal of Consumer Studies, 36*(6), 678–687. doi:10.1111/j.1470-6431.2011.01045.x

Hardin, G. (2009). The Tragedy of the Commons. *Journal of Natural Resources Policy Research, 1*(3), 243–253. doi:10.1080/19390450903037302

Harris, J. (2006). New Paths to Business Value: linking environment, health performance to strategic sourcing. In J. Sarkis (Ed.), *Greening the supply chain* (pp. 39–65). Worcester, MA: Springer. doi:10.1007/1-84628-299-3_3

Hassini, E., Surti, C., & Searcy, C. (2012). A literature review and a case study of sustainable supply chains with a focus on metrics. *International Journal of Production Economics, 140*(1), 69–82. doi:10.1016/j.ijpe.2012.01.042

Hervani, A. A., Helms, M. M., & Sarkis, J. (2005). Performance measurement for green supply chain management. *Benchmarking, 12*(4), 330–353. doi:10.1108/14635770510609015

Hopp, W., & Ebrahim-Khanjari, N. (2012). Trust and information sharing in supply chains. *Production and Operations Management, 21*(3), 444–464. doi:10.1111/j.1937-5956.2011.01284.x

Hutchins, M. J., & Sutherland, J. W. (2008). An exploration of measures of social sustainability and their application to supply chain decisions. *Journal of Cleaner Production, 16*(15), 1688–1698. doi:10.1016/j.jclepro.2008.06.001

IEA. (n.d.). *CO2 Emissions from Fuel Combustion2013*. IEA. Retrieved from http://www.iea.org/statistics

Jacobs, P. (1985). A sustainable society through sustainable development: Towards a regional development strategy for northern Quebec. *Landscape Planning, 12*(3), 267–283. doi:10.1016/0304-3924(85)90006-1

Joshi, G. N., & Dave, M. B. (2012). Green Manufacturing for Greener Tomorrow. In *International Conference on Technology and Business Management.*

Kaplan, R., & Norton, D. (1992). The balanced scorecard: Measures that drive performance. *Harvard Business Review, 70*(1), 71–79. PMID:10119714

Kaya, M., & Özer, Ö. (2009). Quality risk in outsourcing: Noncontractible product quality and private quality cost information. *Naval Research Logistics, 56*(7), 669–685. doi:10.1002/nav.20372

Khosla, A., Prakash, S., & Revi, A. (1986). A transcultural view of sustainable development. The landscape of design. *Landscape and Urban Planning, 13*(C), 401–410. doi:10.1016/0169-2046(86)90057-5

Kolk, A., & Pinkse, J. (2005). Business responses to climate change: Identifying emergent strategies. *California Management Review, 47*(3), 6–20. doi:10.2307/41166304

Lau, K. H. (2011). Benchmarking green logistics performance with a composite index. *Benchmarking, 18*(6), 873–896. doi:10.1108/14635771111180743

Lepoutre, J., & Heene, A. (2006). Investigating the impact of firm size on small business social responsibility: A critical review. *Journal of Business Ethics, 67*(3), 257–273. doi:10.1007/s10551-006-9183-5

Linn, M. W., Sandifer, R., & Stein, S. (1985). Effects of unemployment on mental and physical health. *American Journal of Public Health, 75*(5), 502–506. doi:10.2105/AJPH.75.5.502 PMID:3985238

Lowry, K., & Carpenter, R. A. (1985). Institutionalizing sustainable development: Experiences in five countries. *Environmental Impact Assessment Review, 5*(3), 239–254. doi:10.1016/0195-9255(85)90004-6

Markley, M. J., & Davis, L. (2007). Exploring future competitive advantage through sustainable supply chains. *International Journal of Physical Distribution & Logistics Management, 37*(9), 763–774. doi:10.1108/09600030710840859

McFadden, D. (1978). Cost, Revenue and Profit Functions. In M. Fuss & D. McFadden (Eds.), *Production Economics: A Dual Approach to Theory and Applications*. Elsevier.

Mentzer, J. T., DeWitt, W., Keebler, J. S., Min, S., Nix, N. W., Smith, C. D., & Zacharia, Z. G. (2001). Defining supply chain management. *Journal of Business Logistics, 22*(2), 1–25. doi:10.1002/j.2158-1592.2001.tb00001.x

Mentzer, J. T., Stank, T. P., & Esper, T. L. (2008). Supply chain management and its relationship to logistics, marketing, production, and operations management. *Journal of Business Logistics, 29*(1), 31–46. doi:10.1002/j.2158-1592.2008.tb00067.x

Mossakowski, K. N. (2009). The influence of past unemployment duration on symptoms of depression among young women and men in the United States. *American Journal of Public Health, 99*(10), 1826–1832. doi:10.2105/AJPH.2008.152561 PMID:19696382

Nidumolu, R., Prahalad, C. K., & Rangaswami, M. R. (2009). Why sustainability is now the key driver of innovation. *Harvard Business Review, 87*(9), 56–64.

Overcash, M. (2012). A comparison of reusable and disposable perioperative textiles: Sustainability state-of-the-art 2012. *Anesthesia and Analgesia, 114*(5), 1055–1066. doi:10.1213/ANE.0b013e31824d9cc3 PMID:22492184

Papista, E., & Krystallis, A. (2013). Investigating the types of value and cost of green brands: Proposition of a conceptual framework. *Journal of Business Ethics, 115*(1), 75–92. doi:10.1007/s10551-012-1367-6

Petroleum Times. (1951). New oil jetty is largest in Thames Estuary. *Petroleum Times, 55*(1399), 216.

Ramanathan, U., & Gunasekaran, A. (2014). Supply chain collaboration: Impact of success in long-term partnerships. *International Journal of Production Economics, 147*, 252–259. doi:10.1016/j.ijpe.2012.06.002

Rao, P., & Holt, D. (2005). Do green supply chains lead to competitiveness and economic performance? *International Journal of Operations & Production Management, 25*(9), 898–916. doi:10.1108/01443570510613956

Romer, J. R. (2012). Single-use plastic carryout bags: An icon of waste. *Sustainability, 5*(6), 341–343. doi:10.1089/SUS.2012.9921

Schieberl, J., & Nickles, M. (2014). Outsourcing U.S. jobs abroad: Why? *The International Business & Economics Research Journal, 13*(2), 253.

Searcy, C. (2012). Corporate sustainability performance measurement systems: A review and research agenda. *Journal of Business Ethics, 107*(3), 239–253. doi:10.1007/s10551-011-1038-z

Seuring, S., & Müller, M. (2008). From a literature review to a conceptual framework for sustainable supply chain management. *Journal of Cleaner Production, 16*(15), 1699–1710. doi:10.1016/j.jclepro.2008.04.020

Sheng, E. (2012). Evaluating the cost of sustainability. *WWD, 204*(86), 8.

Silvestre, B. S. (2014). (in press). A Hard Nut to Crack! Implementing Supply Chain Sustainability in an Emerging Economy. *Journal of Cleaner Production*.

Simpson, B. J. K., & Radford, S. K. (2012). Consumer perceptions of sustainability: A free elicitation study. *Journal of Nonprofit & Public Sector Marketing, 24*(4), 272–291. doi:10.1080/10495142.2012.733654

Srivastava, S. K. (2007). Green supply-chain management: A state-of-the-art literature review. *International Journal of Management Reviews, 9*(1), 53–80. doi:10.1111/j.1468-2370.2007.00202.x

Stevens, G. C. (1989). Integrating the supply chain. *International Journal of Physical Distribution & Logistics Management, 19*(8), 3–8. doi:10.1108/EUM0000000000329

Svensson, G. (2002). The theoretical foundation of supply chain management: A functionalist theory of marketing. *International Journal of Physical Distribution & Logistics Management, 32*(9), 734–754. doi:10.1108/09600030210452422

Tapiero, C. S. (2007). Consumers risk and quality control in a collaborative supply chain. *European Journal of Operational Research, 182*(2), 683–694. doi:10.1016/j.ejor.2006.07.039

Taticchi, P., Tonelli, F., & Pasqualino, R. (2013). Performance measurement of sustainable supply chains: A literature review and a research agenda. *International Journal of Productivity and Performance Management, 62*(8), 782–804. doi:10.1108/IJPPM-03-2013-0037

UNFCCC. (n.d.). *Issues - bunker fuels.* UNFCCC. Retrieved on 7 March 2014, from http://unfccc.int/methods/emissions_from_intl_transport/items/1057.php

Varki, S., & Colgate, M. (2001). The role of price perception in an integrated model of behavioral intentions. *Journal of Service Research, 3*(2), 232–240. doi:10.1177/109467050133004

WCED. (1987). *Report of the World Commission on Environment and Development: Our Common Future*. United Nations.

Westing, A. H. (1986). International cooperation for sustainable development and peace. *Environmental Conservation, 13*(2), 97–99. doi:10.1017/S0376892900036663

Woods, L., & Bakshi, B. R. (2014). Reusable vs. disposable cups revisited: Guidance in life cycle comparisons addressing scenario, model, and parameter uncertainties for the US consumer. *International Journal of Life Cycle Assessment, 19*(4), 931–940. doi:10.1007/s11367-013-0697-7

Xu, J., & Wang, H. (2008). Build competitive advantage through the integration of sustainable supply chains. In *Proceedings of 2008 IEEE International Conference on Service Operations and Logistics, and Informatics, IEEE/SOLI 2008*.

Yakovleva, N., Sarkis, J., & Sloan, T. (2012). Sustainable benchmarking of supply chains: The case of the food industry. *International Journal of Production Research, 50*(5), 1297–1317. doi:10.1080/00207543.2011.571926

Yates, L. (2008). Sustainable consumption: The consumer perspective. *Consumer Policy Review, 18*(4), 96–99.

ADDITIONAL READING

Beamon, B. M. (1999). Designing the green supply chain. *Logistics Information Management, 12*(4), 332–342. doi:10.1108/09576059910284159

Beske, P., Land, A., & Seuring, S. (2013). (in press). Sustainable supply chain management practices and dynamic capabilities in the food industry: A critical analysis of the literature. *International Journal of Production Economics*.

Brandenburg, M., & Seuring, S. (2011). Impacts of supply chain management on company value: Benchmarking companies from the fast moving consumer goods industry. *Logistics Research, 3*(4), 233–248. doi:10.1007/s12159-011-0056-7

Cuthbertson, R. (2011). The Need for Sustainable Supply Chain Management. In *Sustainable supply chain management: Practical ideas for moving towards best practice* (pp. 3–13). Heidelberg, Germany: Springer. doi:10.1007/978-3-642-12023-7

Glavic, P. & Lukman, R. (2007). Review of sustainability terms and their definitions, *Journal of Cleaner Production, 15*(18), 1875-1885.

IEA. *CO_2 Emissions from Fuel Combustion2013* retrieved from http://www.iea.org/statistics on March 11, 2014.

Jaber, M. Y. (2009). *Inventory management: Non-classical views*. Boca Raton, FL: CRC Press. doi:10.1201/9781420079982

Jaber, M. Y. (2011). *Learning curves: Theory, models, and applications*. Boca Raton, FL: CRC Press. doi:10.1201/b10957

Rogers, D. S., & Tibben-Lembke, R. S. (1998). Going backwards: reverse logistics trends and practices. Pittsburgh, PA: Reverse Logistics Executive Council (RLEC).

Sarkis, J. (2006). *Greening the supply chain*. Worcester, MA: Springer. doi:10.1007/1-84628-299-3

Searcy, C. (2009). Setting a course in corporate sustainability performance measurement. *Measuring Business Excellence, 13*(3), 49–57. doi:10.1108/13683040910984329

Searcy, C. (2009). *The role of sustainable development indicators in corporate decision-making*, Winnipeg, MB, CAN: International Institute for Sustainable Development Searcy, C. (2011). Updating corporate sustainability performance measurement systems. *Measuring Business Excellence, 15*(2), 44–56. doi:10.1108/13683041111131619

Seuring, S., & Gold, S. (2013). Sustainability management beyond corporate boundaries: From stakeholders to performance. *Journal of Cleaner Production, 56*(2), 1–6. doi:10.1016/j.jclepro.2012.11.033

Srivastava, S. K. (2012). Managerial implications from Indian case studies on e-reverse auctions. *Business Process Management Journal, 18*(3), 513–531. doi:10.1108/14637151211232687

Chapter 2
An Introduction to Sustainable Supply Chain Management and Business Implications

Gowri Vijayan
Universiti Putra Malaysia, Malaysia

Nitty Hirawaty Kamarulzaman
Universiti Putra Malaysia, Malaysia

ABSTRACT

The chapter will explain the concept of sustainability in supply chain management. Sustainability in business operations is fast becoming a vital element to successful global acceptance. It is also one of the major marketing strategies for Points-of-Difference (POD) product positioning. It is necessary to have a basic knowledge of how sustainability is brought about in supply chains in order to truly comprehension the opportunities from it. An understanding of the basic elements to sustainability in supply chain management and global models adopted for implementation of sustainability by corporations is essential. The chapter further focuses on introducing the concept of reverse logistics as sustainability supply chain process and the strategic advantage for industries from adoption of reverse logistics across the supply chain. An insight into United Nations Global Compact program for sustainability across businesses and sustainability governance has been provided.

DOI: 10.4018/978-1-5225-0635-5.ch002

INTRODUCTION

Sustainability in business operations is a challenging issue that corporates have been facing due to growing awareness of environmental issues and cause factors by public and governments. The increasing contribution of business operations such as sourcing, manufacturing and logistics to environmental issues, created a need for optimization of business operations with reduced negative effects. The increasing stakeholder pressures on companies further influence adoption of sustainability by companies. The increasing competition among corporates to show their environment commitment through green supply chains and its marketing, complete supply chain revolution into sustainable is in trend. In order to understand the decision making process behind this change, it is essential to recognize the areas in supply chain wherein the changes are being implemented by corporates. This ground zero knowledge of the business supply chain is essential while planning of operations, especially in a global sense. This is also essential since sustainable changes are not a short-term application, and responsible decisions made might lead to a successful green company in the future.

Learning Objectives

After reading this chapter, readers will be familiar with concepts on:

1. An understanding of sustainability and its ranking.
2. Understand sustainability in supply chain management and its parameters.
3. Familiarize with the green concept of reverse logistics.
4. Explore the UN Global Compact Management for sustainable supply chain management.

SUSTAINABILITY

Sustainability is based on the simple fact that all activities for survival depend on the natural environment. The objective of sustainability is to preserve harmony in the environment, fulfilling the social, economic and economic requirements of the current generation and preserving for future generations. The Brundtland Report of 1987 popularized sustainability and described it to be any development meeting present needs without affecting future generation's capability to survive. The maintenance of practices and factors supporting environment conservation summarize sustainability. This is an initiative taken in order to repair the damages from direct and indirect human actions on the environment. A business is said to be sustainable when it considers sustainability issues in its decision making, provides environment-friendly products or services, have a green supply chain, have made environmental commitment as part of its operations. The sustainable activities emphasize on maximum utilization and minimal wastage. Environmental sustainability involves serious decision-making and provides a balance between environment commitment and interest of the company. It is not a short-term application, and responsible decisions made might lead to a successful green company in the future. Several models of sustainability have been developed over the years. The Brundtland Report depicted sustainability as a three-legged stool of profit, planet and people. Businesses try to cover these three areas while implementing sustainable activities. The Triple Bottom Line (TBL) model of Elkington (1998), collaborating economic, environmental and social dimensions of sustainability and their interdependency, is another accepted study in research circles (Figure 1). This model focuses on holistic development considering the three spheres of social, economic and environmental impacts on society, business and environment. A sustainable approach focuses on understanding the interactions among the environmental, social and economic pillars as a way to measure the impact of business operations. The environmental pillar includes activi-

Figure 1. The three overlapping circles of sustainability
Source: Elkington (1998)

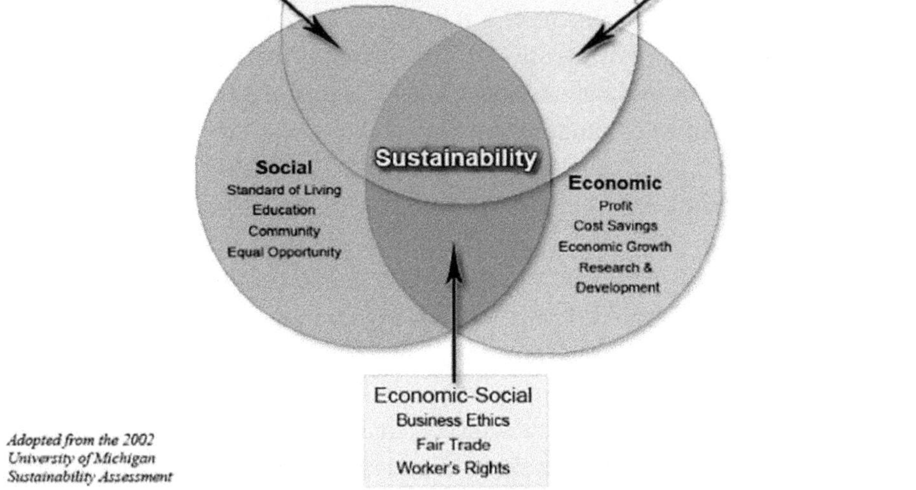

ties like ecosystem services, green engineering, air quality, water quality, stressors, resource integrity, while social pillar includes environmental justice, human health, participation, education, resource security, sustainable communities. The economic pillar consists of activities like incentives, jobs, supply and demand, natural resource accounting, costs and prices. Other researches on sustainability also brought out these three pillars of sustainability. The Triple Top Line (TTL) value concentrates on the same three important requisites for a sustainable business namely financial benefits, work betterment, and social advantages for its stakeholders.

The Three Tiers of Sustainability

The sustainability involvement of an organization can be classified into three groups using the ranking system developed by Future Laboratory in 2008, called Three Tiers of Sustainability. The three tiers are based on the sustainability practices implemented by companies and their commitment to its enhancement. The three tiers are namely:

Tier 1: *The Basics* – The implementation of basic environment friendly measures like switching off lights and PCs when not in use, paper management, and reduction in carbon footprint from daily business activities.

Tier 2: *Thinking Sustainably* – Companies incorporate sustainability into their supply chain operations. This level is attained through corporate assessment of their operations. Supplier management, Life Cycle Management (LCM) and resource optimization belong to this group of sustainable activities. The companies at this stage of sustainability understand the benefits from incorporation of sustainable practices into their supply network.

Tier 3: *The Science of Sustainability* – The creation of an auditing system and benchmarks for development of a framework for governance of Sustainable Supply Chain Management (SSCM). The environmental impact of supply chain agility, flexibility and the costs incurred are explained here. The companies at this level will be committed to sustainability and act as a pushing force to implementation of sustainability standards and regulations across the industrial and government levels.

SUPPLY CHAIN MANAGEMENT (SCM)

A supply chain consists of all parties involved in fulfilment of a customer request, thereby including suppliers, manufacturers, transporters, distributors, retailers, warehouses, and customers themselves. The end objective of every product/service supply chain is to generate profits through customer satisfaction. The integrated effort of all concerned supply chain parties to fulfil this mission is termed as Supply Chain Management (SCM). Globalization and open markets across the globe further necessitated the development of this synchronized effort by supply chain players. The term "supply chain management" was coined in 1982 by Keith Oliver, a consultant for Booz & Company. SCM derives its applicability by combining operations management, logistics, procurement, and information technology (IT) towards development of an integrated and efficient supply chain. According to Lambert *et al.* (1998), SCM can be defined as the "integration of key business processes from end-user through original suppliers that provide products, services, and information that add value for customers and other stakeholders. Mentzer *et al.* (2001) argued that SCM could be defined either as a management philosophy, the implementation part of the philosophy or as a set of process. SCM is thereby described by Mentzer *et al.* (2001) as being a strategic and systematic integration of business functions and strategies, with the intension of long term performance of the business and viability of the supply chain. However, Svensson (2002) defines SCM focusing on the philosophy part of SCM, describing SCM as a philosophy of business that need to address the dual direction flow of resources and information between the players, on a tactical, strategic as well as operational level, from point or origin to point of consumption. The SCM components include supply chain strategy, supply chain enterprise application, supply chain planning, procurement, asset management, product lifecycle management, and logistics. The supply chain processes across these SCM components could be explained in terms of its cycle of movement of products across the players and their marketing outlook. The four main cycles of SCM are namely customer order cycle, replenishment cycle, manufacturing cycle, and procurement cycle. The marketing outlook consists of the push and pull process of marketing. While the pull process of SCM is initiated by a customer order, the push process is triggered by anticipation in customer orders.

SCM has four elements which focus on supply chain value creation and relationship management.

1. **Supply Chain Demand Management:** Demand management focuses on meeting consumer needs over the production process. The increased awareness of consumer expectations and needs, supported by increasing collaboration between supply chain players create increased business opportunities and improved competitiveness of the supply chain network.
2. **Supply Chain Communication**: An effective communication strategy helps in efficiency and productivity enhancement of the supply chain operations, through information sharing between the concerned players. It also makes the supply chain flexible and agile to changing market conditions.

3. **Supply Chain Integration**: The integration of supply chain processes leads to reduced inventory costs, better inventory management among players and improved competitive pricing to customers. The establishment of information networks is critical to the integration across the players in supply chain.
4. **Supply Chain Collaboration**: Collaboration among the supply chain players strengthens relationships, increases efficiency and lead to enhanced business performance.

The efficiency management of SCM across its players requires consideration of four main steps into the strategic planning process, given below:

1. **Plan:** Efficiency in resource utilization requires careful planning of its management across the supply chain. This therein necessitates the development of metrics to monitor the utilization and conversion rate of resources into final output.
2. **Make:** Scheduling activities of production, testing, product packaging and delivery based on necessity is a crucial step in effective management of SCM. This is the most metric centric work of the supply chain, wherein quality levels, production output and labour productivity could be measured for efficiency.
3. **Delivery:** The coordination during receipt of orders, storage, transportation, and delivery stages will help faster movement of goods across the supply chain.
4. **Return:** This is a crucial part of the supply chain necessary in today's green world, but often avoided by many companies. The development of a reverse flow supply chain process is critical to management of return of defective, excess and end-of-life products along the supply chain. The absence of such as system only results in cluttering of the supply chain flow for forward movement of goods.

Studies on impact of supply chain activities of transportation, packaging and warehousing on the environment were conducted to understand SCM's contribution to carbon footprint of a company. SCM operations was found to contribute to nearly 75% of a company's carbon footprint. The sole focus of business on customer satisfaction from reduction of costs and faster delivery further enhance the negative effects of logistics on environment. Globalization, out sourcing and just-in-time inventory management systems have further reduced the control over logistics activities. The complexity of the new supply chains as a result of new markets and opportunities could further damage the environment, unless steps are taken to prevent it. Transparency, integration and coordination between the supply chain players is necessary to streamline these complex supply networks. The pressure for short-term profits often lead to overlook on long term durability of the business. Sustainability is the key to this solution. SCM has a wide range of influence on various stages of the life cycle of materials, and is thereby an excellent platform to initiate sustainability across all products, services and processes that a company is currently engaging in. Any change or chance for optimizing the operations through sustainability needs to start from the top management. In order for the board of members to consider sustainability as a solution, it is necessary to address sustainability in terms of its benefits of resource efficiency. Quantification of long term benefits from sustainability is often difficult and the proposal for sustainability in supply chain falls short of approval. The focus for commitment should thereby be on business volatility and profit generation, for capturing the board's attention. Some of the practices like compulsory standard maintenance by suppliers goes a long way towards ensuring sustainability across the supply chain. Government legislations further ensure acceptance of sustainability by companies. Legislations like the Landfill Tax by the UK government ensure development of proper waste management and disposal systems. Legislation and collaboration between stakeholders are some of the suggested ways to create confidence among companies to follow the sustainable way of business. For that, ample knowledge of sustainability and its incorporation into business operations is essential.

SUSTAINABLE SUPPLY CHAIN MANAGEMENT (SSCM)

Sustainable supply chain management encompasses a set of sustainable logistics practices designed to include environmental parameters into the forward and reverse logistics supply chains. Supply chain sustainability is the management of environmental, social and economic impacts, and the encouragement of good governance practices, throughout the lifecycles of goods and services. The objective of supply chain sustainability is to create, protect and grow long-term environmental, social and economic value for all stakeholders involved in bringing products and services to market. Sustainability is often measured in terms of the triple bottom line of economic, social and environmental impacts, while supply chain sustainability is rated using social, ethical, cultural, and health footprints (SECH). The increasing awareness among customers regarding SECH ratings and its significance, combined with their involvement with NGOs and reaction to unethical companies has forced companies to ensure sustainability in their supply chain. Reducing packaging waste, supplier assessment for sustainable practices, development of sustainable products, reducing carbon footprint, compliance with eco standards and specifications are some of the practices of sustainable supply chain management (SSCM). SSCM integration is necessary into the supply chain activities of product design, procurement, manufacturing, logistics, distribution and returns management. Business sustainability initiatives are often conducted under the title of 'Corporate Sustainability.' Corporate sustainability can be defined as "meeting needs of current stakeholders without compromising ability to meet the needs of the future stakeholders."

Sustainable supply chain management (SSCM) has been widely studied over the years, leading to development of different definitions for it. But, all the definitions of SSCM singularly focus more on the flow of materials and the coordination required for integrity in performance (Payman & Corry, 2013). SSCM is defined as "the strategic, transparent integration and achievement of an organization's social, environmental, and economic goals in the systemic coordination of key inter-organizational business processes for improving the long-term economic performance of the individual company and its supply chains" (Carter & Rogers, 2008). Another definition focuses on the critical difference between SCM and SSCM through the pillars of environmental and social aspects, stating SSCM as SCM with added values of environmental and social/ethical angles (Wittstruck & Teuteberg, 2011). A detailed definition of SSCM by Payman and Corry (2013) based on compilation of previous studies describes SSCM as "the creation of coordinated supply chains through the voluntary integration of economic, environmental, and social considerations with key inter-organizational business systems designed to efficiently and effectively manage the material, information, and capital flows associated with the procurement, production, and distribution of products or services in order to meet stakeholder requirements and improve the profitability, competitiveness, and resilience of the organization over the short and long term." Hassin, Surti, and Searcy (2012) explained SSCM as the management of supply chain operations and resources to maximize profitability while increasing social well-being and decreasing the environmental impacts. Companies often adopt SSCM practices to ensure compliance with sustainability regulations and universal business code of conduct. SSCM aims to create economic, social and environmental values for all the stakeholders involved in that product/service supply chain. Long term viability of business is maintained through adoption of SSCM. This is supported by the framework on triggers to SSCM highlighted by Seuring and Müller (2008) (Figure 2).

The main factors behind SSCM implementation are often government regulation, brand image, innovation, and cost reduction (Ryan, 2010). The need to satisfy the stakeholders and meet its bottom line objective, makes companies walk a fine line in SSCM implementation.

Critical Areas in Sustainable Supply Chain Management

Sustainable supply chain management (SSCM) consists of five basic elements of sustainability metrics, sustainable suppliers, responsible design, socially-responsible

Figure 2. Triggers for sustainable supply chain management
Source: Seuring and Müller (2008)

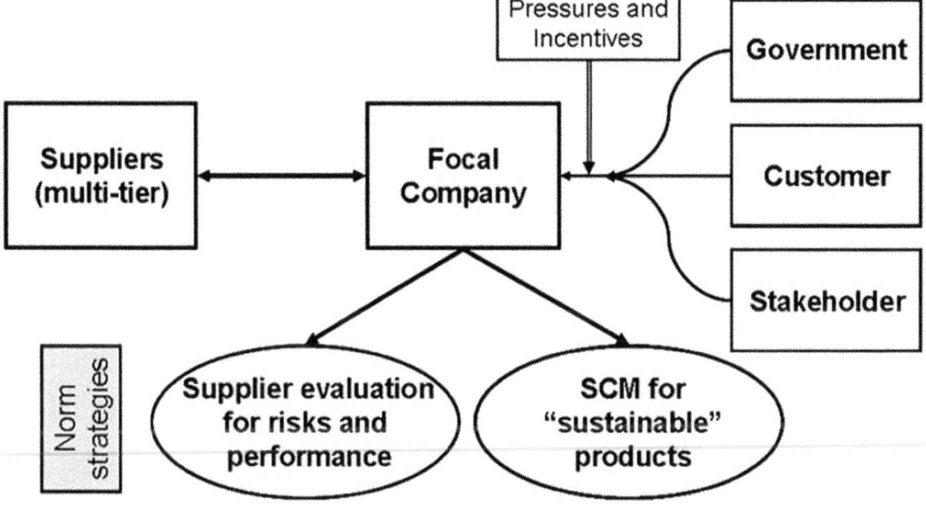

partners and sustainability leadership. Sustainability measurement metrics are essential in order to prove value creation from SSCM, create opportunities to analyze reduction in consumption of resources across the supply chain. Responsible design and sustainable supplier are crucial to sustainable management of products. The presence of socially responsible partners will help businesses in successful implementation of SSCM across the supply stream. Sustainability leadership is essential to plan, execute and monitor the SSCM practices across the supply chain by all the supply chain players.

A logistics network consists of many players and functions. In order to make a logistics network sustainable, all the main functions of manufacturing, transportation, usage and end-of-life of product need to be addressed (Figure 3). Studies on the logistics design have focused on two main approaches: minimization of costs and minimization of environmental impacts across the supply chain. The optimization of these two approaches is essential for acceptance of sustainability by all businesses.

SSCM is mainly practices in the supply chain operational areas of:

- **Sustainable Design:** A sustainable design is aimed to reduce the environmental impact of the product throughout its product life cycle.
- **Sustainable Sourcing:** Sustainable sourcing or procurement deals with the decision making process behind procurement decisions for the company.

Figure 3. Framework for a sustainable logistics network
Source: Adapted from Sheu et al. (2005) and Neto et al. (2008)

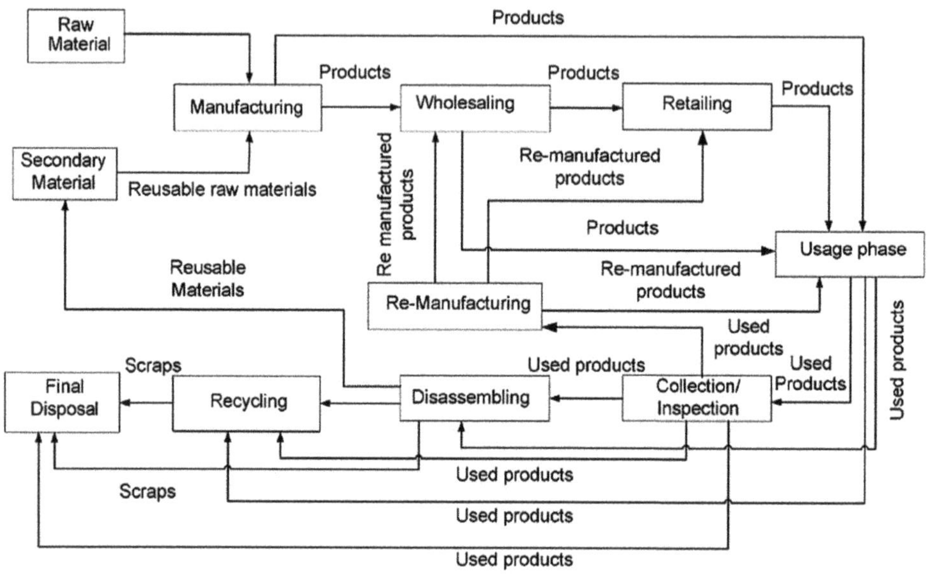

- **Sustainable Manufacturing:** It is aimed to reduce the environmental impact of manufacturing process, by optimization in use of resources, energy and process of production.
- **Green Logistics:** The objective is to meet customer demands, at the same time reduce the negative effects from logistics operations of transportation, warehousing, inventory management and distribution functions on the environment.
- **Reverse Logistics:** Deals with the returns and waste management of materials in the supply chain. It is crucial part of SSCM, managing the backward supply system thereby preventing the cluttering and lagging in forward movement of products.

Reverse Logistics

Reverse logistics is the process of moving products from their typical final destination to another point, in an attempt to recapture some of the value of the product through resale, reverse auction or for the purpose of disposal of the item if not resalable or reusable (Walden, 2005). Other synonyms are Aftermarket logistics, Retrogistics, or Aftermarket supply chain. The reverse supply chain is also a term used in the industry (Reverse Logistics Trends, n.d.). The reverse logistics process depends on nature of returns. Roughly, returns can be classified into three categories namely type of return, source of return, and condition of return (Table 1).

The activities companies generally follow with return goods are given in Table 2.

Returned product types can be listed down generally under the following groups namely close-outs, buy-outs/lifts, job-outs, surplus, defective products, non-defec-

Table 1. Category of returns

Returns Category	Description
Type of Return	The reverse logistics process can be broken into two general areas, depending on whether the reverse flow consists primarily of products, or primarily of packaging. For product returns, a high percentage is represented by customer returns (Rogers and Tibben-Lembke, 1998). In case of packaging, maximum times, it goes for reuse within the stores.
Source of Return	Reverse logistics can include a wide variety of activities and members in its supply chain. This gives a chance for returns to occur from any of the members in the supply chain. Sources of return can include the end-users/consumers, and other members of the distribution channel such as a retailer or distribution center.
Condition of Return	The condition of return influences the company's decision to what to do with the return. If the return goods are in good condition, it can be resold directly or after refurbishment, in a secondary market. If the product cannot be salvaged, then it will have to be destroyed in an environment-friendly manner. Same goes for packaging returns. In this case, often reuse takes place until it cannot be salvaged further.

Source: Authors

Table 2. Reverse logistics activities

Material	Reverse Logistics Activities
Products	• Return to supplier • Resell • Sell via outlet • Salvage • Recondition • Refurbish • Remanufacture • Reclaim materials • Recycle • Landfill
Packaging	• Reuse • Refurbish • Reclaim materials • Recycle • Salvage

Source: Rogers and Tibben-Lembke (1998)

tive products, salvage damaged items, and returns. While the first 4 groups are mainly concerned with retailers and distribution centers, the last 4 includes consumer's returns.

The reasons behind this has been studied on and listed down in Table 3.

The returns policies of most retailers in the name of customer service have caused reverse logistics to become a very visible process. In addition, these policies have created a "rent to own" mentality where the customer does not worry about purchases because if they decide that they do not want or like the product, they can return it. According to Walden (2005) processing or handling of returned items include the following:

1. Processing returned merchandise that is either damaged, seasonal, has to be restocked, is returned for salvage, is returned for rebuild/refurbishment, has been recalled (for any no of reasons), and items that may now be excess to your inventory.
2. Recycling of packaging and packing materials and/or containers. This process is creating a new industry to support returnable/reusable containers and totes.
3. Reconditioning for resale, remanufacturing, refurbishing of items. This is very common in the automobile repair parts industry.
4. Disposal or donation of obsolete items.
5. The recovery of hazardous materials to prevent environmental implications. In some cases hazardous materials can be recycled to further reduce the environmental impacts.

Reverse logistics is an integral part of the product life cycle and corporate procedures that should not be ignored. Reverse logistics programs can aid a company in finding problem areas and patterns of defects, thus creating a way to minimize

Table 3. Reasons behind product returns

Perspective	Reasons
Customer	• Product did not meet customer needs • Difficult for customer to understand and follow product instructions • Defective product • Misuse of liberal return policy by customer
Retailer	• Outdated packaging of product • Seasonal product • Newer version of product available • Product discontinued/Product recall • High product inventory • Bad business

Source: Rogers and Tibben-Lembke (1998)

the amount of returns a company receives. Reverse logistics encompasses areas such as where the products go once they are returned, the recycling process, the relationships between buyers and sellers, and the technology supporting the return process. In order for a company to be competitive in today's market, management must leverage some type of reverse logistics process in place.

Buyers may return items to the seller due to product defects, overages, shipping errors, trade-ins, or other reasons. Return goods handling has been likened to going the wrong way on a one-way street because the great majority of product shipments flow in one direction. Most logistics systems are ill equipped to handle product movement in a reverse channel (Stock *et al.,* 2001). Therefore, when a customer returns a product to a company, management must determine what the item is that is being returned. Frequently when customers make returns, they do not have the original packaging that the product came in when purchased. In these cases, the company must decide what the product is and whether to move it to the next step in the return process. Once the company determines what the item is, they must then determine what kind of condition the item is in and whether or not the item can be resold as new or if the item should be returned to the vendor or supplier. Finally, when a customer or retailer returns the products, the company must decide where the item should be sent. Depending on company logistics policies, the product may be sent to a centralized returns center (CRC) or sent back to the distribution center (DC) (Smith, 2005).

Return of products also may be due to product recalls leading to sudden pull out of products. The reasons behind these sudden pull outs may be quality problems or health threats. This requires the existence of a good reverse supply chain in order to support the pull-back of products. Recall of products to protect consumers from risk of injury or death is a common phenomenon in Japan, USA, UK, Australia, and New Zealand. Special agencies dealing with recalls provide consumers and businesses with detailed information on product safety, recalls and recall alerts through their websites. Cases of pull-out of cars by manufacturers due to failure in brakes of any manufacturing defects have also been seen. Companies like Honda, Toyota, and Hewlett Packard have their own voluntary recalls occasionally, initiated by their headquarters overseas (Consumer, n.d.). Even food industry is not immune to this, especially with their quality specifications. An outbreak of Salmonella forced Cadbury Schweppes to recall more than one million of their chocolate bars in August, 2006, costing Euro 20 million. This resulted in Cadbury's market share in UK falling by 1.1% within 4 weeks on the incidence. The recall cost Cadbury Euro 5million in sales (Guardian, 2006). In September 2008, Cadbury recalled its Chinese-made products over fears it containing contaminated milk. The chocolate giant recalled the products as news of contaminated milk in china leading to deaths reached it. The company, with its immediate recall and presence of mind, considering the safety of its consumers, proved its stance in quality management and efficient supply chain (NBC News, 2008; Daily Mail, 2008). Companies exhibiting strong values and commitment to consumer safety by perfect recall of such products automatically stand apart for their efficient reverse supply chain and gain strong customer loyalty. Reverse logistics also involves removal and disposal of waste materials from the production, distribution, or packaging processes. If waste materials cannot be used to produce other products, they must be disposed of in some manner. Whatever the by-product, the logistics process must effectively and efficiently handle, transport and store it. If the by-products are reusable or recyclable, logistics manages their transportation to remanufacturing or reprocessing locations. Often these activities are out-sourced by the company to various third parties (Stock, 1998).

Sustainable Supply Chain Integration

A successful SSCM requires integration of its upstream and downstream processes and the life cycle management of its products. Life Cycle Management (LCM) aims to lessen the environmental, social and economic impacts of a product during its life cycle and value chain. It is a management system for dissemination of collected information on the social, economic and environmental aspects of the product across its life cycle. LCM involves product information from its stages of product design, manufacture, distribution, consumption/usage, and end-of life. This information is beneficial for the value chain of the product with reference to resource efficiency gains, reduction in production costs, better relationship management, company image and reputation. Voluntary standards, labels, product announcements further help customers in identifying the sustainability of products based on its life cycle assessment.

The proactive integration of social and environmental issues into its operations prevents corporate vulnerability to the fluctuating market situations. This requires the monitoring of operations along upstream of their supply chain, since the blame for the unethical operational practices of suppliers and sub-contractors on labour practices, impact to environment, chemical content, often falls on the company pursuing their services. The current policies and environmental regulations further necessitate downstream players to take preventive steps to avoid such situations. Furthermore, companies have started realizing the relevance of traceability, transparency and accountability of their operations for SSCM. A sustainable system of purchasing policies would help the company to reduce liability risks with greater operating efficiency and also increase the innovative product development. In order to ensure sustainability along with profitability for the upstream and downstream players, the integration of sustainability practices and its monitoring across the supply network is crucial. The first step to sustainability integration along the supply chain is to select companies and suppliers based on their ethical and social code of conduct. Selection of a sustainable upstream player will go miles towards ensuring sustainability across the supply chain. Secondly, the value chain should be examined to ensure suppliers are provided with sufficient resources to fulfil their responsibilities sustainably. The promotion of social responsibility and sustainability across supply chain networks would help in getting necessary support groups for SSCM integrity. A peer network support for development of best methods to sustainability and opportunities will fasten its adoption across the business networks. Lastly, suppliers and customers need to be treated fairly to ensure their commitment to sustainability, cooperation and loyalty. Integration of SSCM is not an easy task, but good relationship management, committed team, good metrics for monitoring operations and a committed top management will go a long way to ensure sustainability and profitability to the business.

Though sustainability and SSCM has received much hype in the business world, its universal acceptance as a core operational strategy has still a long way to go. Some of the common business sustainability issues reported by companies are complexity in metrics for sustainability measurement, lack of motivation of internal resources, lack of better guidelines for better implementation, lack of long term commitment of top management towards environmental and social pillars of sustainability, and compatibility of SSCM practices into the current business format. The focus on bottom line of the company often deviate its top management from focusing on need for sustainability in the supply chain.

THE UNITED NATIONS GLOBAL COMPACT

The United Nations Global Compact is a United Nations initiative to encourage global businesses to adopt sustainable and socially responsible policies, and to report on their implementation. The UN Global Compact has given ten principles of sustainability, for all round SSCM implementation by businesses. The principles act as core values to sustainable development and corporate citizenship practices in the areas of human rights, labour standards, environment, and anti-corruption for the global business community. The United Nations Environment Programme (UNEP) had defined nine major sustainability issues under four categories of human rights, labour rights, environment, and anticorruption. This was the base for the ten principles developed by the Global Compact.

Ten Principles

Human Rights

Principle 1: Businesses should support and respect the protection of internationally proclaimed human rights; and
Principle 2: make sure that they are not complicit in human rights abuses.

Labour

Principle 3: Businesses should uphold the freedom of association and the effective recognition of the right to collective bargaining;
Principle 4: the elimination of all forms of forced and compulsory labour;
Principle 5: the effective abolition of child labour; and
Principle 6: the elimination of discrimination in respect of employment and occupation.

Environment

Principle 7: Businesses should support a precautionary approach to environmental challenges;
Principle 8: undertake initiatives to promote greater environmental responsibility; and
Principle 9: encourage the development and diffusion of environmentally friendly technologies.

Anti-Corruption

Principle 10: Businesses should work against corruption in all its forms. Including extortion and bribery.

The UN Global Compact is the world's largest corporate citizenship initiative with two objectives: Mainstream the ten principles in business activities around the world and catalyze actions in support of broader UN goals, such as the Millennium Development Goals (MDGs). The UN Global Compact is a forum for discussion and a network for communication including governments, companies and labour organizations, whose actions it seeks to influence, and civil society organizations. The UN Global Compact focus on the major environmental issues though their engagement platforms Caring for Climate and CEO Water Mandate. While the Caring for Climate acts as platforms for businesses seeking to showcase their stand on climate change, the CEO Water Mandate assist companies in development of water management policies and practices. In order to assist businesses in successful implementation of SSCM programs, the UN Global Compact in partnership with Duke University and a group of committed participants developed the next generation Environmental Stewardship Strategy and related resource. This was developed to assist the top management in setting up a comprehensive strategy for SSCM. The Environmental Stewardship Strategy is based on UN Global Compact's Management Model, which focuses on continuous improvement across the supply chain.

UN Global Compact Management Model for SSCM

The UN Global Compact Management Model was developed to act as a flexible framework to establishment of SSCM by its signatory companies (Figure 4). It complements the tier 2 sustainability involvement ranking for companies, incorporat-

Figure 4. UN global compact management model
Source: UN Global Compact (2010)

ing sustainability in supply chain. The six steps of the framework namely commit, assess, define, implement, measure, and communicate reflect the decision flow on processes and objectives to SSCM establishment for businesses. In addition, the three core principles of governance, transparency, and engagement critically support successful SSCM across companies.

1. Commit

The first step towards establishment of a corporate sustainability programme for its supply chain, is to understand the external market situation and drivers to business development. This is essential to evaluate the current supply chain issues faced by the company, to evaluate risks and opportunities and to understand the internal support for change. Factors like stakeholder expectations, business strategies, organizational culture, nature of the industry and its carbon footprint all influence the corporate's decision on changes for sustainability in its operations. The most common drivers to business sustainability are managing risks, realizing efficiency and creation of sustainable products.

- **Managing Risks:** Effective implementation of compliance programmes and robust management systems and its monitoring across suppliers, protect companies from potential supply chain delays from unethical production practices of upstream players, governance and environmental practices.
- **Realizing Efficiencies:** Efficiency in resource management not only reduces operational costs of the company, but also reduce carbon footprint on its supply chain, as well as improve relationship management across the players and internal resources. An in-depth understanding of implications of sustainability to business operations and prioritization for improvement will enable companies to harvest the benefits of SSCM effectively.
- **Creating Sustainable Products**: Joint ventures and collaborations between the upstream and downstream members of the supply chain creates ample opportunities for product innovation. There are instances when proper management of sustainability issues have created opportunities for companies, for example markets for energy saving appliances, organic food, and innovative technologies.

Along with identification of business drivers, it is also important to understand the stand of peer companies, the expectations of stakeholders and opportunities for joint ventures into sustainable business development.

- **Benchmarking against Peer Companies:** It is essential to benchmark against one's peer companies in terms of SSCM implementation, internal structure support for SSCM, codes of conduct followed, labour rights, metrics of measurement and reporting practices, to get a comprehensive understanding of current sustainability practices followed in the industry.
- **Understanding Stakeholder Expectations:** Companies should strive to understand its stakeholder expectations, especially considering the role that customers and investors play in implementation of SSCM programmes across industries. The stakeholders are also the biggest support team to companies for evaluation of credibility of available options and approaches to SSCM.
- **Emerging Risks and Opportunities**: Stakeholders are often the first parties to know about market opportunities and issues in the supply chain. Companies who are in regular contact with its stakeholders will be able to take proactive steps to avoid such issues. Also, early identification of issues would help companies assume the leader position in comparison with its peer companies.

The establishment of a vision and objectives is essential for development of a SSCM programme for a company. It provides the necessary directions for strategizing the company commitment. This requires the involvement of top management in its development. The inclusion of potential barriers and risk factors which would affect the success of SSCM implementation also need to be considered while developing the company vision and objectives.

The next logical step to take is to convert the expectations into guidelines for directing the upstream and downstream players of the supply chain. Supplier compliance of sustainability guidelines set by governments should be ensure at the least, to avoid probable environmental and social damage. The establishment of codes of conduct is the next essential step to be taken. Codes of conduct should be made in accordance with expectations on consumers and suppliers. International standards should be consulted before establishing the codes of conduct. The development of a framework for stakeholder action and engagement is essential to ensure SSCM integrity. Some of the environmental policy areas to consider while writing a supplier code of conduct are material toxicity and chemicals, raw material use, recyclability and end of life products, greenhouse gas emissions, energy use, water use and waste water treatment, air pollution, and biodiversity.

2. Assess

Incorporation of sustainability into the entire supply chain is not practical nor possible for first time companies. Instead, companies often tend to focus on sustain their core areas like main suppliers and manufacturing phases which are critical to production. This will not only allow monitoring of performance across the stages, but also ensure that critical areas of the supply chain which require immediate attention are attended to. Companies often find that sub-tier suppliers find sustainability implementation more challenging than other levels of upstream players. It is necessary to consider this while developing and implementing a SSCM programme. Companies can also self-assess themselves on their capability to manage the changing of its supply chain to the sustainable way of business. The definition of company's supply chain is essential to get the true scope of sustainability in the supply chain.

- **Supply Chain Mapping:** Supply chain mapping will help companies to trace back the activities involved in product development and management across the supply chain. This will also give an opportunity to find out the key issues plaguing each section of your supply chain. Mapping of the supply chain requires identification of the main products and services offered by the company, tracking of flow of materials and information for each product, information on sustainability issues of human and labour rights, corruption issues and environmental impacts at every stage of the supply chain.
- **Supply Chain Segmentation:** Segmentation of suppliers will help in determination of resource allocation for sustainable production and distribution along the supply chain. Good segmentation requires balance between risk taking and positive impacts to society and business. Risk calculation for segmentation includes risk to society, risk to business and risk to economic development. The identification and assessment of risks is crucial for successful implementation of SSCM.

3-4. Define and Implement

Relationship management between companies and suppliers is essential to develop a common mindset on implementation of sustainability across the supply chain. It is always better to work with suppliers who share similar thoughts on sustainability vision, strategy and performance. The relationship between companies and existing suppliers must be based on mutual transparency, realistic timelines, continuous improvement, and open communication. Companies can induce sustainability in suppliers through setting expectations, monitoring and evaluation, remediation and capability building, and partnership. Setting expectations is a broad engagement and is about good communication between the companies and their suppliers regarding the company's stand on sustainability, their expectations, code of conduct and contracts. It is the basic level of sustainability relationship between upstream and downstream players in a supply chain. The monitoring and evaluation phase focuses on supplier self-assessment of their sustainability performance, while the remediation and capability building stage addresses supplier focus on performance enhancement and sustainability management. The partnership phase is a deep engagement and allows companies to partner with suppliers to understand the real cause of poor acceptance of sustainability across the production process. The selection of a suitable phase from the above 4 levels is a strategic decision and depends on the extent to which the company wants to commit to sustainability among suppliers.

One of the most challenging barriers to SSCM integration is the commitment issues between sustainability objectives and commercial objectives of a company. It is up to the top management and board of directors to resolve this issue in a way that SSCM implementation guarantees fulfilment of both the commercial and sustainable objectives. The cross functional coordination among managers of different business departments is also necessary for flexibility in implementation and monitoring of SSCM in business. Both the issues need to be resolved before implementation of SSCM across the supply chain.

Collaboration between companies and stakeholders is one of the ways to address sustainability issues across the supply chain. In addition, such collaborations aid small and medium enterprises (SMEs) to contribute towards SSCM. The two main types of corporate collaboration are best practice sharing, and joint standards and implementation. The best practices sharing focuses on sharing of success stories and information found useful for sustainability implementation by companies, while joint standards focus on increasing consistency among corporate expectations and SSCM programmes. While credibility, resource sharing and good relationship with suppliers are added advantages of collaboration, draining of resources, resistance to change and commitment to own company are risks involved in such relationships. Multi-stakeholder partnerships are also becoming common nowadays, with recognition of role of stakeholders in decision-making.

5-6. Measure and Communicate

The establishment of measurement standards and metrics is necessary to complete the assessment of SSCM implementation by a company. This starts from the development of goals across the departments. Goal setting is a collective activity involving heads of all concerned departmental functions. Next, goals need to be developed for supplier performance, and internal performance. This places the platform on which basis the measurement metrics are calculated. Communication of sustainability progress and reporting enhance transparency in the supply chain. It is also the most common way of communication with all the company stakeholders. The main objectives behind the disclosure of sustainability data is to highlight the company's sustainability commitment and performance.

Environmental Stewardship Strategy

Based on the management model, the Environmental Stewardship Strategy outlines an approach to sustainability based on the four pillars of embed, balance, diffuse and translate.

- Leaders *embed* environmental stewardship into all facets of the organization.
- They *balance* short-term targets and long-term goals that are both critical to performance and Environmental Stewardship.
- They *diffuse* best practices throughout value chains and business networks by collaborating and engaging stakeholders.
- They *translate* best practices into processes and practices that are applicable in the diverse geographies in which they operate.

Environmental Stewardship Strategy was designed to be the tier 3 level of involvement of companies into SSCM. It basically involves companies in the core of sustainability pushing government regulations for increased enforcement of sustainability across industries and nations. The combination of business support and policy endorsements create the necessary ripple effect for pushing SSCM across borders.

CONCLUSION

Sustainable Supply Chain Management (SSCM) is at the intersection of productivity and sustainability, providing differentiation in terms of cost savings to material procurement to environmental impact reductions. The future of SSCM is in managing waste industry and procurement. Complete transparency of operations, communication, and cooperation between the small and medium sized enterprises (SMEs), large corporations and governments is necessary for further development in this tier 2 field of sustainability adoption. This will not only help in fast acceptance of sustainability, but also increase the level of innovation in the industry.

REFERENCES

Ahi, P., & Searcy, C. (2013). A comparative literature analysis of definitions for green and sustainable supply chain management. *Journal of Cleaner Production*, *52*, 329–341. doi:10.1016/j.jclepro.2013.02.018

Carter, R. C., & Rogers, D. S. (2008). A framework of sustainable supply chain management: Moving toward new theory. *International Journal of Physical Distribution & Logistics Management*, *38*(5), 360–387. doi:10.1108/09600030810882816

Consumer. (n.d.). *Consumers Association of Penang*. Retrieved from: www.consumer.org.my

Dailymail. (2008). *Cadbury recall all chocolate made in China over fears it may contain contaminated milk*. Retrieved 29 October 2012 from: https://www.dailymail.co.uk/news/article-1063950/cadbury-recall-chocolate-China-fears-contain-contaminated-milk.html

Elkington, J. (1998). *Cannibals with forks- The triple bottom line of the 21st century*. Stoney Creek, CT: New Society.

Hassini, E., Surti, C., & Searcy, C. (2012). A literature review and a case study of sustainable supply chains with a focus on metrics. *International Journal of Production Economics*, *140*(1), 69–82. doi:10.1016/j.ijpe.2012.01.042

Lambert, D. M., Cooper, M. C., & Pagh, J. D. (1998). Supply chain management: Implementation issues and research opportunities. *International Journal of Logistics Management*, *9*(2), 1–19. doi:10.1108/09574099810805807

Mentzer, J. T., DeWitt, W., Keebler, J. S., Min, S., Nix, N. W., Smith, C. D., & Zacharia, Z. G. (2001). Defining supply chain management. *Journal of Business Logistics*, *22*(2), 1–25. doi:10.1002/j.2158-1592.2001.tb00001.x

NBC News. (2008). *Cadbury recalls Chinese-made chocolates*. Retrieved 29 October 2012 from: http://www.msnbc.msn.com/id/26938647/ns/health-food-safety/t/cadbury-recalls-chinese-made-chocolates/

Neto, J. Q. F., Bloemhof-Ruwaard, J. M., van Nunen, J. A. E. E., & Heck, E. V. (2008). Designing and evaluating sustainable logistics networks. *International Journal of Production Economics*, *111*(2), 195–208. doi:10.1016/j.ijpe.2006.10.014

Reverse Logistics Trends. (n.d.). Retrieved from: www.reverselogsiticstrends.com

Rogers, D.S., & Tibben-Lembke, R. S. (1998). *Going backwards: Reverse logistics trends and practices*. Academic Press.

Ryan, P. (2010). *Sustainable Logistics: Towards the development of environmentally conscious supply chains*. (Bachelor of Science Thesis). University of Limerick.

Seuring, S., & Müller, M. (2008). From a literature review to a conceptual framework for sustainable supply chain management. *Journal of Cleaner Production*, *16*(15), 1699–1710. doi:10.1016/j.jclepro.2008.04.020

Sheu, J. B., Chou, Y. H., & Hu, C. C. (2005). An integrated logistics operational model for green-supply chain management. *Transportation Research Part E, Logistics and Transportation Review*, *41*(4), 287–313. doi:10.1016/j.tre.2004.07.001

Smith, A. D. (2005). Reverse logistics programs: Gauging their effects on CRM and online behavior. *Vine*, *33*(3), 166–181. doi:10.1108/03055720510634216

Stock, J. R. (1998). *Development and Implementation of Reverse logistics programs*. Council of Logistics Management.

Stock, J. R., & Lambert, D. M. (2001). *Strategic logistics management* (4th ed.). McGraw-Hill.

Svensson, G. (2002). The theoretical foundation of supply chain management: A functionalist theory of marketing. *International Journal of Physical Distribution & Logistics Management*, *32*(9), 734–754. doi:10.1108/09600030210452422

The Guardian. (2006). *Product recall costs Cadbury Euro 20 million*. Retrieved 29 October 2012 from: http://www.guardian.co.uk/business/2006/aug/02/food.foodanddrink

UN Global Compact. (2010). *Report on UN Global Compact Management Model: Framework for implementation of human rights, labour, environment and anti-corruption*. Retrieved 20 May 2014 from: http://www.unglobalcompact.org/docs/news_events/9.1_news_archives/2010_06_17/UN_Global_Compact_Management_Model.pdf

Walden, J. (2005). *Reverse logistics: Important or irritant?* Retrieved 18 September 2012 from: http://www.honeycombconnect.com/Supply_Chain_Management/document

Wittstruck, D., & Teuteberg, F. (2011). Understanding the success factors of sustainable supply chain management: Empirical evidence from the electrics and electronics industry. *Corporate Social Responsibility and Environmental Management*, *19*(3), 141–158. doi:10.1002/csr.261

Section 2

Measuring the Performance of a Sustainable Supply Chain

Chapter 3
The Development and Analysis of Environmentally Responsible Supply Chain Models

Ehab Bazan
Ryerson University, Canada

Mohamad Y. Jaber
Ryerson University, Canada

ABSTRACT

Inventory management and supply chain management are integral issues and prime areas of concern in production and operations management, and industrial engineering. There is a drive for businesses and organizations to be accountable for their environmental impacts. Guaranteeing environmentally conscious supply chain operations is strongly linked to an organization's sustainability and overall success. The responsible management of forward and return product flows in production and inventory environments is a rapidly increasing requirement. This can be attributed to economic, environmental and/ or regulatory motivations. Mathematical modeling of such systems assists decision-making processes and provided a better understanding of the behavior of such production and inventory environments.

DOI: 10.4018/978-1-5225-0635-5.ch003

INTRODUCTION

Inventory modeling in supply chains is a prime concern for research in production and operations management and industrial engineering. The approach is to have mathematical representations of systems that can be studied and optimized to satisfy dynamic market demands. Most of these inventory models are managed by the classical analysis (profit-maximization/cost-minimization) approach. There has been a push for businesses and organizations to be accountable and responsible for environmental and social impacts of their operations. This drive has been derived from various internal and external stakeholders leading to new regulations being imposed. Such responsibilities have accounted for the introduction and application of various concepts, programs, and efforts (Richards, 1997). Some of these include: environmental management systems, integrated management systems, corporate social responsibility, life cycle assessment, design for environment, pollution prevention, sustainable development, environmental indicators and reporting to name a few.

Applying environmental management concepts to supply chains is becoming known as green supply chain management. Relatively, this concept is considered in its infancy and many individual efforts using a variety of approaches and methodologies are existent throughout the literature. In light of the current environmental responsibilities, interest in industrial environmental performance metrics is increasing. The coupling of such measures and their integration into product procurement/purchasing decisions, investment decisions, and their effect on supply chain environmental performance will only increase environmental awareness and the ability of decision makers to reach balanced judgments and achieve sustainable choices.

This chapter addresses the importance of environmental awareness in firms and organizations, and the integration of environmental consciousness into supply chains as a necessity for sustainability and continuous improvement. This chapter will focus on the mathematical modeling of supply chains highlighting research gaps and providing insights to future work that remains to be done to achieve environmental responsibility.

SUPPLY CHAIN AND SUPPLY CHAIN MANAGEMENT

Supply chains are an important element of any business whether it is manufacturing products or providing a service. It is a direct result of differences and discrepancies between supply and demand throughout the different stages of a business. Assume the following scenario of a vendor who supplies products to a buyer: if the demand is more than what the supplier offers, shortages will occur and may lead to possible back-logs or lost sales. Conversely, if the supply available is more than the buyer's demand, excess inventory may incur additional costs. Expanding, there may be more than two parties involved: e.g. many suppliers, multiple products, numerous warehouses, various distributors, etc. The more parties in a supply chain, the more complex it is and the need for careful management is more evident. For a successful supply chain, it should be efficient and responsive. Furthermore, storage and material handling costs in supply chains can be as high as 50% of a product's indirect operating expenses (Rosenblatt, 1986). Consequently, reducing inventory related costs in a supply chain is a priority. The following will provide readers with some of the concepts and definitions.

Supply chains have been defined as "the alignment of firms that bring products or services to market" (Lambert et al., 1998), or as a chain that "consists of all stages involved, directly or indirectly, in fulfilling a customer request. The supply chain not only includes the manufacturer and suppliers, but also transporters, warehouses, retailers, and customers themselves . . ." (Chopra et al., 2003). It has also been defined as "a network of facilities and distribution options that performs the functions of procurement of materials, transformation of these materials into intermediate and finished products, and the distribution of these finished products to customers." (Ganeshan & Harrison, 1995). Though slightly different, all definitions share a network, active participants in this network, and a goal to bring a service or product to the end customer. It is not uncommon that participants have conflicting objectives, where each participant would like to maximize (minimize) its profit (cost). Accordingly, supply chain management becomes the means to manage these networks.

Mentzer et al. (2001) defines supply chain management as: "the systematic, strategic coordination of the traditional business functions and the tactics across these business functions within a particular company and across businesses within the supply chain, for the purpose of improving the long-term performance of the individual companies and the supply chain as a whole." From this definition, any participant within a supply chain must make individual and collective decisions concerning five functional areas: production, inventory, facility location, transportation, and information. For example, if the business strategy is to compete on the basis of price, the decisions made must optimize the supply chain for minimum cost. However, if the business strategy is to compete on customer service, the decisions must optimize for responsiveness. Examples of the decisions to be made could be what products to be produced, the quantities of each product produced and delivery times, how much inventory to be stored at each stage, the amount to be held as raw-materials, semi-finished, or finished goods, and/or the optimal inventory level and re-order points. Further questions include the location of production and inventory storage facilities, the means of transportation, what data should be collected, the information to be shared and to what degree of accuracy and timeliness.

Essentially, supply chain data synchronization, collaborative planning, forecasting, and replenishment become a necessity for success.

Optimizing a supply chain is the coordination of the decisions along the five functional areas. These coordination policies can be categorized as centralized or decentralized. Decentralized policies involve numerous decisions by each participant within the chain. This often leads to conflict of interest or conflict of individual objectives. However, centralized policies allow a single decision-maker (a team) to manage the supply chain as a whole to achieve an overall objective that is optimal for all participants within the supply chain. Centralized coordination is discussed extensively throughout the literature including various quantitative models. The bulk of this literature considers the economic order quantity model as the base foundation, and the main objective is to optimize the supply chain's combined costs or maximize the overall profit (Jaber & Zolfaghari, 2008; Glock, 2012). Typically when modeling supply chains, the associated costs considered are inventory holding costs, lost sales or shortage costs, order costs and possibly production setup costs if the vendor is a manufacturer.

Equally important to determining the optimal parameters that run the supply chain is the ability to measure its performance. Dr. Peter Drucker's (a notable management consultant) famous quote "what gets measured, gets managed" is often emphasized (Prusak, 2010). Resources, output and flexibility are three types of performance measure that can be used in measuring a supply chain's performance (Beamon, 1999). Generally, there are four measurement categories: customer service, internal efficiency, demand flexibility, and product development (Hugos, 2006). Furthermore, the Supply Chain Council's SCOR model suggests that a system be developed to present data at three levels of details: strategic, tactical (performance metrics), and operational (diagnostic metrics).

As evident, the complexities and depth of each topic in supply chains requires extensive research and investigation. Most of these topics have been economically driven, however there have been recent initiatives to look into both environmental and social aspects.

ENVIRONMENTAL AWARENESS IN SUPPLY CHAINS

According to a joint initiative by the Supply Chain and Logistics Association Canada, and the Retail Council of Canada (2009) in their report on green supply chain management, they state that green supply chain management is the incorporation of environmental thinking into supply chain management. They further state that this "includes introducing technical and innovative processes into materials sourcing and selection, delivery of the final product to consumers, and end-of-life product management." The wished-for result is the improvement of an organization's environmental impact while increasing its efficiency and progress within its supply chain. Examples of green supply chain management practices include (but are not limited to): energy efficiency, reduction of greenhouse gas emissions, water conservation or processing, waste reduction, reduced packaging or increased use of biodegradable packaging, product and packaging recycling, and green procurement practices. The above list can also be used to illustrate environmental benefits stemming from green supply chain management principles. Furthermore, the report also lists business benefits that can be achieved including: distribution efficiency, compliance, distribution cost, customer retention, and differentiate services.

Implementing green supply chain management is not without challenges and pitfalls. To maximize the benefits from productive green supply chain management practices, organizations should align their business targets with their environmental ones, tailor their own roadmaps or implementation plans and make the benefits from such an alignment clear to their supply chain partners. Moreover, an important aspect is the development of clear and appropriate metrics that are universally understood to measure performance.

Generally, the literature classifies green supply chain management into three main categories: the importance of green supply chain management, green design, and green operations (Srivastava, 2007). Green design is fundamentally based on either life cycle analysis/assessment (LCA) or on an environmental conscious design (ECD). Green operations, as presented in the review by Srivastava (2007), consists of three main subcategories: green manufacturing and remanufacturing, reverse logistics and network design, and waste management. Srivastava (2007) further illustrates a timeline of the literature exhibited under each of the categories. However, what is seen to be a necessity is the overlapping of these interdisciplinary topics from these subcategories to produce more effective research agendas resulting in applicable and relevant benefits academically and, more important, practically via the industry.

MODELS FOR GREEN SUPPLY CHAINS

The literature shows an effective supply chain is essential for the success of firms and organizations; however, as stakeholders become more environmentally responsible, the same responsibility is beginning to translate into supply chain management policies.

Lamming and Hampson (1996) investigate the issues present for purchasing and supply chain managers from an environmental perspective. They mention that, traditionally, environmental issues have not been a priority for supply chain decision makers. Their study suggests that supply chain management practices such as vendor assessment, total quality management, lean supply and collaborative supply strategies can benefit from life-cycle analysis, waste management and other environmental management approaches.

Inman (1999) review the impacts and effects on production planning and inventory control from both an academic and industrial perspective. It is evident that environmental concerns have more of an impact on business operations and represent a challenge to be addressed efficiently and effectively. Inman (1999) explores these implications with a focus on the areas of production planning, inventory control, and product distribution.

Angell and Klassen (1999) utilize a focus group of environmental and operations management researchers to generate a framework that helps identify opportunities for integrating environmental issues into the field of operations management. The objective of Walton et al. (1998) is to identify supply chain "environmentally-friendly practices" (EFP). Two topics become apparent: the significance of management's commitment to apply supply chain EFP, and the need to be more proactive rather than just comply with environmental regulations. As a result, supply chain management is evolving and to support this evolution, tools such as life cycle assessment (LCA) are integrated into supply chains (Hagelaar & van der Vorst, 2002). Darnall et al. (2008) examine whether business adopting environmental management systems would be more likely to utilize green supply chain management practices. Their results suggest that both complement each other and that the organizations that have an environmental management system in place are more likely to practice green supply chain management practices and have a higher probability of improving the environment.

Zhu and Sarkis (2004) investigate the relationship between green supply chain management practices and environmental and economic performance. They conclude that the application of green supply chain management practices was more likely to create win-win situations with regards to both environmental and economic performance of the enterprise. Furthermore, they show that the enterprises investigated who have an existing quality management program in place, along with the green practices, demonstrated superior performance. Extending their previous work, Zhu and Sarkis (2006) use an empirical study to investigate the relationship between green supply chain management operational practices and performance amongst early adopters in China. Other empirical studies are performed by Holt and Ghobadian (2009) in the United Kingdom where they indicate the most pressure is on manufacturers to implement green practices, by Jabbour and Jabbour (2009) where their focus was on green practices in supplier selection, and by Thun et al. (2009) for the automobile industry in Germany where they show implementation of green supply chain management yields higher performance. Their conclusion also coincides with the results presented from another empirical research performed by Green Jr et al. (2012).

Seuring and Muller (2008) look into stimulating further research in the field by offering a literature review on sustainable supply chain management from 1994 to 2007. They present a conceptual framework to summarize the research in this field and deduce two strategies for research moving forward: supplier management and risk performance, and supply chain management for sustainable products. They highlight that the research is dominated by green and environmental issues, and that the social issues studied are few and occasional. A "state-of-the-art" literature review was performed by Srivastava (2007) to identify major research work and provide a clear classification of green supply chain management. The purpose of the review was to classify the literature to identify research gaps, issues and potential areas for future research. Srivastava (2007) concludes that green supply chain management can reduce the ecological consequences of business activities without sacrificing performance, quality, reliability, or cost. Other literature reviews were performed by Luthra et al. (2009) to discuss green supply chain management issues, by Ilgin and Gupta (2010) where the review was more focused on issues pertaining to environmentally conscious manufacturing and product recovery, and by Mollenkopf et al. (2010) to investigate research and industry practice regarding simultaneous implementation of green, lean, and global supply chain strategies. Further categorization of green supply chain management and future research directions are presented in the review by Sarkis et al. (2011).

Sustainability is a multidisciplinary topic that has gathered momentum and most recently within economic, business and management fields (Linton et al., 2007). Linton et al. (2007) provide a background review to help understand the current concepts and trends of sustainability in operations management and to present the research opportunities and challenges associated with it. They show how supply chains and sustainability fuse together. As a result, the focus on environmental management and operations is shifted from a more local perspective to a more comprehensive one covering the entire supply chain. They suggest that "supply chains must be explicitly extended to include by-products of the supply chain, to consider the entire lifecycle of the product, and to optimize the product not only from a current cost standpoint but also a total cost standpoint." They further discuss that the total cost must reflect the effects of depleting resources, pollutants, wastes, etc. Linton et al. (2007) conclude that research into the implications of these concerns on the operations and supply chain is a binding necessity. Similar works concerning research challenges for sustainability are presented in the work of Garetti and Taisch (2011).

Examining the importance of inventory planning to the environment in detail is the focus of Bonney and Jaber (2011). They highlight the value of designing inventory systems that echo the needs of the environment and coin this term "responsible inventory systems". Notable in their research are the environmental inventory performance metrics suggested and the extension of a simple economic order quantity (EOQ) model to an "environmental-EOQ" model to account for the "true" cost of an activity, one that includes the environmental costs. True costs in a classical supply chain may be difficult to calculate (Jaber, 2009) which highlights that environmental costs may be of a more daunting task.

Clearly, several research frameworks have been presented, however the need to operationalize these concepts remain open for exploration. Table 1 lists research papers in chronological order and classifies the literature into (marked by "X"): (1) qualitative, (2) quantitative, (3) conceptual framework, (4) surveys, (5) case studies, (6) reviews, (7) research agendas, and (8) performance measures.

Table 1. Summary of the select literature reviewed pertaining to the issues of the environment and supply chain management

Research Article	Qualitative	Quantitative	Conceptual Framework	Modeling	Focus-Groups and/or Surveys	Case Study	Literature Review	Research Opportunities	Performance Measures
Lamming and Hampson (1996)	X		X		X		X	X	
McIntyre et al. (1998)		X	X			X		X	X
Walton et al. (1998)	X				X	X			
Angell and Klassen (1999)	X		X		X		X	X	
Beamon (1999)	X		X				X		X
Inman (1999)	X						X	X	
van Hoek (1999)	X		X					X	
Sarkis (2003)			X	X			X	X	
Zhu and Sarkis (2004)		X		X	X	X			X
Hervani et al. (2005)			X		X	X			X
Kainuma and Tawara (2006)		X		X		X			X
Vachon and Klassen (2006)		X			X				
Zhu and Sarkis (2006)		X			X				
Kumar (2007)		X		X					X
Linton et al. (2007)							X		
Srivastava (2007)							X	X	
Darnall et al. (2008)	X		X		X		X		
Marsillac (2008)	X		X				X	X	
Seuring and Müller (2008)							X	X	
Zhu et al. (2008)		X			X	X			X

continued on following page

Table 1. Continued

Research Article	Qualitative	Quantitative	Conceptual Framework	Modeling	Focus-Groups and/or Surveys	Case Study	Literature Review	Research Opportunities	Performance Measures
Holt and Ghobadian (2009)		X	X		X				
Jabbour and Jabbour (2009)	X					X			
Luthra et al. (2009)		X	X	X					
Thun et al. (2009)		X	X		X				
Benjaafar et al. (2010)		X		X			X		
Chaabane et al. (2010)		X		X					
El Saadany and Jaber (2010)		X		X					
Ilgin and Gupta (2010)	X						X		
Mollenkopf et al. (2010)		X	X						
Bonney and Jaber (2011)	X		X				X	X	
Bonney and Jaber (in-press)	X		X				X	X	
El Saadany et al. (2011)		X	X	X				X	X
Faruk et al. (2011)	X		X		X	X			
Garetti and Taisch (2011)			X				X		
Hua et al. (2011)		X		X					
Sarkis et al. (2011)		X							
Wahab et al. (2011)		X		X					
Battini et al. (2012)		X		X					
Green Jr et al. (2012)		X		X	X				
Jaber et al. (2013)		X		X					

Earlier works (from 1996 to 2003) listed in Table 1 focused on conceptual ideas and designs and suggested frameworks for future research opportunities. Later papers (from 2004 onwards) tried to quantify the concepts and frameworks develop in earlier works and introduce performance metrics to help assess and evaluate the environmental performance of supply chains. These studies were mainly based on case studies and focus groups. Subsequently, from 2009 onwards there have been considerable efforts to quantify the environmental performance of a supply chain, but these studies focused, only, on measuring carbon emissions (carbon footprints) and energy consumption. Research works considering a more holistic view of the supply chain and its environmental issues are more qualitative in nature and mathematically modeling their behavior remains a challenge.

REVERSE LOGISTICS: A FIRST STEP

There are numerous definitions presented for reverse logistics. Some of these define reverse logistics in a similar fashion to that of a traditional (forward) supply chain. If a traditional supply chain is one that involves raw material acquisition, manufacturing, and distribution of products to the customer, then reverse logistics is the reverse flow of products from the customer that involves the collection, inspection, disassembly, and distribution of used products to the point where recycling, remanufacturing, repair and recovery is possible with the purpose of recapturing value or appropriate disposal (Rogers & Tibben-Lembke, 1999; Dowlatshahi, 2000; Bei & Linyan, 2005).

Reverse logistics is not a newly introduced concept. The economic benefits of reusing products and materials have been previously applied (Fleischmann et al. 1997). Moreover, environmental concerns and newly introduced legislations have added motivation to the existing economic benefits leading to increase in the development of reverse logistics activities (Fleischmann, 2001). Some of these environmental concerns include scarce availability of landfill sites, declining raw material and energy resources, damages to the ozone, etc. These environmental concerns coupled with the environmental legislations allowed businesses and organization to extend developments in increasing product life, the ability to recycle more of the products and even reduce greenhouse-gas (GHG) emissions to the air (Bei & Linyan, 2005; Gülsün et al, 2006; Bonney & Jaber, 2011).

Van Hoek (1999) looks into reverse logistics as a research point that can be expanded to other research areas in green supply chain management. The perspective is to understand the impacts of some business practices on the environment and the entire supply chain. The focus of this study was to look at challenges of research on green supply chain management to lower the "ecological footprint" of supply chains. Of their findings was a categorization of green approaches in the supply chain research field. The impact of upstream and downstream integration of green practices along the supply chain is explored by Vachon and Klassen (2006). Reverse-logistics mathematical models such as El Saadany and Jaber (2010) do consider disposal and remanufacturing. Utilizing reverse logistics to manage returned products coincides with the goals of being environmentally responsible by reducing or minimizing environmental impacts. Reverse logistics and green supply chains overlap and have shared commonalities as explored by Marsillac (2008). Marsillac (2008) emphasises these shared processes and aims and concludes by suggesting that the integration of green supply chains and reverse logistics into a comprehensive system will improve and contribute beyond the individual successes of each separately. The dissimilarity between green supply chains and reverse logistics, it that the first focuses on the economic benefits stemming from the recovery activities whereas the latter shares these with environmental benefits (De Brito & Dekker, 2003; Bei & Linyan, 2005).

In a nutshell, reverse logistics can be considered a sub-set of closed-loop supply chains which in turn are a sub-set of green supply chains which are part of a more encompassing supply chain referred to as a sustainable supply chain. This is depicted in Figure 1.

More on green supply chains shall be discussed in the upcoming section. What is important is that there is growing concern regarding the management of product return flows. These growing concerns stem from both economic and environmental motives. Inventory models for reverse logistics should be extended to encompass both these economic and environmental aspects. The literature shows that there is plenty of expansion possible in this area (Bazan et al., 2016).

Figure 1. Reverse logistics, closed-loop, green and sustainable supply chains

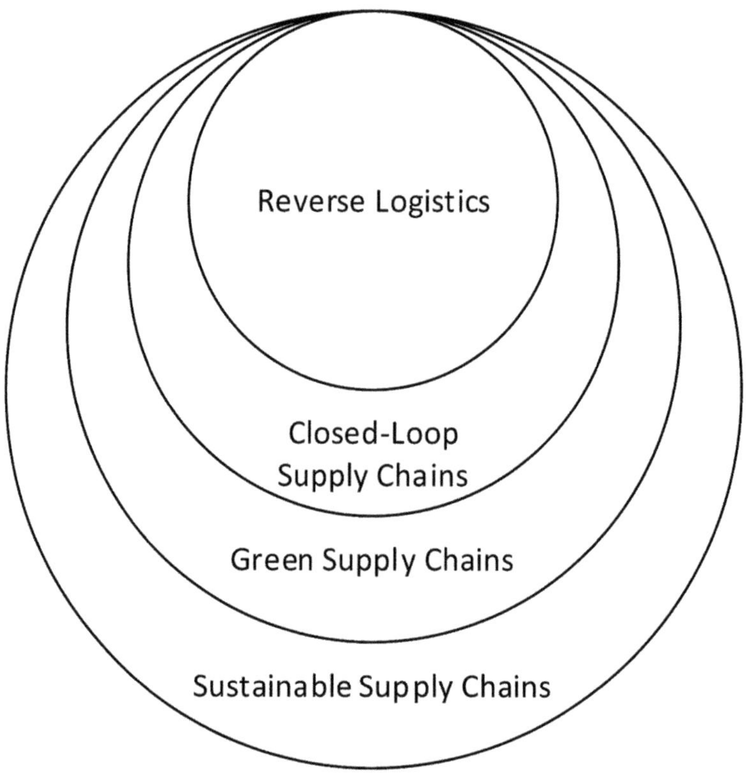

MODELING OF GREEN SUPPLY CHAINS

The necessity for incorporating environmental thinking into supply chain management is presented in Bonney and Jaber (2011). Numerous environmental problems stem from the production and transportation of goods across a supply chain. Examples include greenhouse-gas (carbon) emissions due to transportation and manufacturing processes, other air emissions, fuel consumed for transportation, depletion of natural resources and raw material, energy consumed for manufacturing, scrapping (solid waste) and biodegradability of products and packaging, water usage, chemical and toxic/hazardous waste, thermal pollution and noise. The world's resources are finite, whether it be material for manufacturing, or fuel or energy sources, or even clean air and a pristine ozone layer. The risk of having unsustainable practices along a supply chain can hinder the environment tremendously. The question remains, can inventory be planned to help mitigate and even better, help the environment. Storage locations, shipped batch sizes, the number of orders and the means of transportation are some questions that can be addressed when modeling supply chains and can have considerable impact on the environment. With the intention of operating supply chains to protect and sustain the environment, performance measures need be introduced that consider the inter-relationships between inventory and the environment. Generally, the environmental performance measures should push towards environmentally-sound activities and stray away decisions that are harmful to the environment (Bonney & Jaber, 2011). For example, activities such as reuse or remanufacturing are more desirable than recycling as less energy is involved and less waste generated.

The amalgamation of environmental interests into the organizational and inter-organizational practices of a supply chain including that of reverse logistics is how Sarkis et al. (2011) define green supply chain management. There are over twenty other definitions for green supply chain management as presented in Ahi and Searcy (2013). It is clear from Ahi and Searcy (2013) that there are overlaps in the definitions provided, but the common theme is the integration of the environmental concerns into the supply chain activities and processes with the purpose of saving resources and reducing both emissions and wastes. It is clear that green supply chain management is a requirement for a more comprehensive sustainable supply chain management approach. The 'greening' of activities have several benefits including saving natural resources, lowering energy costs, and avoiding unnecessary waste and pollution (Zhu & Sarkis, 2004; Ferretti et al., 2007).

An environmental decision making tool was constructed by McIntyre et al. (1998) for the integrated supply chain at Xerox Ltd. They show how they provide a measure of environmental performance for the whole supply chain as well as for the individual functional elements within the chain. Preliminary results of their work show that the working-life of a product is what causes the biggest environmental impact. Of significant importance in their research is the "environmental common denominator" approach that has been implemented. They assume that all processes or functions have three main environmental issues, namely: the amount of energy consumed, the materials used, and the pollutants emitted, that comprise the environmental common denominators. The integrated supply chain presented consists of the following seven functions: acquire, assemble, distribute, install, working-life, remove, and asset-recovery. In their opinion, their decision tool helps in devising a practical approach to achieve a win-win scenario where the "right-product" can be delivered to the "right-customer" at the "right-time" whilst minimizing the associated environmental impacts.

According to Beamon (1999), the traditional structure of supply chains ought to be extended to include mechanisms for product recovery and new associated performance measures. Their work discusses and investigates the environmental factors leading to the development of an extended environmental (green) supply chain. Beamon (1999) develops a general procedure towards achieving and maintaining a green supply chain.

Faruk et al. (2001) develop a management tool for analyzing, mapping, and managing environmental impacts along supply chains. In their research, Faruk et al. (2001) assess environmental impacts as either "guidance hierarchies" or "categories of emission stress". The guidance hierarchies are classified as: ecosystem disturbance, material types, energy sources, solid waste, shipping distance, and mode of transport. The categories of emission stress relate to emissions of air and water only and are classified into two classes where class 1 consists of global climate change, ozone depletion, and toxicity to water, and class 2 consists of acidification, nutrification, and photochemical smog formation. Faruk et al. (2001) consider the following stages: material acquisition, preproduction, production, use, distribution, and disposal. In the assessment matrix presented the guidance hierarchies and categories of emission stress are listed against these stages with regards to material inputs, energy use, solid waste, liquid emissions, atmospheric emissions, and intersite contexts.

Sarkis (2003) presents a decision framework to aid in managerial analysis and decision making in the area of green supply chain management. It suggests structuring and modeling the elements of a supply chain network and applying an analytical network process to the problem. The advantage of the technique presented is the ability to provide decision makers with the flexibility to identify and integrate the inter-dependencies present in a supply chain network, while considering the network's environmental and economic characteristics.

Hervani et al. (2005) integrate research in supply chain management, environmental management and performance management into a single framework that allows for evaluation and review based on the following categories: inputs, outputs, controls and tools. They present a list of selected metrics and measures of environmental performance, each of which has strategic, tactical and operational managerial implications. Even though the indicators are many, the predicament remains in which to use, how to measure it, and when to measure it.

Kainuma and Tawara (2006) propose a multi attribute utility theory method to assess supply chains that they consider to be a "lean and green supply chain" method. The objective is to extend the supply chain to include reuse and recycling throughout the life cycle of products and services in the chain.

Energy or fuel consumed due to transport and storage, carbon-dioxide emissions due to transport and storage, and financial cost of operating a supply chain (excluding the production steps) are three metrics used by Kumar (2007) when presenting a model to analyse a two-level supply chain. Kumar (2007) uses this model and studies the energy usage, emissions and cost for various industries. In general, the model shows opportunities for improving energy and emissions footprints of supply chains.

Zhu et al. (2008) looks into how to evaluate green supply chain management practices and their implementation amongst manufacturers in the Chinese industry. Their study was performed through an empirical investigation where the data collected is tested against two measurement models for green supply chain management. The first list of measurable factors includes internal environmental management, green purchasing, cooperation with customers, eco-design and investment recovery each of which may have several measurement items. The second list of measurement items is for performance outcomes; namely environmental and economic performance, again each of which is measured with several measures.

Benjaafar et al. (2010) use a simple model to illustrate how carbon emissions could be integrated into procurement, production and inventory management operational decisions. Their approach is to modify a traditional model to support decision making by associating carbon emissions parameters with the various decision variables. The main use of the model is to show the extent where carbon emissions can be reduced by addressing operational adjustments. Also focusing on carbon management strategies under the carbon emission trading mechanism is the works of Chaabane et al. (2010) and Hua et al. (2011). Fixed and variable emission costs are considered in a two-level supply chain model presented by Wahab et al. (2011). Jaber et al. (2013) also consider a two-level supply chain and present a model that jointly minimise the costs related to inventory and green-house-gas emissions costs when penalties for exceeding emissions limits are considered. They consider different emissions trading schemes.

Bonney and Jaber (2013) present an input-output activity matrix (IOAM) to analyze manufacturing and logistics systems and its performance. This IOAM illustrated in their work shows how each tier of a logistics chain may be analyzed to improve both its economic and environmental performance. Bonney and Jaber (2013) suggest that using the IOAM linking the economic and environmental aspects of a system can be a successful analysis. What they propose is a framework that has not yet been operationalized.

El Saadany et al. (2011) consider a coordinated two-level supply chain and seek to fill a gap in the literature by modeling the supply chain and developing an analytical decision model that can explore the performance when product, process, and environmental quality features are considered. Results show that investing to reduce environmental costs will improve environmental performance and increase profits. In the work of El Saadany et al. (2011) is a list of numerous quality environmental measures of a supply chain that are categorized under the following: product-based elements, manufacturing-based elements, product working life, operations-based elements, and finally green image and perceived quality.

Battini et al. (2014) explore the integration of economic and environmental objectives within a traditional EOQ model and propose a sustainable EOQ model. The approach requires a complete analysis from the beginning of the purchase order to the end of the product life at the buyer's facility. To identify all these environmental impacts arising during the life time of the purchase order a life cycle assessment (LCA) approach is applied. Essentially, Battini et al. (2014) consider the environmental inputs to be materials and energy, and the outputs to be air, water and solid emissions at each stage of the life cycle for the purchase order. They compute these inputs and outputs in order to apply them to the EOQ theory relating to transportation costs and external costs.

Principally, there are three thoughts raised. First, the notion that the majority of literature has been qualitative in nature, but nevertheless, present important concepts in implementing green supply chain management and suggesting future research streams. There are exceptions where quantitative research is shown but these are restricted to inventory models incorporating carbon emissions and/or energy consumption. Second, there is research that highlights the progression of reverse logistics to green supply chain management. However, again the mathematical models present for reverse logistics do not encompass all environmental factors associated with the impacts. Third, with the establishing of environmental performance measures there are several works where frameworks have been suggested to analyze supply chain systems, but such frameworks have not been operationalized. Furthermore, performance measures presented in the literature have not been applied in a supply chain context.

SUGGESTED RESEARCH DIRECTIONS

The question moving forward is what is there to be done to enhance the current inventory and supply chain models as they exist now to be more 'green', that is, to be more environmentally responsible? The answer to this question is seen to be the combination of three main (see Figure 2) advancements: system analysis, environmental performance measures, and supply chain modeling.

Each of these three are inter-related where each feed off the other advancements. The ability to analyze a current system and determine the scope of the study will allow for performance metrics to be developed accordingly. Also, advancements in supply chain modeling techniques will allow for more to be considered when studying a system as it can be modeled and taken into account. The remainder of this section will discuss each of these required advancements.

System Analysis

It can be argued that one of the most important steps is defining the system scope to be analyzed (Mumford, 1985). Defining the system scope and requirements allows specialists to then determine a set of outputs against which the performance can be measured. One must determine what the system objective is and consequently

Figure 2. The triangle of environmentally responsible supply chains

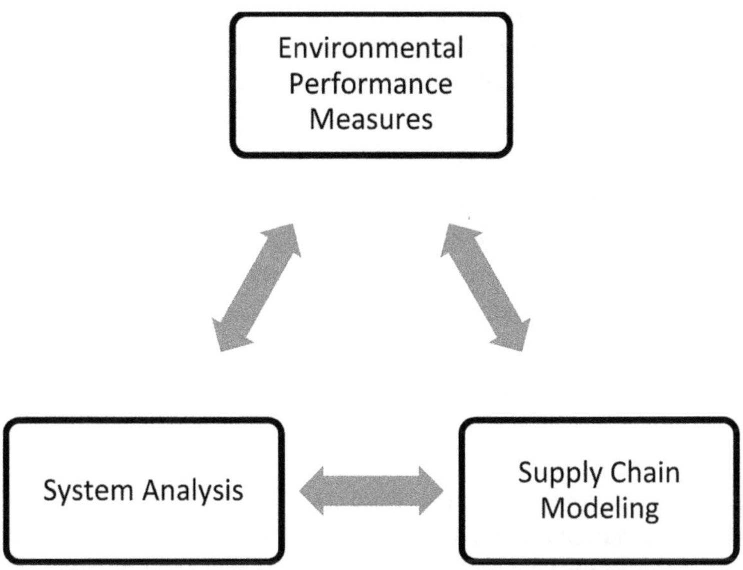

look at all implications, limitations, and constraints that may hinder achieving this system objective. Doing so will create a drive to better utilize given resources and enhance developments to increase and expand on opportunities.

Outlining the chain and scope of study involves in determining the participants to be considered. Is the study involved considering the extraction of raw materials, the manufacturing of components, sub-assemblies and assemblies, the distribution centers and warehouses involved, and/or the use of the product by the customer? Is it considering the energy involved for manufacturing, the fuel used during transportation, the energy used for storage, the greenhouse gases emitted from production and/or transportation, the disposal of products and their biodegradability, etc…? Each of these questions targets numerous factors to be considered. Moreover, it highlights the necessity to consider the type of industry and product manufactured involved. For example, if we are considering automobiles or photocopy machines, we may have leased products which allow for a continuous return flow of products that may be reused/resold after brief inspection and cleaning.

Let us assume a general manufacturing case, where a manufacturer acquires materials and components to manufacture and assemble a single product, which is later delivered to retailers before being sold or leased to the end customer for use. At the end of the lease term or if the product has reached its end life, the end customer may return the product to the retailer or just scrap the product. The returned product will be inspected and upon doing so decide if the product as a whole or certain components or assemblies may be reused, or those that require some remanufacturing (which could be done at the remanufacturing facility), and for the remaining components determine if some material may be recycled or if it is to be scrapped. Figure 3 depicts this assumed scenario with solid arrows showing movement of material and products forward and dotted lines showing the return flow of product from the end customer.

Figure 3 is only a simplified version of what actually happens in the real world. Performing a life-cycle assessment (LCA) for the product is considered a well-established procedure to obtain a better picture of what occurs from extracting the raw material to the use of the product to either its disposal or its return back to the manufacturing stages for possible repair. The LCA is considered a broad and comprehensive method that makes it best suitable for environmental analysis (Keoleian, 1993). With such level of detail it becomes clear that the LCA will form itself to each type of industry covering the various details that make that industry unique. Major studies consider the automotive industry, the chemical industry, the electronics industry, and the pulp and paper industry (National Research Council (US) Committee on Industrial Environmental Performance Metrics, 1999). This is clearly evident when we consider the associated environmental performance measures. The idea of integrating LCA into supply chains is one that is gathering mo-

Figure 3. An example for the forward and return flow of products, materials and components for a single manufactured product

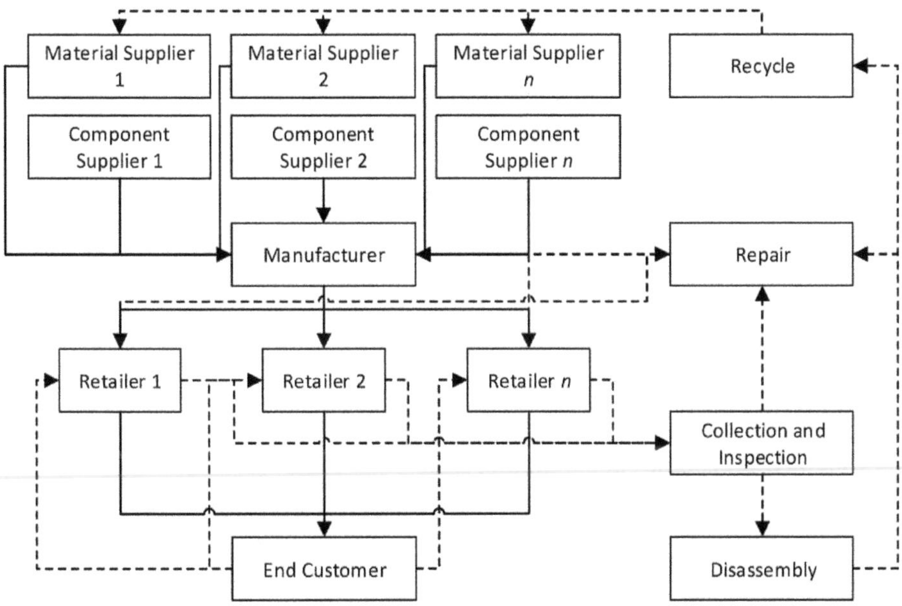

mentum showing great promise, however, guidelines for this integration are not as evident as the need. Some efforts to present primary points that can guide this process are presented in the work of Hagelaar and van der Vorst (2002). They also conclude in their work that if care is not taken to cater the study to a specific industry's chain, the implementation of the LCA and the realized results will not be as successful.

With the LCA being such a comprehensive tool that goes into considerable detail, the question becomes determining the limitations of the study. Are we including all participants depicted in Figure 3 or can we ignore some from the scope of the study? The boundaries of the supply chain considered needs to be set. For example, in Figure 3, the extraction of raw material and its transportation to warehouse and supply locations have been ignored. Setting boundaries is important especially in the early stages of modeling the environmental concerns in supply chains. Just as inventory modeling has evolved from the basic EOQ model, one must take a gradual approach to the integration of environmental concerns by considering a simple chain structure to understand and develop understanding of specific behaviors.

One of the tools that could facilitate the modeling of environmentally responsible supply chains is the framework proposed by Bonney and Jaber (2013). They suggest an input-output analysis approach that can be utilized to encompass the environmental factors associated with the supply chain. Such a task is daunting and requires efforts, but the input-output analysis provides a reasonable framework for a complicated task. Inputs including budgets, energy, materials, water, etc., are transformed throughout the chain to desired outputs, the products, and undesired outputs which include waste and other environmental impacts. The input-output approach seeks to link the required inputs with the desired outputs. This can help facilitate a more environmental conscious approach to supply chain modeling.

Figure 4 is a simplified version of Figure 3 with inputs and outputs from each stage. It is clear that outputs from certain stages could be inputs to the following. Some assumptions are made with Figure 4, such as having repaired items and recycled materials as good as new (which may not necessarily be the case for all industries/products), or that painting occurs only at the manufacturing facility constituting possible chemical waste not present at other facilities. What is clear in Figure 4 is that energy use, GHG emissions and waste disposal are present at almost every facility. It becomes apparent that their inclusion in the mathematical modeling of supply chains be manifested. Another evident observation is the amount of transportation required for the return flow of goods from the end customer. There is a massive space for improvement and coordination for the reverse flow that is behind in development from classical/forward supply chain coordination policies.

Numerous studies show a general hierarchy for recovery options of manufactured goods (Carter & Ellram, 1998; Steven, 2004; El Saadany, 2009). The first option is always the reduction of all possible resources required whether it be water, natural minerals and materials, fuel, etc. The second option involves prolonging the life of the product and its reusability. The third option is to perform minor repairs. The next options available include remanufacturing, refurbishing, recycling, disposal with energy recovery, and finally waste disposal in landfill sites. This hierarchy is summarized in Figure 5. From Figure 4 and the hierarchy presented in Figure 5, it becomes obvious that in order to best achieve environmental results, the energy used and the fuel required for transportation must be reduced to a minimum, and if possible be acquired from 'green' or renewable sources. The second focus is on increasing the lifetime of the product with the customer. The focus can then be shifted to having more items repaired as opposed to recycled, and finally minimize the waste disposal as much as possible.

The aforementioned example just briefly touches upon the environmental concerns in supply chains. One limitation in the example presented is that environmental impact of the use of the product with the end customer is neglected. For example, if this product was an automobile, GHG emissions and fuel consumption are major environmental issues. One must design for better fuel efficiency, alternative fuel and reduced or eliminated GHG emissions. A more detailed input-output analysis can lead to discussions on Design for Environment and the rest of the Design for X

Figure 4. An example input-output flowchart for a manufactured item with forward and return material flow

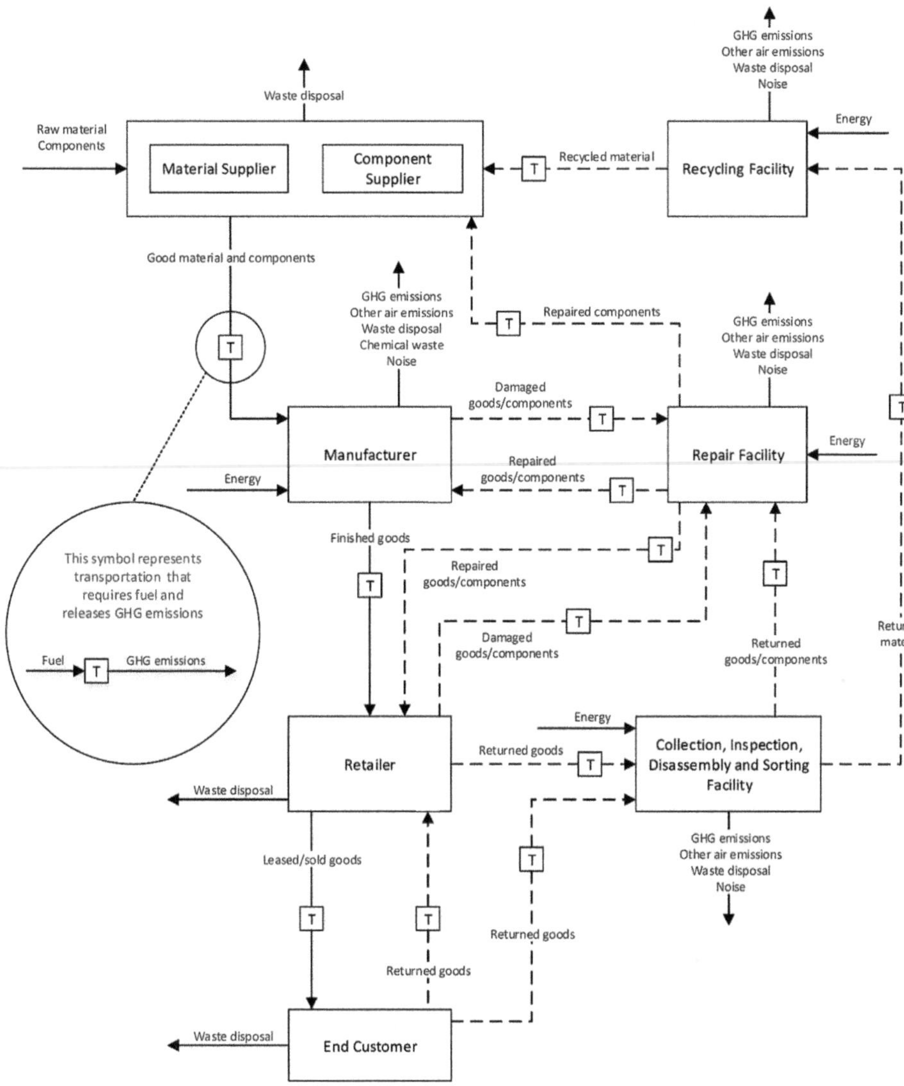

categories. Design for Environment covers material selection which can help extend product life and its biodegradability when disposing; Design for Manufacturing entails spending less time and energy for manufacturing, Design for Disassembly facilitates disassembly operations allowing minimal efforts for inspection, cleaning and repair operations (Kuo et al, 2001; Hauschild et al., 2004). The linking of these outputs to the inputs and the associated investment costs is vital to having an envi-

Figure 5. General hierarchy of recovery operations

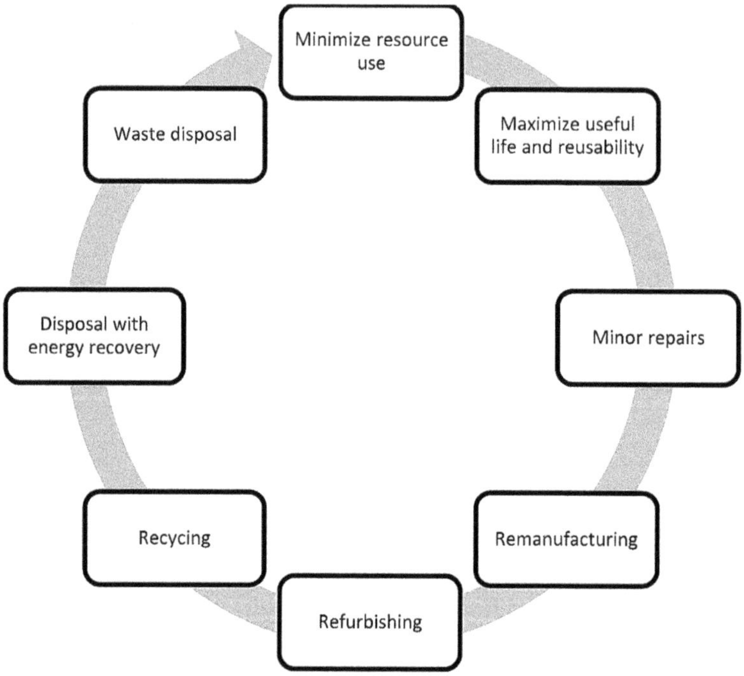

ronmental responsible model. There are huge challenges pertaining to cost accounting as they are not as sophisticated to capture the complex relationships presented.

Relationships between investment costs and the projected results may be implemented and their effect studied regarding inventory policies. The combination of both LCA and the framework suggested for the input-output analysis is seen as a strategic development that should be considered to tackle this problem. The questions remain: what complexity and level of detail is to be considered and at what cost?

Environmental Performance Measures

In order to effectively combine the efforts of LCA and the input-output analysis with the objective of modeling environmentally responsible supply chains, performance measures become mandatory. Dr. Peter Drucker's quote "what gets measured, gets managed" presented in Section 2 is significant and should resonate with those attempting to quantify the environmental impacts. Generally, the approach is to quantify the costs associated with all activities and then optimize by minimizing the total cost associated with the processes in the supply chain considered. However, as previously discussed, the cost of certain effects cannot necessarily be accurately estimated. The cost of GHG emissions is assumed by having a fixed cost for each ton emitted and can be found in the works of Jaber et al. (2013) and Zanoni et al. (2014). They further incorporate additional penalty costs GHG emission costs as outlined by the European Union. Energy costs can be obtained per kilowatt hour used, but how much energy is exactly used for the manufacturing or recovery operation? How much energy is used for storage operations? What are the costs for repair and remanufacturing which may vary from one returned item to another depending on the damage caused or its deterioration? Undoubtedly, the majority of the associated costs are estimates, judgements and proposed penalty costs enforced by legislative and governing bodies. The need for a more accurate performance measure other than just cost is necessary.

Just as presented in Section 7.1 each industry will have its own set of environmental performance measures, with some commonalities (National Research Council (US) Committee on Industrial Environmental Performance Metrics, 1999). The Global Reporting Initiative (GRI) has a set of guidelines for sustainability reporting which include providing indicators for environmental factors. However, these are merely guidelines and suggestions for businesses and organizations to consider and adapt to their specific industry (www.globalreporting.org). The work of El Saadany et al. (2011) discuss environmental performance measures for a supply chain. They use these performance measures to develop a model to investigate the performance of a supply chain considering financial, quality and environmental concerns. El Saadany et al. (2011) suggest an aggregated quality measure for all the performance measures listed in their work and assigning weights develop an environmental quality cost function. They seek to optimize a value for the aggregate quality measure that minimizes the overall system cost.

The main issue for performance measures is to establish measures that can be gathered timely, accurately and consistently so that appropriate and realistic targets can be achieved.

Supply Chain Modeling Implementation

The final stage is the combination of the results of the system analysis and the suggested performance measures. Modeling assumptions are required to facilitate the translation of the real-world to the mathematical model. Consequently, the assumptions in the developed model must first be validated and then the math verified. For example, a general assumption for many reverse logistics and closed-loop supply chain models is that reworked products are as good as new. This may be the case for some products, but for some instances, this may not be valid. A repaired engine will never be as efficient as a brand new one and consequently this assumption is no longer valid. The model must account for different quality products and multiple market demands for each product quality (e.g., new and used engines). A suggested top-level approach for developing and modeling is presented in Figure 6.

Generally, mathematical modeling of supply chains (based on the EOQ and joint economic lot size, JELS, settings) are a single objective function, mainly linear programming or mixed-integer programming problems (Jaber & Zolfaghari, 2008; Glock, 2012). The objective being either minimizing costs or maximizing profits. With the incorporation of environmental performance measures formulating a single-objective optimization problem may not be possible unless all measures have the same unit which is most likely not the case. Furthermore, there may be conflicting objectives; for example, reducing the amount of energy consumed may result in higher GHG emissions or excess production waste. The solution efforts must then be directed towards techniques for multi-objective optimization formulation. Such techniques are discussed in the works of (Fonseca & Fleming, 1995; Van Veldhuizen & Lamont, 2000; Jin et al., 2001). The overall goal is to treat both the environmental and financial concerns as the objective as opposed to having a financial objective with environmental constraints.

Figure 6. Suggested flowchart for modeling and optimizing environmentally responsible supply chains

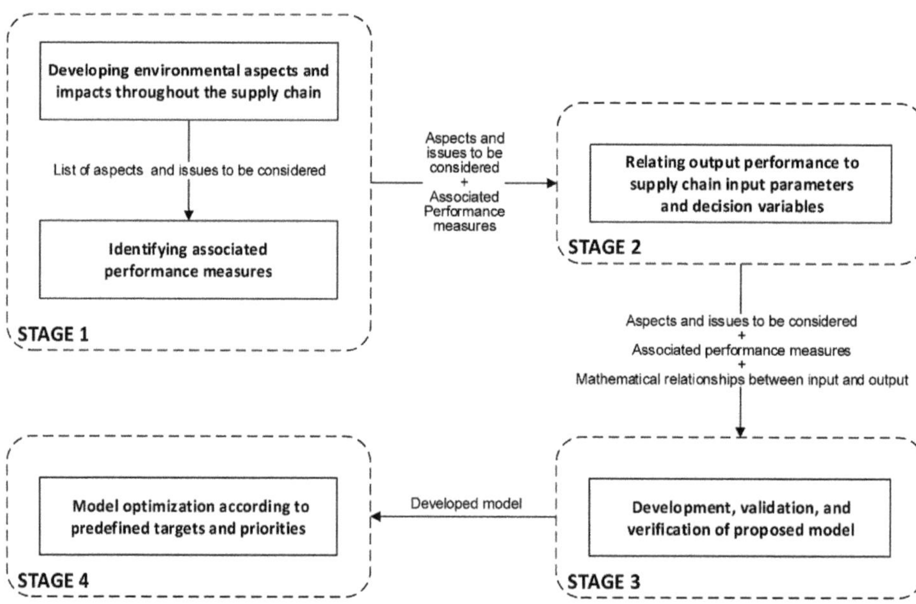

SUMMARY

The importance of modeling supply chain models that are environmentally responsible is an issue that requires immediate attention. Natural resources are finite and the impact of current industry supply chains are detrimental to the environment. This chapter summarizes the current efforts in the inventory modeling of 'green' supply chain and shows a growing need for supply chain models to expand beyond modeling GHG emissions.

Recommendations presented include addressing the need for appropriate environmental performance measures in a supply chain context and using these measures to expand on existing reverse logistic inventory models and closed-loop supply chain models covering both economic and environmental aspects. Some environmental issues of immediate concern are the amounts of energy used, material used (and disposed of), and air emissions released to the environment.

Concluding, the integration of LCA with input-output analyses are suggested for a more comprehensive analysis of supply chains with the intent of relating system inputs with desired and undesired output. The issue of environmental performance measures is in its infancy and require advancements in order to capture the true costs and effects of a supply chain on the environment. With current research at a relatively early stage, numerous challenges remain to quantify and measure environmental issues. The integration of environmental factors is pushing inventory and supply chain modeling towards multi-objective modelling techniques. It is evident that in order to achieve progress towards a more sustainable operation both researchers and practitioners alike need to collaborate and tackle all three phases of environmentally responsible supply chains: system analysis, performance metrics, and finally valid mathematical modeling of supply chains.

REFERENCES

Ahi, P., & Searcy, C. (2013). A comparative literature analysis of definitions for green and sustainable supply chain management. *Journal of Cleaner Production*, *52*, 329–341. doi:10.1016/j.jclepro.2013.02.018

Angell, L. C., & Klassen, R. D. (1999). Integrating environmental issues into the mainstream: An agenda for research in operations management. *Journal of Operations Management*, *17*(5), 575–598. doi:10.1016/S0272-6963(99)00006-6

Battini, D., Persona, A., & Sgarbossa, F. (2014). A sustainable EOQ model: Theoretical formulation and applications. *International Journal of Production Economics*, *149*, 145–153. doi:10.1016/j.ijpe.2013.06.026

Bazan, E., Jaber, M. Y., & Zanoni, S. (2016). A review of mathematical inventory models for reverse logistics and the future of its modeling: An environmental perspective. *Applied Mathematical Modelling, 40*(5), 4151–4178.

Beamon, B. M. (1999). Designing the green supply chain. *Logistics Information Management, 12*(4), 332–342. doi:10.1108/09576059910284159

Bei, W., & Linyan, S. (2005). A review of reverse logistics. *Applied Sciences, 7*, 16–29.

Benjaafar, S., Li, Y., & Daskin, M. (2010). Carbon footprint and the management of supply chains: Insights from simple models. *IEEE Transactions on Automation Science and Engineering, 10*(1), 99–116. doi:10.1109/TASE.2012.2203304

Bonney, M., & Jaber, M. Y. (2011). Environmentally responsible inventory models: Non-classical models for a non-classical era. *International Journal of Production Economics, 133*(1), 43–53. doi:10.1016/j.ijpe.2009.10.033

Bonney, M., & Jaber, M. Y. (2013). Developing an input–output activity matrix (IOAM) for environmental and economic analysis of manufacturing systems and logistics chains. *International Journal of Production Economics, 143*(2), 369–377. doi:10.1016/j.ijpe.2011.12.016

Carter, C. R., & Ellram, L. M. (1998). Reverse logistics: A review of the literature and framework for future investigation. *Journal of Business Logistics, 19*(1), 85–102.

Chaabane, A., Ramudhin, A., & Paquet, M. (2012). Design of sustainable supply chains under the emission trading scheme. *International Journal of Production Economics, 135*(1), 37–49. doi:10.1016/j.ijpe.2010.10.025

Chopra, S., & Peter, M. (2003). *'Supply chain* (2nd ed.). Upper Saddle River, NJ: Prentice-Hall.

Darnall, N., Jolley, G. J., & Handfield, R. (2008). Environmental management systems and green supply chain management: Complements for sustainability? *Business Strategy and the Environment, 17*(1), 30–45. doi:10.1002/bse.557

De Brito, M. P., & Dekker, R. (2003). A Framework for Reverse Logistics [R]. ERIM Report Series[-LIS]. *Research Management*, ERS-2003–ERS-2045.

Dowlatshahi, S. (2000). Developing a theory of reverse logistics. *Interfaces, 30*(3), 143–155. doi:10.1287/inte.30.3.143.11670

El Saadany, A. M. (2009). *Inventory management in reverse logistics with imperfect production, learning, lost sales, subassemblies, and price/quality considerations*. Toronto: Ryerson University.

El Saadany, A. M. A., & Jaber, M. Y. (2010). A production/remanufacturing inventory model with price and quality dependant return rate. *Computers & Industrial Engineering*, *58*(3), 352–362. doi:10.1016/j.cie.2009.01.017

El Saadany, A. M. A., Jaber, M. Y., & Bonney, M. (2011). Environmental performance measures for supply chains. *Management Research Review*, *34*(11), 1202–1221. doi:10.1108/01409171111178756

Faruk, A. C., Lamming, R. C., Cousins, P. D., & Bowen, F. E. (2001). Analyzing, mapping, and managing environmental impacts along supply chains. *Journal of Industrial Ecology*, *5*(2), 13–36. doi:10.1162/108819801528301l4

Ferretti, I., Zanoni, S., Zavanella, L., & Diana, A. (2007). Greening the aluminium supply chain. *International Journal of Production Economics*, *108*(1), 236–245. doi:10.1016/j.ijpe.2006.12.037

Fleischmann, M. (2001). *Reverse logistics network structures and design (No. ERS-2001-52-LIS)*. ERIM Report Series Research in Management.

Fleischmann, M., Bloemhof-Ruwaard, J. M., Dekker, R., Van der Laan, E., Van Nunen, J. A., & Van Wassenhove, L. N. (1997). Quantitative models for reverse logistics: A review. *European Journal of Operational Research*, *103*(1), 1–17. doi:10.1016/S0377-2217(97)00230-0

Fonseca, C. M., & Fleming, P. J. (1995). An overview of evolutionary algorithms in multiobjective optimization. *Evolutionary Computation*, *3*(1), 1–16. doi:10.1162/evco.1995.3.1.1

Ganeshan, R., & Harrison, T. (1995). *An introduction to supply chain management*. University Park, PA: Penn State University.

Garetti, M., & Taisch, M. (2012). Sustainable manufacturing: Trends and research challenges. *Production Planning and Control*, *2*(2-3), 83–104. doi:10.1080/09537287.2011.591619

Glock, C. H. (2012). The joint economic lot size problem: A review. *International Journal of Production Economics*, *135*(2), 671–686. doi:10.1016/j.ijpe.2011.10.026

Green, K. W. Jr, Zelbst, P. J., Meacham, J., & Bhadauria, V. S. (2012). 'Green supply chain management practices: Impact on performance', Supply Chain Management. *International Journal (Toronto, Ont.)*, *17*(30), 290–305.

Gülsün, B., Bıyık, G., & Özgen, D. (2006). *Reverse logistics: A survey*. 10th International Research/Expert Conference TMT 2006, Barcelona-Lloret de Mar, Spain.

Hagelaar, G. J. L. F., & van der Vorst, J. G. A. J. (2002). Environmental supply chain management: Using life cycle assessment to structure supply chains. *International Food and Agribusiness Management Review*, *4*(4), 399–412. doi:10.1016/S1096-7508(02)00068-X

Hauschild, M. Z., Jeswiet, J., & Alting, L. (2004). Design for environment—do we get the focus right? *CIRP Annals-Manufacturing Technology*, *53*(1), 1–4. doi:10.1016/S0007-8506(07)60631-3

Hervani, A. A., Helms, M. M., & Sarkis, J. (2005). 'Performance measurement for green supply chain management', Benchmarking. *International Journal (Toronto, Ont.)*, *12*(4), 330–353.

Holt, D., & Ghobadian, A. (2009). An empirical study of green supply chain management practices amongst UK manufacturers. *Journal of Manufacturing Technology Management*, *20*(7), 933–956. doi:10.1108/17410380910984212

Hua, G., Cheng, T. C. E., & Wang, S. (2011). Managing carbon footprints in inventory management. *International Journal of Production Economics*, *132*(2), 178–185. doi:10.1016/j.ijpe.2011.03.024

Hugos, M. H. (2006). *Essentials of supply chain management* (2nd ed.). John Wiley and Sons.

Ilgin, M. A., & Gupta, S. M. (2010). Environmentally conscious manufacturing and product recovery (ECMPRO): A review of the state of the art. *Journal of Environmental Management*, *91*(3), 563–591. doi:10.1016/j.jenvman.2009.09.037 PMID:19853369

Inman, R. A. (1999). Environmental management: New challenges for production and inventory managers. *Production and Inventory Management Journal*, *40*(3), 46–49.

Jabbour, A. B. L., & Jabbour, C. J. (2009). Are supplier selection criteria going green? Case studies of companies in Brazil. *Industrial Management & Data Systems*, *109*(4), 477–495. doi:10.1108/02635570910948623

Jaber, M. Y. (Ed.). (2009). *Inventory management: non-classical views*. CRC Press. doi:10.1201/9781420079982

Jaber, M. Y., Glock, C. H., & El Saadany, A. M. A. (2013). Supply chain coordination with emissions reduction incentives. *International Journal of Production Research, 51*(1), 69–82. doi:10.1080/00207543.2011.651656

Jaber, M.Y., & Zolfaghari, S. (2008). Quantitative models for centralised supply chain coordination. In *Supply Chain Theory and Applications*. Vienna, Austria: I-Tech Education and Publishing.

Jin, Y., Olhofer, M., & Sendhoff, B. (2001). *Dynamic weighted aggregation for evolutionary multi-objective optimization: Why does it work and how?*. Academic Press.

Kainuma, Y., & Tawara, N. (2006). A multiple attribute utility theory approach to lean and green supply chain management. *International Journal of Production Economics, 101*(1), 99–108. doi:10.1016/j.ijpe.2005.05.010

Keoleian, G.A. (1993). The application of life cycle assessment to design. *Environmental Life Cycle Assessment and its Applications, 1*(3-4), 143 – 149.

Kumar, V. (2007). Analyzing and optimizing the environmental performance of supply chains. In *Proceedings of ACEEE Summer Study on Energy Efficiency in Industry T*. Surya Technologies, Inc. Retrieved from www.suryatech.com

Kuo, T. C., Huang, S. H., & Zhang, H. C. (2001). Design for manufacture and design for 'X': Concepts, applications, and perspectives. *Computers & Industrial Engineering, 41*(3), 241–260. doi:10.1016/S0360-8352(01)00045-6

Lambert, D. M., Stock, J. R., & Ellram, L. M. (1998). *Fundamentals of logistics management*. Boston, MA: Irwin/McGraw-Hill.

Lamming, R., & Hampson, J. (1996). The environment as a supply chain management issue. *British Journal of Management, 7*(s1), s45–s62. doi:10.1111/j.1467-8551.1996.tb00147.x

Linton, J. D., Klassen, R., & Jayaraman, V. (2007). Sustainable supply chains: An introduction. *Journal of Operations Management, 25*(6), 1075–1082. doi:10.1016/j.jom.2007.01.012

Luthra, S., Kumar, V., Kumar, S., & Haleem, A. (2009). 'Green supply chain management issues: A literature review approach', Journal of Information. *Knowledge and Research in Mechanical Engineering, 1*(1), 12–20.

Marsillac, E. L. (2008). Environmental impacts on reverse logistics and green supply chains: Similarities and integration. *International Journal of Logistics Systems and Management*, *4*(2), 422.

McIntyre, K., Smith, H., Henham, A., & Pretlove, J. (1998). 'Environmental performance indicators for integrated supply chains: The case of Xerox Ltd', Supply Chain Management. *International Journal (Toronto, Ont.)*, *3*(3), 149–156.

Mentzer, J. T., DeWitt, W., Keebler, J. S., Min, S., Nix, N. W., Smith, C. D., & Zacharia, Z. G. (2001). Defining Supply Chain Management. *Journal of Business Logistics*, *22*(2), 1–25. doi:10.1002/j.2158-1592.2001.tb00001.x

Mollenkopf, D., Stolze, H., Tate, W. L., & Ueltschy, M. (2010). Green, lean, and global supply chains. *International Journal of Physical Distribution & Logistics Management*, *40*(1), 14–41.

Mumford, E. (1985). Defining system requirements to meet business needs: A case study example. *The Computer Journal*, *28*(2), 97–104. doi:10.1093/comjnl/28.2.97

National Research Council (US). (1999). Committee on Industrial Environmental Performance Metrics. Industrial Environmental Performance Metrics: Challenges and Opportunities. National Academy Press.

Prusak, L. (2010). What Can't Be Measured. *Harvard Business Review Blog Network*. Retrieved 16 January 2013 from: http://blogs.hbr.org/davenport/2010/10/what_cant_be_measured.html

Richards, D. J. (1997). *The industrial green game: Implications for environmental design and management*. National Academy of Engineering.

Rogers, D. S., & Tibben-Lembke, R. S. (1999). *Going backwards: reverse logistics trends and practices* (Vol. 2). Pittsburgh, PA: Reverse Logistics Executive Council.

Rosenblatt, M. J. (1986). The dynamics of plant layout. *Management Science*, *32*(1), 76–86. doi:10.1287/mnsc.32.1.76

Sarkis, J. (2003). A strategic decision framework for green supply chain management. *Journal of Cleaner Production*, *11*(4), 397–409. doi:10.1016/S0959-6526(02)00062-8

Sarkis, J., Zhu, Q., & Lai, K. H. (2011). An organizational theoretic review of green supply chain management literature. *International Journal of Production Economics*, *130*(1), 1–15. doi:10.1016/j.ijpe.2010.11.010

Seuring, S., & Müller, M. (2008). From a literature review to a conceptual framework for sustainable supply chain management. *Journal of Cleaner Production, 16*(15), 1699–1710. doi:10.1016/j.jclepro.2008.04.020

Srivastava, S. K. (2007). Green supply-chain management: A state-of-the-art literature review. *International Journal of Management Reviews, 9*(1), 53–80. doi:10.1111/j.1468-2370.2007.00202.x

Steven, M. (2004). Networks in reverse logistics. In H. Dyckhoff, R. Lackes, & J. Reese (Eds.), Supply Chain Management and Reverse Logistics (pp. 163–180). Berlin: Springer-Verlag. doi:10.1007/978-3-540-24815-6_8

Supply Chain and Logistics Association Canada, Retail Council of Canada. (2009). *Green Supply Chain Management: Retail Chains and Consumer Product Goods – A Canadian Perspective*. Industry Canada.

Thun, J. H., & Müller, A. (2010). An empirical analysis of green supply chain management in the German automotive industry. *Business Strategy and the Environment, 19*(2), 119–132.

Vachon, S., & Klassen, R. D. (2006). Extending green practices across the supply chain: The impact of upstream and downstream integration. *International Journal of Operations & Production Management, 26*(7), 795–821. doi:10.1108/01443570610672248

Van Hoek, R. I. (1999). From reversed logistics to green supply chains. *Supply Chain Management International Journal (Toronto, Ont.), 4*(3), 129–135.

Van Veldhuizen, D. A., & Lamont, G. B. (2000). Multiobjective evolutionary algorithms: Analyzing the state-of-the-art. *Evolutionary Computation, 8*(2), 125–147. doi:10.1162/106365600568158 PMID:10843518

Wahab, M. I. M., Mamun, S. M. H., & Ongkunaruk, P. (2011). EOQ models for a coordinated two-level international supply chain considering imperfect items and environmental impact. *International Journal of Production Economics, 134*(1), 151–158. doi:10.1016/j.ijpe.2011.06.008

Walton, S. V., Handfield, R. B., & Melnyk, S. A. (1998). The green supply chain: Integrating suppliers into environmental management processes. *International Journal of Purchasing and Materials Management, 34*(2), 2–11. doi:10.1111/j.1745-493X.1998.tb00042.x

Zanoni, S., Mazzoldi, L., & Jaber, M. Y. (2014). Vendor-managed inventory with consignment stock agreement for single vendor–single buyer under the emission-trading scheme. *International Journal of Production Research*, *52*(1), 20–31. doi: 10.1080/00207543.2013.812812

Zhu, Q., & Sarkis, J. (2004). Relationships between operational practices and performance among early adopters of green supply chain management practices in Chinese manufacturing enterprises. *Journal of Operations Management*, *22*(3), 265–289. doi:10.1016/j.jom.2004.01.005

Zhu, Q., & Sarkis, J. (2006). An inter-sectoral comparison of green supply chain management in China: Drivers and practices. *Journal of Cleaner Production*, *14*(5), 472–486. doi:10.1016/j.jclepro.2005.01.003

Zhu, Q., Sarkis, J., & Lai, K. H. (2008). Confirmation of a measurement model for green supply chain management practices implementation. *International Journal of Production Economics*, *111*(2), 261–273. doi:10.1016/j.ijpe.2006.11.029

Chapter 4
Assessing the Green Supply Chain Management for the United Arab Emirates Construction Industry

Sreejith Balasubramanian
University of Wollongong – Dubai, UAE

Balan Sundarakani
University of Wollongong – Dubai, UAE

ABSTRACT

The United Arab Emirates (UAE) construction industry has been witnessing an unprecedented growth in the last two decades. As a result, the sector is facing a major challenge of reducing the carbon footprint and thus creating major concern for the governments and the environmental agencies in the UAE. The chapter discusses the role of Green Supply Chain Management (GSCM) in achieving sustainability in the UAE construction industry. The various stages of GSCM applicable to the construction industry are critically assessed along with green performance measures in achieving environmental, economic and operational performance. Finally, the chapter provides a list of recommendations that could be used by practitioners and policy makers in implementing and measuring sustainability practices in the construction industry.

DOI: 10.4018/978-1-5225-0635-5.ch004

INTRODUCTION

Environmental pollution and climate change are the main challenges faced by mankind in the 21st century. Increasing carbon emissions from the various stationary and non-stationary facilities are the main cause of global warming and rise in sea level. Hence the carbon footprint quantification, analysis and reduction are considered as the key to identification and mitigation of emissions in every industry. One of the main industries from a sustainability perspective is the construction sector. According to United Nations Environmental Program (UNEP, 2007) report, the construction sector is responsible for 30% of the world greenhouse gas (GHG) emissions. The increased growth in construction industry has led to increased GHG emissions due to higher energy consumption, embodied emission from the production of steel, cement etc., and from the combustion of petrol and diesel from related transportation from construction activities. In addition, construction industry waste makes a significant percentage of the total waste generated in many cities around the world (Li and Zhang, 2012). Hence, there is an urgent need to reduce the environmental impacts of the sector, especially in developing economies like United Arab Emirates (UAE), where the construction is the main contributor of environmental emissions.

Background to the United Arab Emirates (UAE) Construction Sector

During the last decade, UAE has witnessed an unprecedented construction boom, in which the sector on average had grown 10-12 precent, and is the major contributor of environmental pollution. As of 2010, the total construction projects in UAE had an estimated worth of $958bn and is expected to grow by 9.6% annually between 2010 and 2014 (Deloitte Middle East, 2010). According to the 'UAE construction Industry Outlook (2012)' report, UAE accounted for nearly 20.3% of total construction industry in the Gulf Cooperation Council (GCC) having outpaced Saudi Arabia in 2008 in becoming the region's largest construction market. Even during the global financial recession (2009-2012), the sector managed to record strong growth and contributed approximately 8% to the country's GDP in 2009 to over 14% in 2012. Unfortunately, this unprecedented construction boom had largely contributed to UAE becoming one of the top countries in terms of per capita carbon emissions. UAE was ranked 4th, 6th and 1st place in terms of per capita carbon emission in the year 2006, 2008 and 2010 respectively (World Bank Report, 2012), with per capita carbon emission estimated at 22.6 metric tons in 2010. As a result, there is increasing pressure on UAE by global environmental bodies to cut down carbon emissions and per capita energy spending.

Green Initiatives in the United Arab Emirates (UAE) Construction Sector

Due to growing international pressure, UAE government has taken number of green initiatives to promote sustainability in the build environment. For e.g., Abu Dhabi Urban planning Council in 2010 introduced 'Pearl Rating System' a framework for sustainable design and construction in which consideration and documentation of sustainability aspects of new buildings is mandatory. Dubai Electricity and Water Authority (DEWA, 2011) announced the green building guidelines which covers aspects such as site selection, usage of construction materials, waste management and indoor environment quality. The World Energy Forum 2012 and World Future Energy Summit 2013 were hosted by UAE in a bid to create awareness and promote sustainable living in the UAE. These are already some evidences of the green projects in the UAE construction sector. For example, Masdar City is the world's first planned carbon neutral city located in the city of Abu Dhabi. However, these ad-hoc government initiatives and projects alone are not sufficient to green the UAE construction sector. To green the UAE construction sector, the sector needs to be looked at from a holistic perspective. The concept of GSCM, which offers a holistic approach, can be applied for greening the UAE construction sector.

Evolution of Green Supply Chain Management (GSCM)

From a supply chain standpoint of view to ensure future generations will have resources, we need to look into the supply chain and understand its implications on the environment. The environmental impacts of supply chain process have been an issue for many years. Traditionally supply chain management focused primarily on time, quality and cost. Supply chain management (SCM) has been viewed as process where raw materials are converted into final products, then delivered to the end customer (Beamon, 1999). However, in the mid 1990's as SCM evolved, practices such as just-in-time (JIT), lean supply chain, total quality management (TQM), are adopted by organizations to enhance the operational efficiency of the supply chain. However, increasing concern of environmental issues has led to number of independent studies on eco-friendly design (Chandra, 1991), life cycle analysis (Penev, 1995), waste management (Roy & Whelan, 1992), green manufacturing (Van & Salomon, 1997; Guide & Srivastava, 1998), green transportation (Gendreau & Dejax, 1993) and reverse logistics (Kelly & Silver, 1989; Pohlen & Farris, 1992) and together played a pivotal role in the evolution of GSCM, to counter the environmental impacts of supply chain processes in an organization. Having said, GSCM is not the only term used by researchers and practitioners. A review of recent literature in sustainability shows that supply chain environmental management (SCEM), sustainable supply chain management (SSCM), environmental supply chain management (ESCM) are used and often interchangeably. However, a closer investigation revealed that related concepts such as SSCM and ESCM are broader and incorporates the social dimension of sustainability, while GSCM primarily focuses on the environmental dimension along with operational and economic dimensions. The scope of the book chapter is narrowed to GSCM to focus on the environmental issues of construction. According to (Srivastava, 2007), GSCM can be defined as ''Integrating environmental thinking into supply chain management, including product design, material sourcing and selection, manufacturing processes, delivery of the final product to the consumers as well as end-of-life management of the product after its useful life'' (p.12)

Importance of Green Supply Chain Management (GSCM) in the United Arab Emirates (UAE) Construction Sector

The environmental considerations are under looked and often neglected by firms in the UAE construction sector. A typical UAE construction supply chain is shown in Figure 1. Typically the construction project is awarded to a main contractor by the client/developer. Habitually, neither the main contractor nor the suppliers are involved in the planning or design stage of the construction supply chain. Developer hires the consultant and architect for the master planning and design and then hires the main contractor who will be responsible for the entire construction and commissioning of the building. The potential of GSCM in the UAE construction industry is enormous. An end to end GSCM can reduce emission across all stages

Figure 1. A typical UAE construction supply network

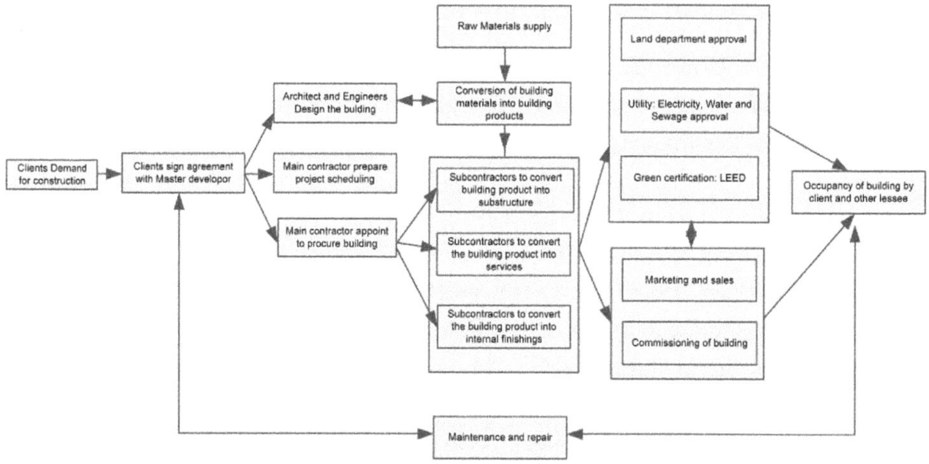

of the supply chain. Support of all the stakeholders is required to implement GSCM. For instance, developer requirement for green buildings will be conceptualized by the architect in the form of green building design. The requirements of the green building design is tendered and awarded to the main contractor, who takes the responsibility to implement the green project. The main contractor should possess the sufficient knowledge of GSCM to implement as per tender specification. Moreover, the other activities undertaken by the main contractors such as project scheduling can forecast the operational improvement resulting from implementing GSCM. However, the integration of GSCM into the construction sector is a complex undertaking. This chapter attempts to integrate the various aspects of GSCM in the UAE construction sector.

Methodology

Due to the limited number studies in green supply chain management (GSCM) in construction, a systematic evaluation of existing literature GSCM from other sectors which has witnessed considerable research in GSCM over the last decade was executed to synthesise the various GSCM aspects relevant to the United Arab Emirates (UAE) construction industry. A comprehensive understanding of the UAE construction sector is achieved through industry and government reports, published company annual reports and newsletters along with number of discussions with industry experts. The findings from the literature collated the various aspects of GSCM that is necessary for the implementation of GSCM for the UAE construction sector. The various GSCM aspects identified include factors that influence construction firms to implement GSCM practices (drivers), the various green supply chain stages for greening the construction sector, environmental, economic and operational benefits of GSCM and challenges faced by the sector in implementing GSCM. The rest of chapter discusses the various aspects of GSCM in detail.

POTENTIAL DRIVING FACTORS TO THE ADOPTION OF GREEN SUPPLY CHAIN MANAGEMENT (GSCM) IN THE UNITED ARAB EMIRATES (UAE)

Driving factors of GSCM adoption for firms include both internal and external factors.

Internal Drivers

Internal drivers are forces that arise within the organization. The various internal drivers in the adoption of GSCM are as follows;

Commitment to Improve Environmental Performance

Environmental performance is considered as a major agenda in the company policy Similar to corporate social responsibility, corporate environmental responsibility is considered by firms to reduce the adverse environmental impacts of their activities (Simpson et al., 2007). However, the extend of commitment depends on the environmental values of the founder or owner (Wycherley, 1999) and support and involvement of top management (Min & Galle, 2001).

Desire to Achieve Economic Benefits

Commitment to GSCM may not be driven by environmental responsibility or regulatory compliance. Increasing number for firms has identified it as a potential management strategy to achieve economic benefits. Previous studies have identified that green practices can lead to cost reduction and savings (Dhanda & Hill, 2005; Fassoula, 2005; Richey et al., 2005).

Improve Corporate Brand Image

Adopting GSCM can improve the corporate brand image resulting in increased market share by winning new projects (Francesco & Fabio, 2010). Recent studies show that firms are using GSCM as a springboard to reach environmental conscious customers (Faisal, 2012). Wycherley (1999) identified the benefits of improved corporate brand image in the case of Body Shop International, one of the early adopters of GSCM.

To Achieve Competitive Advantage

A proactive GSCM strategy can be used by firms to differentiate themselves from their competitors (Rao & Holt 2005). Competitive advantage can be in the form of cost reductions, preferential treatment and subsidies from government and for securing stakeholder approval in environmental projects (Faisal, 2012). To extend of competitive advantage will be more in the region where GSCM is still at infancy stage (Francesco & Fabio, 2010).

Improving Supply Chain Efficiency

In additional to improvement in cost reduction and brand image, organizations are prompted to implement GSCM practices to improve supply chain efficiency. Previous studies such as Lippman (2001) and Lee (2008) identified that adoption of GSCM practices leads to reduction in cycle time and cost, and improved quality. In addition, it can lead to decreased inventory level and improved capacity utilization (Zhu et al., 2005; Zhu et al., 2008).

Table 1 summarizes the various internal drivers of GSCM and its relevance to the United Arab Emirates (UAE) construction sector.

External Drivers

External drivers are external pressures faced by the firms to implement GSCM. The various external drivers in the adoption of GSCM are as follows;

Government Regulation and Legislation

In GSCM, the role of government and legislation plays a vital role. This is considered as the mother of all drivers and leaves no option for a firm but to comply with it or exit the market. Numerous studies have identified government regulation and legislation as the main driver of GSCM (Beamon, 1999; Hall, 2001; Min & Galle, 2001; Zhu & Sarkis, 2006; Walker et al., 2008; Schneiderman; 2009),

Client Pressure

Environmental commitment of client/developer can drive firms to implement GSCM practices. For instance in the construction sector, environmental commitment of the developer can influence contractors, which in turn can influence sub-contractors and suppliers. Hence the pressure is typically exerted from top to bottom in the supply

Table 1. Internal drivers of GSCM relevant to UAE construction sector

Drivers of GSCM	Implications to UAE Construction Sector
1. Commitment to improve environmental performance	Incorporating sustainability strategy as a part of organizational strategy
2. Achieve economic benefits	Substantial opportunities to reduce construction costs through reduction in energy, water, materials and through waste management and recycling
3. Improve corporate brand image and hence winning new contacts.	Improve corporate brand image and hence attract foreign investments and buyers.
4. Proactive approach to achieve competitive advantage	Differentiation for companies in the highly competitive UAE construction market
5. Improving supply chain efficiency.	Improve construction productivity, reduce project time, improve quality in the fast paced UAE construction sector

chain (Walton et al., 1998; Handfield et al., 1997; Hall, 2001). Similarly, studies have shown that large clients can influence small and medium stakeholders in the supply chain to implement GSCM practices (Chen, 2010; Lee, 2008).

Consumer Pressure

Consumer behaviour studies in 1990's found that consumers are increasingly aware of the environmental impact of the products they consume (Roberts, 1996). The individual consumer's belief that they can contribute to environmental causes have led to environmental purchases (Berry & Rondinelli, 1998; Buckholz, 1991). Berns et al. (2009) found that consumer concern about sustainability is a significant driver for firms to adopt GSCM. Lo & Leung (2000) identified the increasing environmental awareness of young customers.

Table 2 summarizes the various external drivers of GSCM and its relevance to the United Arab Emirates (UAE) construction sector.

PROPOSED GREEN SUPPLY CHAIN MANGEMENT (GSCM) STAGES APPLICABLE TO THE UNITED ARAB EMIRATES (UAE) CONSTRUCTION INDUSTRY

Reduction in carbon emission can be achieved from all functional stages of the construction supply chain such as design, purchasing and construction. Therefore for greening the construction sector, it is necessary to green the various phases of the integrated supply chain. The proposed stages identified from the literature include green design, green purchasing, green distribution and green construction (offsite and onsite) and waste management. However, extend of implementation of these supply chain phases depends on the internal and external driving forces as explained in detail in the previous section. Depending on the influence of these drivers, GSCM strategy is adopted organizationally and is reflected in the firm's vision and mission. However, prior to implementation of supply chain practices

Table 2. External drivers of GSCM relevant to UAE construction Industry

Drivers of GSCM	Implications to the UAE Construction Industry
1. Government regulation and legislation	Compliance to green design guidelines and regulation
2. Client pressure	To meet the environmental requirements in the tender documentation
3. Consumer Pressure	Customer desire for green buildings

such as green design and green construction, firms have to ensure they have the necessary internal environment such as environmental management systems (EMS) and ISO 14001 implementation to efficiently implement supply chain practices. Organizations also have to conduct cost benefit analysis on the benefits and costs associated with the implementation of GSCM practices. Performance measurement post implementation will enable organizations to re-assess the cost-benefit analysis for continuous improvement.

Figure 2 shows the designed green supply chain applicable to the United Arab Emirates (UAE) construction sector. Figure 2 is developed by integrating the various GSCM stages indentified from the literature to the UAE construction supply chain as identified in Figure 1.

The various GSCM practices are explained in detail in the following section.

Internal Environmental Management (IEM)

IEM is the practice of developing green supply chain management as a strategic organizational imperative through commitment and support from senior and mid-level managers (Zhu et al., 2008). IEM is a prerequisite for all the GSCM practices and should be part of the overall strategy for the successful implementation of GSCM practices (Murray, 2000). Earlier studies have also shown that management commitment to environmental strategy is necessary to implement GSCM (Byrne & Deeb, 1993; Klassen & McLaughlin, 1993). However, it was Zhu et al., (2008) who defined IEM and validated the factors with IEM.

Figure 2. Green supply chain stages applicable to the UAE construction sector

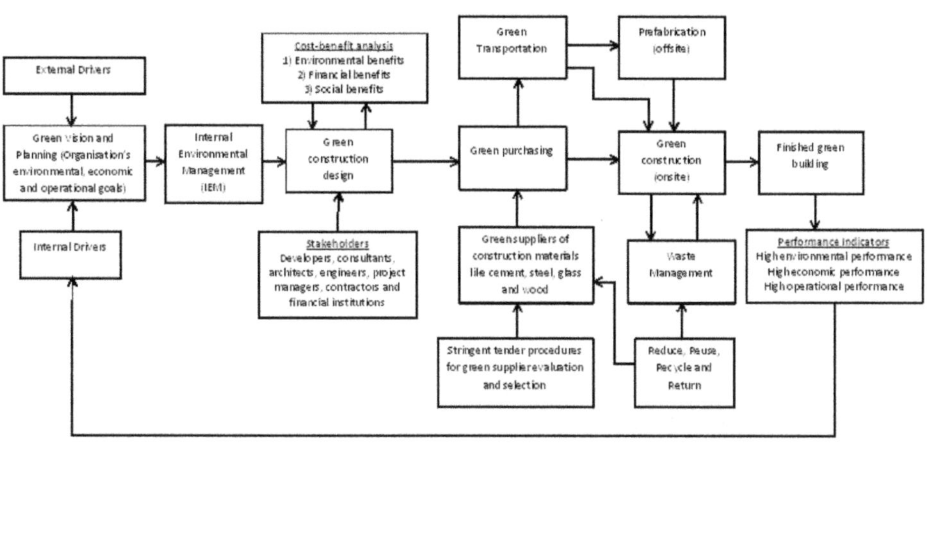

The identified factors for IEM are:

- Cooperation, commitment and communication between top management, senior and middle managers for GSCM.
- Implementation of EMS and ISO14001 certification.
- Environmental compliance auditing programs.
- Cross-functional integration for environmental improvement.

Green Design

Green design refers to environmental conscious design of products and services (Fiksel, 1996). It considers design issues associated with carbon emission, waste management and environmental pollutions. Srivastava (2007) identified life cycle analysis (LCA) is critical in green design concept. Green design includes low material consumption design, selection of environmentally friendly materials, and substitution of hazardous material (Gungor & Gupta, 1999; Tibben-Lembke, 2002). However recent literatures incorporate design of products for reuse and recycle (Zhu et al., 2008). Green-design requires that manufacturers design products that minimize consumption of materials and energy, that facilitate the reuse, recycle, and recovery of component materials and parts, and that avoids or reduces the use of hazardous products within the manufacturing process (Zhu et al., 2008).

Identified factors for green design are:

- Design of products for reduced consumption of material energy.
- Design of products for reuse, recycle, and recovery of material and/or component parts.
- Selection of environmentally-friendly raw materials with low embodied energy.
- Substitution of environmentally hazardous materials.

For the construction sector, implementing green building design requires attention from the conceptual stage of the building design. It could utilize innovative technology and architectural concepts to incorporate design features to reduce material use, promote low energy consumption materials, promote recycled materials such as recycled concrete, wood, steel and plastic, promote end of life recovery of materials, promote alternative source of energy like solar or wind energy, site selection for construction, efficient floor space utilization, energy efficient lighting and cooling technology, reduction in the use of hazardous materials, use of onsite water and solid waste treatment, proper ventilation, promote usage of precast or offsite manufactured materials in controlled environment to reduce onsite construction etc. An effective green building design shall consider the construction phase, operational phase, maintenance and renovation phase and end of life demolition and recovery.

Some of the relevant practices in green building design are:

- Shape and orientation of the building.
- Reducing primary heat losses through the building envelope.
- Reduced cooling loads.
- Introducing energy saving measures and energy creation systems.

Green Purchasing

Green purchasing is important aspect of GSCM. It addresses the requirement of green design. Green purchasing in GSCM is more than mere supplier selection. It involves stringent supplier environmental assessment criteria, environmental audit of suppliers, information and knowledge sharing with suppliers such as trainings and seminars to develop their own environmental programs such as ISO 14001 to improve environmental performance of suppliers (Min & Galle, 1997; Rao, 2004; Rao & Holt, 2005; Zhu et al, 2005; 2007; 2008)

Identified practices in green purchasing relevant to the construction sector are:

- Suppliers environmental assessment as prequalification criteria.
- Guiding suppliers to set up their own environmental programs such as Suppliers' ISO14000 certification.
- Environmental audit for suppliers' internal management.
- Providing design specification to suppliers that include environmental specifications for purchases.
- Bringing together suppliers in the same industry to share their know-how and problems.
- Urging and pressuring suppliers to take environmental actions.
- Information and knowledge sharing to suppliers.

Construction firms in UAE should monitor and evaluate the environmental performance of suppliers. In addition, sharing design specifications to suppliers in advance and urging suppliers to consider environmental actions can improve the environmental performance of the supply chain. Construction firms can also cooperate with suppliers for the purpose of developing products that are environmentally sustainable.

Green Distribution

Green distribution typically consists of green packaging and green transportation. Both the factors are interconnected as green packaging have a direct impact on transportation. Green packaging involves reducing the size, shape and weight of packaging, and the use of environmentally friendly materials. It can be defined as a way to package products that minimizes impact to the environment (Kassaye, 2001). In addition, green packaging is the easiest and foremost practice an organization can adopt in GSCM (Shi & Kane, 1996). By switching to green packaging, organizations will allow their products to gain public image (Kassaye, 2001). Ho et al. (2009) identified green packaging can reduce material usage, improve space utilization and reduce the handling time. Green transportation is the use of environmentally-friendly transportation system such as type and mode of transport such as sea, land and air transport and use of alternative fuel vehicles.

The identified factors for green distribution are:

- Reduced size of packaging.
- Use of environmental friendly packaging materials.
- Better shape for space utilization and handling.
- Use of environmentally friendly mode of transport options.
- Use of alternative fuel options.

Green Construction

Green construction means using sustainable principles and technology in the construction process such as cleaner technologies and automation in the construction process. The other green construction practices include on site usage of energy efficient vehicles and machineries, offsite fabrication of materials in a controlled environment to reduce construction waste and energy usage and use of project management techniques in the construction process to achieve efficiency in the construction process. In addition, having skilled labourers increases the efficiency of the construction process and enables reduction in construction waste. Further, practices such as onsite segregation of construction waste, re-use and recycling can further improve the performance of onsite construction.

Waste Management

It is an important aspect of GSCM. In the UK, construction and demolition waste contributed to 47.6 million tonnes in (UK Gov, 2010). In the US according to EPA (2003), 170 million tonnes of waste have been generated. The UAE is not far behind and is considered as the biggest producers of per capita construction waste in the world behind US. The construction industry in the UAE contributes to 75% of the total waste in the country (UAE Interact, 2007). The construction waste generated in the UAE in 2007 is 27.7 million tonnes compared to 10.6 million tonnes in 2006 (Alkhafaf, 2008). The figures indicates construction waste minimization is very important to reduce emissions and in promoting sustainable development. The Building Research Establishment (1978) defined waste as the 'difference between material ordered and those placed for fixing on building projects'. However waste generation from construction activities is much complex and therefore understanding the source of waste is important to minimize it. Research by (Rounce, 1998; Ekanayake & Ofori, 2000; Keys et al., 2000; Ekanayake & Ofori, 2004; El-Sayegh, 2008) identified improper design as the major cause of construction waste. The labour intensive nature of the construction activity implies, unskilled labourers employed in the construction contribute to excessive waste generation. Lingard et al. (2000) defined the origins of construction waste as procurement, handling, operations and culture.

GREEN PERFORMANCE MEASURES

GSCM management strives to achieve environmental, economic and operational performance. Although environmental performance is the important aspect of GSCM, the latter two are equally important to achieve business performance and act as intrinsic motivators for firms for adopting GSCM and to stay competitive in the industry. However in most cases, the extend of environmental, economic and operational performance achieved is not measured. Hence, performance measurement is an important aspect of GSCM for continuous improvement, to justify investment in GSCM, to publish the measures in annual reports and to benchmark with peers and industry average.

Environmental Performance Measures

Zhu et al. (2008) defined environmental performance in GSCM as the ability to reduce air emissions, effluent waste, and solid wastes and the ability to decrease consumption of hazardous and toxic materials. The key environmental performance indicators relevant to construction sector identified from the literature and its implication to United Arab Emirates (UAE) construction sector are given in Table 3.

Table 3. Environmental performance measures relevant to UAE construction industry

Environmental Performance	Implications to UAE Construction Industry
1. Reduction in energy consumption	Reduction in electricity consumption, reduction in fuel consumption, reduction in embodied energy of materials etc.
2. Reduction in raw materials	Reduction in raw material consumption such as steel, glass, cement etc.
3. Reduction in solid and effluent waste	Reduction in solid and effluent waste during construction site preparation and onsite construction activities.
4. Reduction in air emissions	Reduction in greenhouse gas emission (GHG) due to onsite construction activities and transportation
5. Reduction in water consumptions	Reduction in water consumption during the construction process
6. Reduction in consumption of hazardous/ harmful/toxic materials	Reduction in the use of hazardous materials like lead and sulphate based paints and adhesives
7. Decrease in frequency of environmental accidents	Reduction in environmental accidents like fire, breakage of underground pipes, leakage of hazardous materials etc.

Source: (Zhu et al., 2005; 2007; 2008)

Economic Performance

Economic performance is important for organizations adopting GSCM. From a management perspective, economic performance represents yield of the profit. For evaluating economic performance in GSCM, this study identified direct benefits such as the decrease of cost for energy consumption, material purchasing, waste treatment, and waste disposal (Zhu et al., 2007; 2008). However, GSCM economic performance also include indirect long term benefits such as new market opportunities, increase product price, increased profit margin, increased sales and market share and improved corporate image (Rao, 2002; Rao & Holt, 2005). The key environmental performance indicators and its implication to UAE construction sector are given in Table 4.

Operational Performance

Operational performance is the foundation of the traditional supply chain. It is imperative that GSCM adoption should not compromise the operational performance of the supply chain. Operational performance of GSCM is identical to operational performance measurement of any supply chain and includes overall efficiency of the supply chain, quality improvement, productivity improvement and asset utilization. Studies by (Ru-Jen et al., 2011; Sang et al., 2012) indicates GSCM can improve cycle time and reduce project duration along with increase in value to customer. The key operational performance indicators are given in Table 5.

Table 4. Economic performance measures relevant to UAE construction industry

Economic Performance	Implications to UAE Construction Industry
1. Increase in sales and market share	Increase in the sales of green building projects, Increase in market share through entry to new foreign markets and new environmental conscious buyers.
2. Increase in product price	Higher rent or selling price for buildings
3. Increase in profit margin	Overall increase in profit from green projects.
5. Cost savings - Decrease of cost for energy consumption - Decrease of fee for waste treatment - Decrease of fee for waste discharge - Decrease of fine for environmental accidents - Decrease of cost for materials purchasing	Overall reduction in cost in the construction supply chain through reduction in energy, material consumption and waste. Reduction in fines from environmental accidents.

Source: Rao (2002), Rao and Holt (2005), Ru-Jen et al. (2011), Zhu et al. (2008)

Table 5. Operational performance measures relevant to UAE construction industry

Operational Performance	Implications to UAE Construction Industry
1. Improvement in overall efficiency of the supply chain	Measuring the improvement in the overall efficiency of the construction supply chain
2. Improvement in quality	Improvement in the quality of the construction process
3. Improvement in productivity	Improvement in the productivity of the construction process
4. Improvement in cycle time	Reduction in transportation time for inbound and outbound logistics
5. Reduction in project completion time	Reduction in the construction project handover time.
6. Increased value to customer	Return on investments for the customer through reduced operational cost, reduction, improved indoor air quality, improved safety etc.
7. Improved asset utilization	Optimization of construction resources, reduction in inventory etc.

Source: Rao (2002), Rao and Holt (2005), Ru-Jen et al. (2011), Sang et al. (2012)

POTENTIAL GREEN SUPPLY CHAIN MANAGEMENT (GSCM) CHALLENGES PERCEIVED BY THE CONSTRUCTION INDUSTRY IN THE UNITED ARAB EMIRATES (UAE)

Although there are factors prompting organizations to adopt GSCM, the implementation of GSCM is not easy as firms have to overcome number of challenges. The challenges would be much higher for developing economies like UAE, where GSCM is still at infancy stage. Identifying and overcoming the challenges are key for the sector to overcome the status quo and to implement GSCM. Similar to drivers, challenges faced by firms can be classified as internal and external challenges.

Internal Challenges

Internal challenges are challenges faced within the organization. The identified internal challenges relevant to the construction sector are;

Lack of Knowledge, Training, and Awareness of GSCM

Studies have acknowledged that lack of knowledge and training (Cooper et al., 2000; Bowen et al., 2001), lack of experience (Yu Lin & Hui, 2008) are all important challenges to overcome for the adoption of GSCM. Rettab and Ben (2008) identified lack of knowledge as the primary inhibitor of organizations in UAE for adopting GSCM. Balasubramanian (2012) identified lack of knowledge and experience as an important inhibitor for GSCM in the UAE construction industry.

Lack of Top Management Commitment

Lack of top management support and commitment is one of the main challenges of GSCM. Zhu and Sarkis (2007) identified the importance of top management in the implementation of GSCM practices. Similarly, Ravi & Shankar (2005) mentioned the importance of continuous support from top management for organization wide implementation of GSCM strategy. Balasubramanian (2012) identified lack of top management support as one of the challenges faced by the UAE construction sector.

Fear of High Initial Investment

Many organizations carry the conception that huge investment is required for implementing GSCM practices. In addition, companies are aware for the fact that implementation of GSCM will not deliver immediate financial benefits, and hence many companies are reluctant to invest on GSCM (Nidumolu et al., 2009). For instance, green materials tend to be more expensive and increase the product's overall cost (Koplin et al, 2007; Hoffman, 2008). Several studies have identified high cost as the major challenge faced by organizations (Min & Galle, 2001; Alkhidir & Zailani, 2009).

Lack of Environmental Management Systems, ISO 14001, and Information Technology

An efficient environmental management system (EMS) is necessary for implementing GSCM. EMS promotes communication and knowledge sharing between clients, designers, suppliers and contractors. Balasubramanian (2012) identified lack of EMS leads to lack of understanding among stakeholders. EMS certified with ISO 14001 is very important for firms moving towards sustainability. Toshi et al. (2011) identified a positive relationship between Japanese firm's ISO 14001 certification and GSCM practices and found that ISO 14001 significantly promotes GSCM activities. Studies have found that it is difficult to implement GSCM without a proper environmental management system (Stevels, 2002). Similarly, information technology supports collaborative supply chain processes and enhances supply chain performance (Rogers et al., 1998). An advanced information and technology system is necessary for supporting the GSCM during various stages of product life cycle. It can be useful for product development programs encompassing the design for environment, recovery and reuse. Efficient information systems are needed for tracking and tracing the returns of product, linking with the previous sales (Ravi & Shankar, 2005). It is required to handle information's flows associated with both forward and backward flow of materials and other resources to manage green supply chain efficiently (AlKhidir & Zailani, 2009). Construction sector being traditionally a backward industry with poor IT infrastructure vis-à-vis other sectors, is one of the reasons for its poor environmental performance.

Lack of Sustainable Organizational Culture

Sustainable culture within the organisation is important for GSCM. Long term sustainable vision, informal and improved internal communication enable organizations to adopt green practices (Yu Lin & Hui Ho, 2008). Training and awareness are important for achieving successful implementation of GSCM in any organization (Ravi & Shankar, 2005). Management may encourage employees to learn green information and may provide rewards for green employees (Hsu & Hu, 2008). This is relevant for the UAE construction sector, considering the fact that UAE is relatively new to the concept of green.

Resistance to Change

Adopting and implementing GSCM involves change in the normal business processes, re-aligning the responsibilities, learning new technologies like ISO 14001, etc. Organizations and employees resist these changes and new technology adoption.

Previous studies have identified in their research 'resistance to change' as a major challenge for environmental innovation such as GSCM (Hosseini, 2007; Hsu et al., 2008; Digalwar et al., 2004). The identified internal challenges and its relevance to the UAE constructions sector are provided in Table 6.

Table 6 summarizes the various internal challenges of GSCM and its implication to the UAE construction industry.

External Challenges

External challenges are outside industry challenges faced by the organization. The identified external challenges relevant to the construction sector are;

Lack of Government Support

Lack of government support can hinder the adoption of GSCM. There is increasing need for government to provide industry friendly policies to promote GSCM rather than mandating firms to adopt GSCM. Government should facilitate a win-win situation of GSCM adopters by providing incentives, preferential treatment, best practices awards, and recognition for sustainable leaders and organizations. Lack of these can discourage firms in adopting GSCM. Presently in the UAE construction sector there is little support from government. Numerous studies have highlighted the importance for government support for GSCM (Hosseini, 2007; Hsu et al., 2008; Mudgal et al., 2009; Mudgal et al., 2010; Srivastava, 2007)

Table 6. Internal GSCM challenges relevant to UAE construction sector

Internal Challenges of GSCM	Implications to UAE Construction Industry
1. Lack of knowledge and awareness of GSCM	Government and the industry should conduct GSCM trainings and awareness sessions, conferences and seminars to improve knowledge and awareness of stakeholders in the construction
2. Lack of top management commitment	Government and the industry should recognize leaders in GSCM and would encourage GSCM commitment among other industry leaders
3. Fear of high investment cost.	Perceived high initial investment in the construction sector for green practices
5. Lack of EMS and IT infrastructure	Government regulations and leading clients should make it mandatory for all construction firms to implement EMS and to have a standard IT infrastructure requirements
7. Lack of sustainable organizational culture	Top management and HR department should create a sustainable culture to adopt the concept of GSCM at all employee levels

Lack of Skilled Professionals

GSCM requires skilled professionals both inside the organization as well as throughout the supply chain. Skilled professionals can adapt to the changes quickly, learn new technologies, share ideas and solve environmental problems (Yu Lin & Hui Ho, 2008). Balasubramanian (2012) identified lack of skilled professionals as an important inhibitor of GSCM in the UAE construction sector.

Lack of Trust and Commitment of Suppliers

Trust and commitment of suppliers play an important role in GSCM. Previous studies have identified lack of commitment and trust as the major challenges facing firms to implement GSCM practices (Wycherley, 1999; Walker et al., 2008). Lack of trust and commitment of suppliers is also a challenge for the construction sector, because of the one off nature of the construction projects.

Perceived Lack of Demand for Sustainable Solutions

Customers demand for low prices. Considering the financial crisis of 2008 and afterwards, there is lack of demand for sustainable solutions as customer are not willing to pay more for sustainable solutions. Orsato (2006) identified customer desire for low prices can lead to lack of sustainable practices. Balasubramanian (2012) identified the lack of demand as a major inhibitor to the adoption of GSCM in the UAE construction sector.

Table 7 summarizes the various external challenges of GSCM and its implication to the UAE construction industry.

Table 7. External GSCM challenges relevant to UAE construction sector

Internal Challenges of GSCM	Implications to UAE Construction Industry
1. Lack of government support.	Government should provide support in the form of preferential treatment, incentives, subsidies and best practices award to encourage adoption of GSCM
2. Lack of skilled professionals.	Organizations should attract skilled green professionals from around the world and retain them to implement GSCM
3.Lack of trust and commitment of suppliers	Construction stakeholders should focus on having long term relationships with suppliers
4. Perceived lack of demand for sustainable solutions.	Construction industry should clearly articulate the benefits of green buildings and hence customers will be willing to pay higher for green buildings

The various aspects of GSCM relevant to the construction sector is discussed. Although, more empirical investigation is required to obtain more insights on the various aspects discussed, the chapter provides an initial investigation into the potential GSCM aspects that is relevant for the UAE construction sector.

KEY TAKE AWAY POINTS FOR PRACTITIONERS

The following questions are important for practitioners while implementing GSCM.

Does Adoption of GSCM Improve Economic Performance?

The social responsibility of any business is to be profitable. Economic performance is important for organizations to sustain in the competitive environment. Before adopting GSCM, organizations much do a cost-benefit analysis to understand the financial benefits of GSCM along with achieving environmental and operational benefits. Analysis such as life cycle assessment (LCA), total cost of ownership (TCO) etc. are key to the implementation of GSCM. If the answer is 'Yes' then it's a win-win situation for the environment and business. Most cases organizations having sustainable vision often settle for trade-off between environmental and economic performance.

Is Profitability the Only Concern of SCM?

This may be the case in the past; however the present and future supply chain needs to fulfil the requirements or regulations of governments, NGO or sometimes customers. So achieving environmental performance and social performance is important to sustain business. So organizations need to implement environmental measures to stay competitive in the business, hence profitability is not the only concern of SCM. Implementing GSCM is no longer an option but a must in the future.

Are You Measuring the Environmental, Economic, and Operational Performance of the Supply Chain?

Measuring the performance of the supply chain is important in closing the loop of the supply chain. Implementing GSCM is a continuous process and it requires performance measures to understand the weak links in the supply chain in order to improve the performance. Key performance indicators are important for the successful implementation of GSCM. Green balanced score measures would probably an alternative method organisation can use to measure their Return on Assets (ROA).

Which Organizational Strategy Would You Choose? Proactive or Reactive Strategy?

Proactive organizations will have strong sustainable vision with environmental leaders having the confidence to adopt GSCM. These organizations often have strong financial and human resource capabilities. Proactive organizations will try to achieve first mover benefits, preferential treatment and government incentives and also try to improve the operational efficiency of the supply chain. Hence these companies have long term vision to improve their economic and environmental performance. They will also incorporate necessary internal environmental management changes to implement GSCM practices.

Reactive organizations respond to government rules and regulations, compliance to client specifications, stakeholder pressure and competition to improve their environmental performance. The implement GSCM to achieve minimum compliance with no long term vision. From a government perspective, a balance should be kept to promote proactive GSCM practices such as incentives and subsidies and at the same time tighten the rules and regulations to ensure organizations consider GSCM in their organizational strategy.

For implementing GSCM, integrating GSCM practices with construction industry practices is the best way to move forward. Similarly different industry requires different GSCM approaches. All the stakeholders should be incorporated and study should focus on the end to end supply chain.

CONCLUSION

As the construction sector grows, we are depleting more and more natural resources and in the process harming the environment. It is important that we look into sustainable management practices such as GSCM to reduce the environment impact of the construction sector. Considering the fact that GSCM in construction is still at infancy, present study will significantly contribute to filling the knowledge gap of GSCM in the United Arab Emirates (UAE) constructions sector and for construction sector in general. Some of the significant contributions of the study include identifying the potential drivers and challenges faced by the construction industry in the UAE and in general for implementing GSCM. The various stages in the supply chain for reducing the environmental impacts is identified such as design, purchasing, transportation and onsite and offsite construction. The importance of having an internal environmental management (IEM) for implementing GSCM is identified. Finally, performance measures namely environmental, economic and operational is identified to assess the performance improvement post GSCM implementation. However, considering the fact that present study is largely derived from the literature, the various aspects identified are more indicative than conclusive. Hence, more studies and empirical research is required to validate the various aspects identified in the study. Future studies can also focus on identifying the difference in GSCM implementation among the various construction stakeholders such as developers, consultants, architects, main contractor, subcontractor, and suppliers. Stakeholder analysis is important to ensure all stakeholders are coherent in implementing GSCM. Further, future studies can use case studies on large companies that have implemented GSCM practices to find evidence on the improvement in performance. Such empirical evidence is critical for imparting confidence among construction stakeholders to engage in GSCM.

REFERENCES

Balasubramanian, S. (2012). A Hierarchical Framework of Barriers to Green Supply Chain Management in the Construction Sector. *Journal of Sustainable Development*, *5*(10), 1913-9063.

Beamon, M. (1999). Designing the green Supply Chain. *Logistics Information Management*, *12*(4), 332–342. doi:10.1108/09576059910284159

Ben Brik, A., Mellahi, K., & Rettab, B. (2013). Drivers of Green Supply Chain in Emerging Economies. *Thunderbird International Business Review*, *55*(2), 123–136. doi:10.1002/tie.21531

Berns, M., Townend, A., Khayat, Z., Balagopal, B., Reeves, M., Hopkins, M. & Kruschwitz, N. (2009). The business of sustainability.What it means to managers now. *MIT Sloan Management Review*, 20–26.

Berry, M. A., & Rondinelli, D. A. (1998). Proactive corporate environment management: A new industrial revolution. *The Academy of Management Executive*, *12*(2), 38–50.

Blackburn, J. D., Guide, V. D. R. Jr, Souza, G. C., & Van Wassenhove, L. N. (2004). Reverse Supply Chains for Commercial Returns. *California Management Review*, *46*(2), 6–22. doi:10.2307/41166207

Buckholz, R. (1991). Corporate responsibility and the good society: From economics to ecology. *Business Horizons*, *34*(4), 19–31. doi:10.1016/0007-6813(91)90003-E

Byrne, P. M., & Deeb, A. (1993). Logistics must meet the 'green' challenge. *Transportation & Distribution, 34*(2), 33–37.

Carter, C. R., & Dresner, M. (2001). Purchasing's Role in Environmental Management: Cross-Functional Development of Grounded Theory. *Journal of Supply Chain Management, 37*(2), 12–27. doi:10.1111/j.1745-493X.2001.tb00102.x

Chan, R. Y. K., & Lau, L. B. Y. (2001). Explaining green purchasing behaviour: A cross-cultural study on American and Chinese consumers. *Journal of International Consumer Marketing, 14*(2/3), 9–41.

Dhanda, K. K., & Hill, R. P. (2005). The role of information technology and systems in reverse logistics: A case study. *International Journal of Technology Management, 31*(1/2), 140–151. doi:10.1504/IJTM.2005.006628

Digalwar, A. K., & Metri, B. A. (2004). Performance Measurement Framework for World Class Manufacturing. *International Journal of Applied Management and Technology, 3*(2), 83–101.

Drumwright, M. (1994). Socially Responsible Organisational Buying: Environmental Concern as a Non-economic Buying Criterion. *Journal of Marketing, 58*(3), 1–19. doi:10.2307/1252307

Dubai Electricity and Water Authority. (2011). *Green Buildings Regulations and Specifications.* Retrieved on October 12,2012, from http://www.dewa.gov.ae/images/greenbuilding_eng.pdf

Fassoula, E. (2005). Reverse logistics as a means of reducing the cost of quality. *Total Quality Management and Business Excellence, 16*(5), 631–643. doi:10.1080/14783360500077575

Ferguson, M. E., & Toktay, L. B. (2006). The Effect of Competition on Recovery Strategies. *Production and Operations Management, 15*(3), 351–368. doi:10.1111/j.1937-5956.2006.tb00250.x

Francesco, T., & Fabio, I. (2010). Shadows and lights of GSCM (Green Supply Chain Management): Determinants and effects of these practices based on a multinational study. *Journal of Cleaner Production, 18*(10), 953–962.

González-Benito, J., & González-Benito, O. (2005). A study of the motivations for the environmental transformation of companies. *Industrial Marketing Management, 34*(5), 462–475. doi:10.1016/j.indmarman.2004.08.005

Gonzalo, F., & Fernando, R. (2010). A methodology to identify sustainability indicators in construction project management—Application to infrastructure projects in Spain. *Ecological Indicators*, *10*(6), 1193–1201. doi:10.1016/j.ecolind.2010.04.009

Gorane, S. J., & Ravi, K. (2013). Modelling the SCM enablers: An integrated ISM-fuzzy MICMAC approach. *Asia Pacific Journal of Marketing and Logistics*, *25*(2), 263–286. doi:10.1108/13555851311314059

Gungor, A., & Gupta, S. M. (1999). Issues in environmentally conscious manufacturing and product recovery: A survey. *Computers & Industrial Engineering*, *36*(4), 811–853. doi:10.1016/S0360-8352(99)00167-9

Ho, J.C., & Shalishali, M.K., Tseng, T.L. & Ang, D.S. (2009). Opportunities in green supply chain management. *The Coastal Business Journal*, *8*(1), 18–31.

Hoffman, A. J. (2008). From Heresy to Dogma: An Institutional History of Corporate Environmentalism. Stanford Business Books.

Holt, D., & Ghobadian, A. (2009). An empirical study of green supply chain management practices amongst UK manufacturers. *Journal of Manufacturing Technology Management*, *20*(7), 933–956. doi:10.1108/17410380910984212

Hosseini, A. (2007). Identification of Green Management of system's factors: - A Conceptualized Model. *International Journal of Management Science and Engineering Management*, *2*(3), 221–228.

Hsu, C. W., & Hu, A. H. (2008). Green Supply Chain Management in the Electronic Industry. *International Journal of Science and Technology*, *5*(2), 205–216.

Jonatan, P., & Marcel, D. (2009). Overcoming Barriers to Sustainability: An Explanation of Residential Builders Reluctance to Adopt Clean Technologies. *Business Strategy and the Environment*, *18*(8), 515–527. doi:10.1002/bse.615

Kam, B., Smyrnios, G.C. & Walker, R.H. (2003). Strategic business operations, freight transport and eco-efficiency: a conceptual model. *Greening the Supply Chain, Greenleaf, Sheffield*

Klassen, R. D., & McLaughlin, C. P. (1993). TQM and environmental excellence in manufacturing. *Industrial Management & Data Systems*, *93*(6), 14–22. doi:10.1108/02635579310040924

Lee, S. (2008). Drivers for the participation of small and medium-sized suppliers in green supply chain initiatives. *Supply Chain Management: An International Journal*, *13*(3), 185–198. doi:10.1108/13598540810871235

Lee, S., & Klassen, R. D. (2008). Drivers and enablers that foster environmental management capabilities in small- and medium-sized suppliers in supply chains. *Production and Operations Management*, *17*(6), 573–586. doi:10.3401/poms.1080.0063

Lippman, S. (2001). Supply Chain Environmental Management. *Environmental Quality Management*, *11*(2), 11–14. doi:10.1002/tqem.1301

Lo, C. W., & Leung, S. W. (2000). Environmental agency and public opinion in Guangzhou: The limits of a popular approach to environmental governance. *The China Quarterly*, *163*, 677–704. doi:10.1017/S0305741000014612

Maxwell, D., & Van der Vorst, R. (2003). Developing sustainable products and services'. *Journal of Cleaner Production*, *11*(8), 883–895. doi:10.1016/S0959-6526(02)00164-6

Min, H., & Galle, W. (1997). Green purchasing strategies: Trends and implications. *International Journal of Purchasing and Materials*, *4*, 410–417.

Min, H., & Galle, W. P. (2001). Green purchasing practices of US firms'. *International Journal of Operations & Production Management*, *21*(9), 1222–1238. doi:10.1108/EUM0000000005923

Mudgal, R. K., Shankar, R., Talib, P., & Raj, T. (2009). Greening the supply chain practices: An Indian perspective of enablers' relationships. *International Journal of Advanced Operations Management*, *1*(2/3), 151–176. doi:10.1504/IJAOM.2009.030671

Nidumolu, R., Prahalad, C. K., & Rangaswami, M. R. (2009). Why Sustainability is now the key driver of innovation. *Harvard Business Review*, *87*(9), 56–64.

Ninlawan, C., Seksan, P., Tossapol, K., & Pilada, W. (2010). The implementation of Green Supply Chain Management Practices in Electronic Industry. *Proceedings of the International MultiConference of Engineers and Computer Scientists.*

Orsato, R. (2006). Competitive environmental strategies: When does it pay to be green? *California Management Review*, *48*(2), 127–143. doi:10.2307/41166341

Paulraj, A. (2009). Environmental motivations: A classification scheme and its impact on environmental strategies and practices. *Business Strategy and the Environment*, *18*(7), 453–468. doi:10.1002/bse.612

Rao, P. (2002). Greening the supply chain: A new initiative in South East Asia. *International Journal of Operations & Production Management*, *22*(6), 632–655. doi:10.1108/01443570210427668

Rao, P., & Holt, D. (2005). Do green supply chains lead to competitiveness and economic performance? *International Journal of Operations & Production Management, 25*(9), 898–916. doi:10.1108/01443570510613956

Ravi, V. & Shankar, R. (2005). Analysis of interactions among the barriers of reverse logistics. *International Journal of Technological Forecasting and Social Change, 72*(8), 1011-1029.

Rettab, B., & Ben Brik, A. (2008). *Green Supply Chain in Dubai. Dubai Chamber Centre for Responsible Business, Dubai.* Retrieved October 14, 2012, from http://www.dubaichamber.com/wp.../GREEN-SUPPLY-CHAIN-IN-DUBAI.pdf

Richey, R. G., Stefan, E. G., & Patricia, J. D. (2005). The role of resource commitment and innovation in reverse logistics performance. *International Journal of Physical Distribution & Logistics Management, 35*(4), 233–257. doi:10.1108/09600030510599913

Roberts, J. (1996). Green consumers in the 1990s: Profile and implications for advertising. *Journal of Business Research, 36*(3), 217–231. doi:10.1016/0148-2963(95)00150-6

Ru-Jen, L., Rong-Huei, C., & Thi-Hang, N. (2011). Green supply chain management performance in automobile manufacturing industry under uncertainty. *Procedia: Social and Behavioral Sciences, 25*, 233–245. doi:10.1016/j.sbspro.2011.10.544

Sang, M. L., Sung, T. K., & Donghyun, C. (2012). Green supply chain management and organizational performance. *Industrial Management & Data Systems, 112*(8), 1148–1180. doi:10.1108/02635571211264609

Sarkis, J. (2003). A strategic decision framework for green supply chain management. *Journal of Cleaner Production, 11*(4), 397–409. doi:10.1016/S0959-6526(02)00062-8

Sharma, S., & Vredenburg, H. (1998). Proactive corporate environmental strategy and the development of competitively valuable organizational capabilities. *Strategic Management Journal, 19*(8), 729–753. doi:10.1002/(SICI)1097-0266(199808)19:8<729::AID-SMJ967>3.0.CO;2-4

Simpson, D., Power, D., & Samson, D. (2007). Greening the automotive supply chain: A relationship perspective. *International Journal of Operations & Production Management, 27*(1), 28–48. doi:10.1108/01443570710714529

Sourani, A., & Sohail, M. (2011). Barriers to addressing sustainable construction in public procurement strategies. *Engineering Sustainability, 164*(4), 229–237. doi:10.1680/ensu.2011.164.4.229

Srivastava, S. K. (2007). Green supply chain management: A state-of-the-art literature review. *International Journal of Management Reviews*, *9*(1), 53–80. doi:10.1111/j.1468-2370.2007.00202.x

Tibben-Lembke, R. S. (2002). Life after death: Reverse logistics and the product life cycle. *International Journal of Physical Distribution & Logistics Management*, *32*(3), 223–244. doi:10.1108/09600030210426548

Tseng, M. (2013). Modeling sustainable production indicators with linguistic preferences. *Journal of Cleaner Production*, *40*, 46–56. doi:10.1016/j.jclepro.2010.11.019

United Nations Environmental Program. (2007). *Building Standards and Climate Change: Status, Challenges and Opportunities*. Retrieved October 30, 2012, from http://unep.fr/shared/publications/.../DTIx0916xPA-BuildingsClimate.pdf

Vachon, S. (2007). Green supply chain practices and the selection of environmental technologies. *International Journal of Production Research*, *45*(18/19), 4357–4379. doi:10.1080/00207540701440303

Walker, H., Di Sisto, L., & McBain, D. (2008). Drivers and barriers to environmental supply chain management practices: Lessons from the public and private sectors. *Journal of Purchasing and Supply Management*, *14*(1), 69–85. doi:10.1016/j.pursup.2008.01.007

Wang, G., Wang, Y., & Zhao, T. (2008). Analysis of interactions among the barriers to energy saving in China. *Energy Policy*, *36*(6), 1879–1889. doi:10.1016/j.enpol.2008.02.006

WBR. (2009). *World Bank Report of United Arab Emirates*. Retrieved November 12, 2012, from http://data.worldbank.org/country/united-arab-emirates

Wycherley, I. (1999). Greening supply chains: The case of the Body Shop International. *Business Strategy and the Environment*, *8*(2), 120–127. doi:10.1002/(SICI)1099-0836(199903/04)8:2<120::AID-BSE188>3.0.CO;2-X

Yu Lin, C., & Hui Ho, Y. (2008). An Empirical Study on Logistics services provider, intention to adopt Green Innovations. *Journal of Technology. Management and Innovation*, *3*(1), 17–26.

Zhu, Q., & Sarkis, J. (2006). An inter-sectoral comparison of green supply chain management in China: Drivers and practices. *Journal of Cleaner Production*, *14*(5), 472–486. doi:10.1016/j.jclepro.2005.01.003

Zhu, Q., Sarkis, J., & Kee-hung, L. (2007). Green supply chain management: Pressures, practices and performance within the Chinese automobile industry. *Journal of Cleaner Production*, *15*(11-12), 1041–1052. doi:10.1016/j.jclepro.2006.05.021

Zhu, Q., Sarkis, J., & Lai, K. (2008). Confirmation of a measurement model for green supply chain management practices implementation. *International Journal of Production Economics*, *111*(2), 261–273. doi:10.1016/j.ijpe.2006.11.029

Chapter 5
A Waste Elimination Process:
An Approach for Lean and Sustainable Manufacturing Systems

Sherif Mostafa
University of South Australia, Australia

Jantanee Dumrak
University of South Australia, Australia

ABSTRACT

The continuous improvement using waste elimination has been emphasized as the most important task of modern organizations. Lean manufacturing system has proved its capability to eliminate waste and produce environmental gains. The eight types of waste identified under lean have positive relations to green waste. In this chapter, a waste elimination process is suggested as an approach for lean sustainable outcomes. The process structure contains three consecutive phases: waste documentation, waste analysis, and waste removal. The techniques and tools of each phase are inclusively discussed.

DOI: 10.4018/978-1-5225-0635-5.ch005

INTRODUCTION

Manufacturing sector is a vital and significant sector contributing to economic development of a country. The connectivity between manufacturing sector and other sectors such as agricultural and services allows a chain impact to be predicted if any ineffective and inefficient activities are performed by the manufacturing sector. The concept of sustainability has played a crucial role in manufacturing to maximize resources utilization in delivering products without causing environmental impact. The concept emphasizes on eco-efficient production processes and hazard-free workplace environment. A key focus on attaining sustainability in manufacturing is on an effective waste removal within the production processes. Eight types of production wastes have been identified within the lean manufacturing system. The occurrence of these waste types combined can be transformed into green waste or environmental footprint (Khan, Jaber, & Glock, 2012). It has been confirmed that the waste removal using lean can generate environmental gains (Moreira, Alves, & Sousa, 2010). Sustainability in lean application, therefore, supports long-term environmental and production improvement.

This chapter introduces a waste removal process as a technique for sustaining lean manufacturing to result in gaining green environment. The following sections of this chapter entail background of lean manufacturing and sustainability, waste identification tools, waste elimination process and future directions towards green and sustainable manufacturing systems. All sections reflect the chapter objectives which are:

- To identify the commonalities between lean and sustainability within manufacturing context. This will lead to an understanding of interactions between lean and green manufacturing.
- To review the identification tools used for locating waste or non-value added activities inside the manufacturing supply chain.
- To propose a waste elimination process as a lean sustainable approach. This will include three consecutive phases: waste documentation, waste analysis and waste elimination.

BACKGROUND

Continuous improvement of organization features is a guarantee for surviving in the highly competitive environment. Ohno (1988) emphasizes that continuous improvement must never stop. The activities that customers are willing to pay for are identified as value added activities. Whereas, non-value added activities are considered as waste from customers' view. According to Monden (1998), three types of activities are generally implemented within an organization. These three categories are: non value-adding (NVA), necessary but non value-adding (NNVA), and Value adding (VA). The NVA activities are obvious waste that should be entirely eliminated. The NNVA activities may be wasteful but are necessary under the current operating procedures. These activities include walking long distances to pick up parts and transferring tool from one hand to another. The VA activities encompass any activities required to change or process the raw materials or semi-finished products toward what customer wants (Hines & Rich, 1997).

According to Melton (2005), waste is defined as an activity in a process that does not add value to product/service from customer view. Waste elimination is the next step after defining the end user value. However, Melton has overlooked at the components of the process. In general, a process does not contain only activities, but also inputs, tools, and outputs. Therefore, waste elimination should have a broader spectrum rather than just focusing on activities that do not produce value. Seven major types of waste in manufacturing and business processes were identified in the Toyota Production System (TPS) (Ohno, 1988) consisting of overproduction, waiting, unnecessary transport, incorrect processing, excess inventory, unnecessary movement and defects. In addition to the seven types of waste, Womack and Jones (2003) identify the eighth waste type. This type of waste is related to unused employee creativity to improve the processes and practices (Hicks, 2007; Liker, 2004). Recently, the environmental footprints of an organization are one of the competition criteria added to cost, service, quality and lead time (Khan et al., 2012). Therefore, the environmental waste is considered as ninth waste type. The environmental waste includes any activities that could harm human health or environment such as the excessive of substances released to air, water, or land (Dombrowski, Mielke, & Schulze, 2012; Gehin, Zwolinski, & Brissaud, 2008). To facilitate waste reducing/eliminating, five principles of lean are established to facilitate waste reducing/eliminating.

The lean principles are defining the customer value, mapping the value stream, pulling the value from the customer, pursuing perfection by continuous identification and eliminating wastes. They are applied through lean tools. The lean tools include Kaizen, 5S, Production Preparation Process (3P), Total Productive Maintenance (TPM), Six Sigma, and Kanban. The successful application of the lean principles makes an organization become greener (Mohammed & Sadique, 2011; Vinodh, Arvind, & Somanaathan, 2011) as Fliedner (2008) identifies some of the lean sustainable benefits including reduction in material usage, energy consumption, hazardous waste and water usage. As a result, lean principles lead to enhancing sustainability practice of an organization.

This chapter introduces a waste elimination process to attain lean sustainability. The process consists of six consecutive steps including mapping the value stream, understanding/educating the nine types of waste, identifying the waste types in an organization value stream, conducting waste analysis, ranking the waste using waste elimination template, and selecting of appropriate lean tools for waste elimination. The techniques used for undertaking are included in each step. For instance, a simulation or Radio Frequency Identification (RFID) is used for mapping the current and future state of the value stream. Waste analysis employed the cause and effect analysis is used to explore the root causes of each waste. The waste elimination template is for ranking each waste type on the root causes basis. Afterwards, a score for each waste is calculated from the cost of removal, ease of removal, impact of removal on other types of waste, occurrence of waste and detection mechanisms of the waste. A Quality Function Deployment (QFD) method is used to associate the waste root causes with the lean manufacturing tools.

THE LEAN MANUFACTURING CONCEPT AND SUSTAINABILITY

Definitions of Lean Manufacturing

After the Second World War, Japanese manufacturers suffered vast shortages of materials as well as financial and human resources. These conditions resulted in the origin of lean manufacturing system. Toyota Corporation has been known as the system's originator. In 1980, the term lean manufacturing was used by the authors of the international motor vehicle program (IMVP) carried out by Massachusetts Institute of Technology (MIT) to describe the approach initially developed in Japanese auto manufacturing industry especially Toyota Production System (TPS). The term lean means less of everything compared to the mass manufacturing system (Bruun & Mefford, 2004). Lean aims to maximize the end-user values (i.e. provide good quality products at low price) by eliminating all forms of wastes from a supply chain (Drake, Lee, & Hussain, 2013). The lean manufacturing system is simply an improvement and a repackage of the Just-In-Time (JIT) or 'stockless' production (Narasimhan, Swink, & Kim, 2006). It is argued that lean goes beyond JIT in the waste elimination. Whilst JIT only focuses on eliminating one type of waste (inventory), the lean attacks other types of waste in addition (Holweg, 2007; Mehta & Shah, 2005). Lean is viewed as a systematic approach employed for efficient use of resources by identifying and eliminating waste non-value-added activities (Mostafa, Dumrak, & Soltan, 2013). Shah and Ward (2007) defines lean as a collection of practices that work together synergistically to create a streamlined high quality system that produces finished products at the pace of the customer demand with little or no waste.

Lean Principles

Lean manufacturing is based on five key principles that provide a simple structure to construct a detailed roadmap for lean implementation (Haque & James-moore, 2004). Womack and Jones (2003) identify the five basic principles behind lean thinking as follows:

- **Specify Value**: Value is anything that a customer will pay for. Value represents the customer requirements for product functionality, quality, and delivery. Besides, the relationship of these three criteria to market price. Organizations should specify what does/does not create value from the customer perspective (B. Singh, Garg, & Sharma, 2011).
- **Identify the Value Stream**: Value stream is defined as a collection of the value-added as well as NVA activities that are required to bring a product, product family, or service starting from raw material and ending with the customer (Sullivan, McDonald, & Van Aken, 2002). The value stream highlights what is against value (waste) and to be eliminated.
- **Make the Value to Flow**: An uninterrupted value flow is created by actions that result in sequential completion of all activities in the most efficient manner, from its initiation to its completion (Haque, Badr & James-moore 2004).
- **Let Customer Pull the Value**: This lean principle emphasis on customer to pull the entire value-creating process. The information on customer demand flow from downstream to upstream and therefore, start produce the material flow as contrast to information to end in the hand of customer (Womack & Jones, 2003).
- **Pursue Perfection**: Continually removing the NVA activities while offering a product to achieve customer requirements (Womack & Jones, 2003).

Value and Waste Concept

Lean manufacturing focuses on end-user value. The customer value is any activity, process or operation that changes the product from one form to another in order to get it closer to the customer's specifications. Precisely defining the customer value is the first principle. As the value is not specified correctly, this can result in delivering an unsatisfied or wrong product or service. It is considered as a great waste for the organization. The value should be expressed in term of a specific product that meets the customer's needs at a specific price at a specific time. The success or failure of a supply chain is determined in the market place through the end-user value. The end-user value of a product or service has many metrics that can be aggregated such as service level, quality, cost, and lead-time (Naylor, Naim, & Berry, 1999) as shown in Figure 1.

After defining the value, the next step is to identify and eliminate the waste in the entire value stream. Waste is any activity that adds costs or time to the product without adding any value from the customer perspective. The nature of waste depends on the activities of the organization. Waste may be identified in three major types: unobvious waste, less obvious waste, and obvious waste. Hopp and Spearman (2004) give examples of the obvious waste as inventory, unneeded processes, excessive setup times, unreliable machines, and rework. They argue that less obvious waste is associated with variability in process times, delivery times, yield rates,

Figure 1. Total value metric
(Modified from Naylor et al., 1999, p. 109)

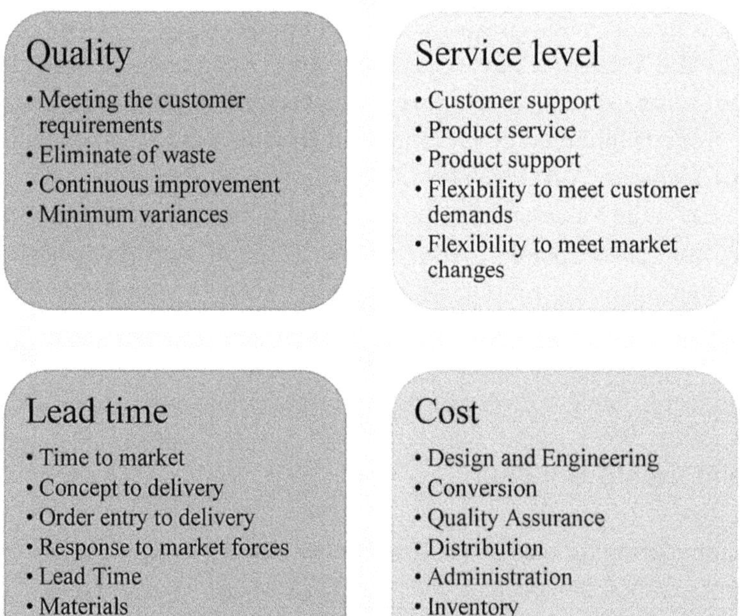

staffing level, and demand rates that create buffering costs. This type of waste also includes anything in the system that is not absolutely regular and predictable exhibits variability.

Lean operations eliminate obvious waste and reduce less obvious waste. The operations can replace expensive buffers (such as inventory) with less expensive ones (such as capacity). Figure 2 displays the effect of lean on the processes included within the supply chain for producing a product. The value added activities include all activities that transform material or information to produce products or services that customers willing to pay for. The non-value added (NVA) activities can be divided into necessary and unnecessary activities. The necessary NVA activities involve activities that create no value but are required by current technology, current thinking, process limitations, or current process. The necessary NVA should be reduced from the value stream. The unnecessary NVA activities are pure waste that must be eliminated.

The waste types within the manufacturing context have been explained by some researchers (Hicks, 2007; Imai, 1986; Ohno, 1988; Uygun & Straub, 2012; Womack & Jones, 2003). Toyota Production System (TPS) identified seven traditional waste types. Womack and Jones (2003) extended one more type of waste. Therefore, waste can be classified into eights types. Khan et al. (2012) stated that the environmental footprints of an organization is added as one of the competition criteria (cost, service, quality and lead time). Therefore, the environmental waste is considered as the ninth type of waste. It includes any activities that could harm human health or environment such as pollutants released to air, water, or land (Dombrowski, Mielke, & Schulze, 2012). Figure 3 demonstrates the nine types of waste. A brief discussion of each waste type is provided in the Table 1.

Figure 2. Supply chain leaning

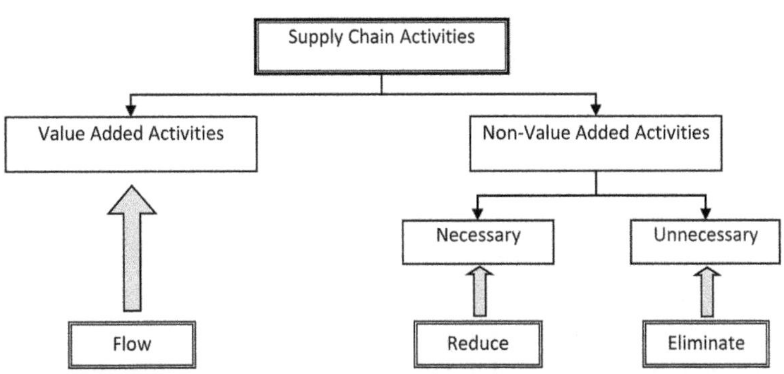

Figure 3. Development of waste types

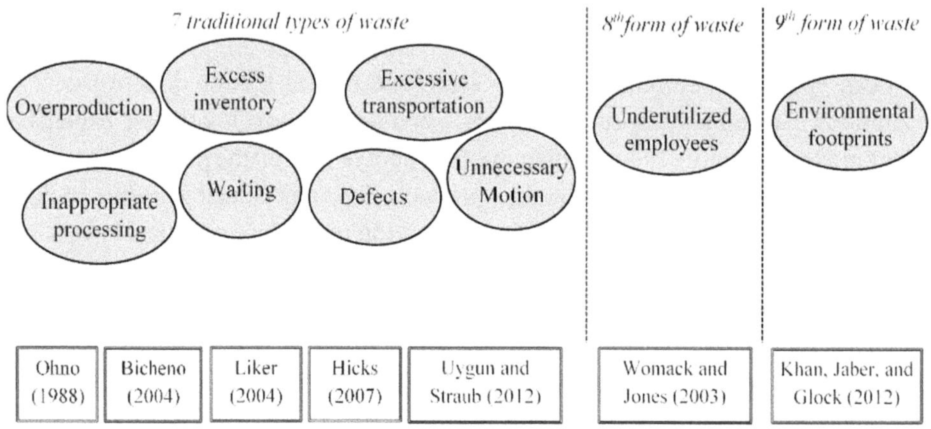

Sustainability

Sustainability is the interchangeable term to sustainable development. The sustainable development refers to meeting the needs of the present generation without compromising the ability of future generations to meet their own needs (R. K. Singh, Murty, Gupta, & Dikshit, 2009). Semantically, sustainability is to maintain a state of a system indeterminately at a certain level. The sustainability was previously understood as the quality of maintaining the environment. The focus was only on the environmental aspects (Amini & Bienstock, 2014). However, the present concept is called the triple bottom line (TBL) of economic, social and environmental performance. The TBL is also known as profit, planet and people (Porter & Kramer, 2006). This means an organization should be operated to sustain long-term economic benefits without scarifying short-term behaviour that is socially and environmentally wasteful. In manufacturing, the sustainability aims at producing future products that are completely recyclable, manufacturing processes with no detrimental impacts on the environment, and disassembly of the products at the end of its functional life (Molamohamadi & Ismail, 2013).

Understanding the linkage between sustainability and manufacturing has been developed from design for environment (DfE) concept (Nagel & Tomiyama, 2004) towards green supply chain management (GSCM), sustainable supply chain management (SSCM) and green factory (Gold, Seuring, & Beske, 2010; Green, Zelbst, Meacham, & Bhadauria, 2012; Rohrmus, Döricht, Weinert, Holst, & Kiener, 2013). The conventional DfE approach was used to design the products and manufacturing system to be friendly environmental. This approach was applied through the life cycle thinking to implement a number of design practices within the product life cycle.

Table 1. Environmental impacts of lean wastes (modified from EPA, 2009, pp. 32-33)

Waste Type	Examples	Environmental Impacts
Overproduction	Producing items for which there are no orders	• More raw materials and energy consumed in making the unnecessary products • Extra products may spoil or become obsolete requiring disposal • Extra hazardous materials used result in extra emissions, waste disposal, worker exposure, etc.
Excess inventory	Excess raw material, work-in-process (WIP), or finished goods	• More packaging to store work-in-process (WIP) • Waste from deterioration or damage to stored WIP • More materials needed to replace damaged WIP • More energy used to heat, cool, and light inventory space.
Defects	Scrap, rework, replacement production, inspection	• Raw materials and energy consumed in making defective products • Defective components require recycling or disposal • More space required for rework and repair, increasing energy use for heating, cooling, and lighting.
Excessive Transportation	Unnecessary movement of materials in and out the storage rooms between production stages	• More energy use for transport • Emissions from transport • More space required for WIP movement, increasing lighting, heating, and cooling demand and energy use.
Unnecessary Motion	Human motion that are unnecessary or straining, carrying WIP long distances, transport	• More packaging required to protect components during movement • Damage and spills during transport • Transportation of hazardous materials requires special packaging to prevent risk during accidents
Inappropriate processing	More parts, process steps, or time than necessary to meet customer needs	• More parts and raw materials consumed per unit of production • Unnecessary processing increases wastes, energy use, and emissions
Waiting	Stock-outs, lot processing delays, equipment downtime, capacity bottlenecks	• Potential material spoilage or component damage causing waste • Wasted energy from heating, cooling, and lighting during production downtime
Underutilization of employees potentials	Lost time, ideas, skills, improvements, and suggestions from employees	• Fewer suggestions of pollution prevention • Fewer waste minimization opportunities

The practices comprised the utilization and elimination phase of the product. The practices included energy use reduction, design for re-usability, substance reduction, and recyclability. Afterwards, the sustainability thinking as expanded to cover the supply chain management (SCM).

Generally, GSCM and SSCM concepts have been used to describe the connection between sustainability and SCM (Ashby, Leat, & Hudson-Smith, 2012). Ahi and Searcy (2013) defined GSCM as an integration of environmental thinking into supply chain management. It includes product design, material sourcing and selection, manufacturing processes, delivery of the final product to the consumers, and end-of-life management of the product. SSCM is defined as a degree to which an organization strategically collaborates with its supply chain stakeholder's and collaboratively manages intra- and inter-organization processes for sustainability (Wolf & Seuring, 2010). The SSCM is the extension of GSCM (Brandenburg, Govindan, Sarkis, & Seuring, 2014). The definition of GSCM is quite similar to the SSCM, but it excludes the integration of social and economic aspects (Ahi & Searcy, 2013). The most recent green factory term which emphasizes on the equilibrium between the people and environment using a green cycle is introduced by Rohrmus et al. (2013). The green factory transforms input materials to final products using the advantages of carbon, hydrogen, oxygen as energy sources in its production processes.

Commonalities between Lean and Sustainability

Several researchers and environmental experts have recognized the compatibility between lean and sustainability (Gort, 2008; Larson & Greenwood, 2004). Lean and sustainability require more leadership than financial investment and this would be succeeded if an organization management walks the talk. This means lean and sustainability can be assured if what is stated in the implementation reports (the talks) representing the actual of the organization (the walk). Adopting lean or sustainability is a journey rather than short-span projects (Langenwalter, 2007).

Hitherto, the drive for waste removal is considered as the common link between lean and sustainability. Environmental waste is obviously different from the eight types of lean waste, but it is related as presented in Table 1. Therefore, large environmental gains can be achieved by implementing lean. For instance, less inappropriate processing and more efficient transportation results in lower emissions. In addition, less storage and inventory space from the creation of right-sized production unit results in reduced materials, land, and energy consumed (EPA, 2007).

Moreira et al. (2010) highlighted a positive impact of lean on eco-efficiency of production systems. Tice, Ahouse, and Larson (2005) mentioned that there is a link between waste removal using lean concept and for environmental improvement. The environmental waste (e.g. CO2 emissions and environmental risks) is not explicitly included in the eight wastes (shown in Table 1). However, the concept of lean manufacturing naturally reduces some of the environmental impacts due to its inherent waste removal focus. In conclusion, waste removal is an important aspect for both lean and environmental improvements as it adds more value to the customer and the organization.

Waste Identification Tools

Waste identification refers to detect the location of NVA activities within a supply chain. It is considered as the first step in lean manufacturing implementation (Mostafa et al., 2013). Lean manufacturing consists of tools/practices that are able to identify the sources of waste in an organization. Value stream mapping (VSM) is considered as a significant lean tool for visualizing the information and materials flow (Mike Rother & Shook, 2003). Through the literature, VSM tools can be classified into two main categories: traditional and dynamic VSM (Dotoli, Fanti, Iacobellis, & Rotunno, 2012; Ramadan, Zelong, & Noche, 2012; Schmidtke, Heiser, & Hinrichsen, 2014; Solding & Gullander, 2009). A discussion of each category is provided as following.

Traditional Value Stream Mapping (VSM)

The VSM is a key lean tool. VSM is used for visualizing and analyzing the current state information and material flows for a product or service (Braglia, Carmignani & Zammori 2006). This tool is also used in designing and improving the future state of the value stream (Rother & Shook 1999; Seth & Gupta 2005). VSM can be used for highlighting several deficiencies in the process (Rother & Shook 1999). It represents the value stream using a predefined set of standard icons (Sullivan, McDonald & Van Aken 2002). The VSM icons, firstly introduced by Rother and Shook (1999), fall into three categories: material, information and general icons (Figure 4). To portray the entire value stream, the function of each icon must be established. Then the interrelations among icons are determined (Figure 5). VSM has been employed in many organizations for planning and starting the lean transforma-

Figure 4. Value stream mapping icons
(Source: Edraw max Software)

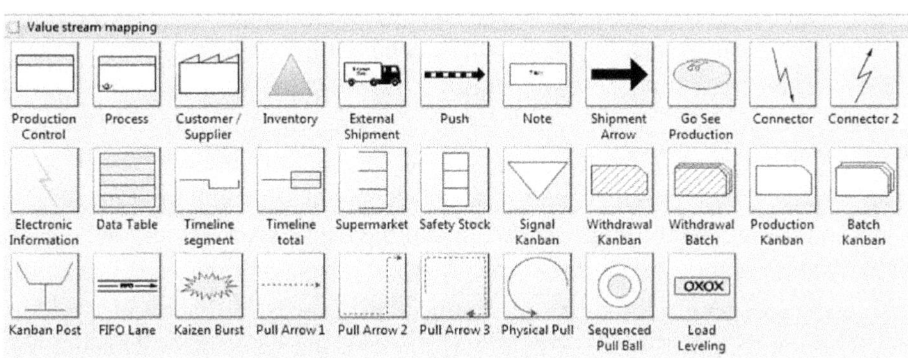

Figure 5. An example of value steam mapping
(Source: Edraw max software)

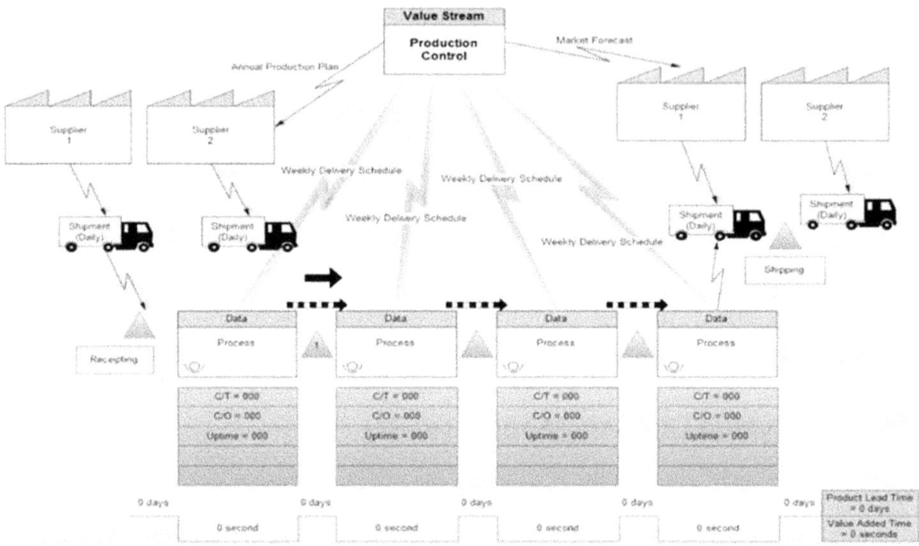

tion (Ramesh, Prasad & Srinivas 2008). The current-state map is primarily used to assist a lean team in understanding an entire system and identifying opportunities for cost savings. The future-state map represents an ideal state of the system with the value added processes and waste eliminating improvements implemented (Lian & Van Landeghem, 2002).

VSM tools become popular from the work of M. Rother and Shook (1999) as the tools were developed to understand the interdependence of one function, department or the whole unit, and for capturing a holistic view about the value stream (Seth & Gupta, 2005). The complexity of current manufacturing processes is the driver behind the development of various VSM tools. Hines and Rich (1997) introduced seven VSM tools: process activity mapping, supply-chain response matrix, production variety funnel, quality filter mapping, demand amplification mapping, decision point analysis and physical structure mapping.

Despite the continuous development of VSM tools, most tools do not capture the impact of variability and the dynamic interactions within the process (Ramesh, Prasad, & Srinivas, 2008). Moreover, creating the current- and future-state maps tends to be carried out using a manual-based technique. This will limit the level of details and the number of different versions of the maps (Lian & Van Landeghem, 2002). Additionally, drawing these maps by hand, displaying the map details, and making changes within the maps are cumbersome processes that require a lot of time (Gurumurthy & Kodali, 2011). Nevertheless, the work of McDonald, Van

Aken, and Rentes (2002) pointed out that modelling tools such as simulation can be used as supplements to VSM traditional tools. Powell and Skjelstad (2012) apply Radio Frequency identification (RFID) to map the internal flows of material and information. It is suggested that RFID can be used to map the entire value stream in real-time, which simplifies the distribution and replenishment plans.

Dynamic Value Stream Mapping (DVSM)

RFID-Based VSM

The development of RFID technology in supply chains was based on the limitations of the barcoding technology used in the service sector (Rawal, 2009). Each item has a unique code that is made up of a series of wide and narrow parallel lines and spaces (Jain, 2014). The code comprises the item information including manufacturer, retailer, price, and weigh. It allows product tracking universally (Chen et al., 2013). However, in some cases scanning and reading the barcodes is ineffective due to environmental conditions such as temperature, dirt, or hazardous contamination. The radio frequency identification (RFID) technology has been used in these cases. The RFID technology has been considered as an important enabler of efficient and effective organizations (Angeles, 2013). Previous studies focused on associating RFID with the sustainability and environmental issues (Lu, Chang, & Liao, 2013), logistics and warehouse management (Chen et al., 2013)and healthcare supply chains (Kumar & Rahman, 2014). In the manufacturing sectors, RFID technology has been applied for tracking and tracing products. It reduces the recycle times and stock levels, increases product visibility, and updating the supply chain with real-time information for making key decisions (Leung, Cheung, & Chu, 2014; Saremi & Taghizadeh, 2013). The RFID system consists of a tag, a reader and antennas, and a computer equipped with a middleware program, in which business rules are configured to automate some decisions (see Figure 6).

Figure 6. RFID system structure

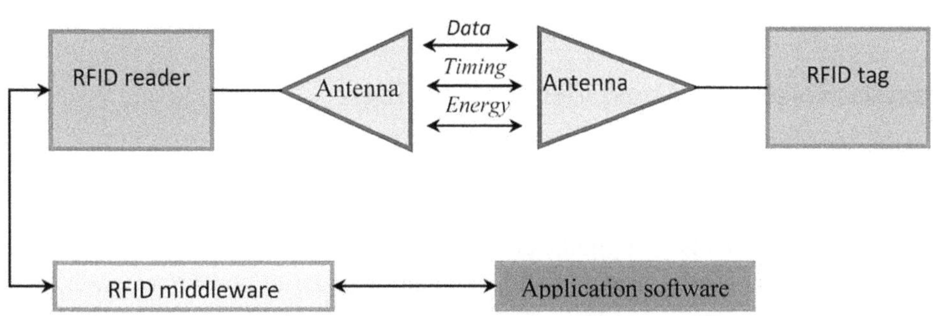

The RFID tag consists of a microchip connected with an antenna and attached to a product. The antenna enables the chip to transmit identification information it contains to a reader antenna through radio frequencies (RF). The reader converts the frequencies into digital information that software systems can use for processing. Normally, the reader sends information to the application software: the tag identification, the reader's identification, and the time the tag was read. By knowing which readers are in which locations, organizations can identify where the product is, as well as what it is, and by tracking the tag data by time, it allows organizations to investigate where the product has been.

RFID technology has been incorporated with lean for maximizing the customer value through eliminating all NVA activities (Ilin, Simić, & Simić, 2012). In the lean supply chain, all components connected with RFID tags. Therefore, the RFID technology is oriented to create the VSM based on the data from all RFID tags. The data is analyzed in order to evaluate the processes. Regardless of the RFID benefits, various challenges including cost of the RFID technology, impacts of environmental conditions on the tag reliability and limited application to some products (e.g. metal and fluid) (Kaur, Sandhu, Mohan, & Sandhu, 2011) still remain.

Simulation-Based VSM

In comparison to the traditional VSM, the integration of the simulation within the traditional VSM facilitates analysis on more complex manufacturing systems (Solding & Gullander, 2009). The discrete event simulation (DES) is used for analyzing and designing advanced manufacturing systems. DES is based on developing a model that mimics the real system in a virtual environment (Rohana, Nooririnah, Isa, Kamat, & Mehad, 2013). Simulation was applied in the study of Abdulmalek and Rajgopal (2007) in supporting VSM to visualize the current state map and validate the future state of steel manufacturing company. The steps of using simulation-based VSM are including production family selection, data acquisition of the system, develop the current sate map, develop and run the simulation model, and designing the future value stream map (Dotoli et al., 2012; Schmidtke et al., 2014). Several limitations of using simulation based VSM include programing and simulation knowledge, high investment costs, and need for extra data and information for developing the model (Solding & Gullander, 2009).

Issues Related to Waste Identification Tools

Traditional and dynamic VSM are used for visualizing a supply chain. The classical VSM has been complemented with RFID and simulation to map complex processes and family of products (Schmidtke et al., 2014). This is called DVSM. Simulation and RFID technologies provide real time data of material and information flow within the supply chain (El-Haik & Al-Aomar, 2006). However, several factors affect the use of DVSM. These factors include high setup cost and complexities of using RFID and simulation technologies. Moreover, it might be difficult to use these technologies for visualizing the environmental waste. Furthermore, DVSM could not provide waste analysis and selection the appropriate lean tools for the waste removal process. Waste identification is not enough for sustainable gains from applying lean in an organization. Waste identification is considered as the grounded stage that should be followed by waste analysis and waste removal stages using lean tools. The sustainability of lean improvements in this chapter can be achieved by adopting the waste elimination process which comprises three phases: waste documentation and acknowledgments, waste analysis and waste removal. Each phase is comprehensively discussed in the following section.

LEAN AND SUSTAINABILITY

Sustainability can be defined as achieving economic growth and industrialization without environmental damages (Keiner, 2006). The concept of sustainability covers three dimensions: economic, social and environment. Lean successful implementation has proved that it contributes to sustainability (Herrmann, Thiede, Stehr, & Bergmann, 2008). Figure 7 demonstrates the sustainability dimension grounded in the lean philosophy. Lean achieves the economic sustainability by improving and sustaining profits margins. The social sustainability is associated with lean as the lean concept emphasizes on human resources within lean culture, health and safety conditions in a workplace (EPA, 2003). From a sustainability perspective, all environmental impacts are considered as waste and resulted from poor manufacturing system. Lean philosophy is a powerful tool to enhance the environmental sustainability (Martínez-Jurado & Moyano-Fuentes, 2013; Torielli, Abrahams, Smillie, & Voigt, 2011).

Lean and Green (Environmental Sustainability)

The US Environmental Protection Agency (EPA) highlighted the correlation between lean and Green concepts. Lean can serve as a catalyst for green path in an organization by optimizing the use of the available resources (EPA, 2003). Previous research confirmed that the lean organizations are transcending to green concept. Bergmiller (2006) indicated that lean organizations are significantly greener than the non-lean organizations. Dües, Tan, and Lim (2013) found that lean is beneficial for green practice in a supply chain. Womack and Jones (2003) mentioned that lean thinking must be green as it reduces the amount of energy and wasted resources required to produce a given product. King and Lenox (2001) asserted that lean production is

Figure 7. Lean philosophy as a sustainable approach

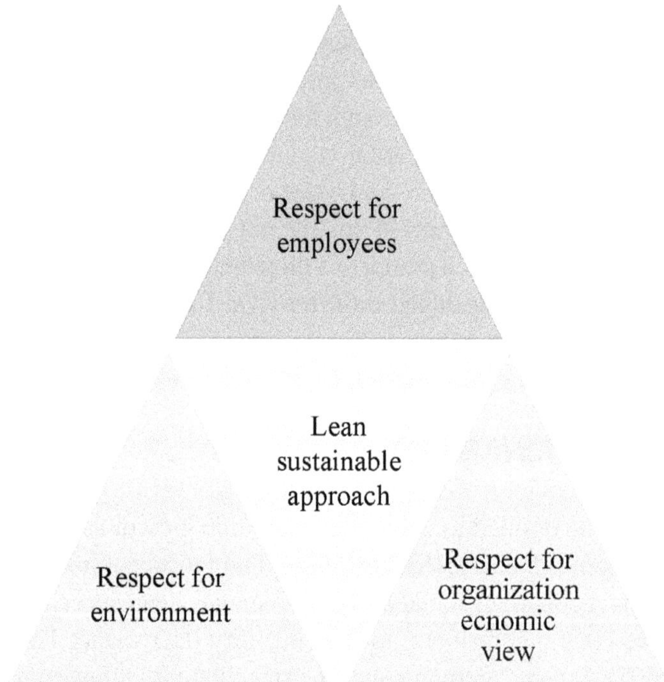

complementary to environmental performance. They believed that implementation of lean will reduce the overall cost of pollution prevention by decreasing the source wasting in the organizations. The study of Rothenberg, Pil, and Maxwell (2001) concluded that production plants managed by lean yield higher green results.

The common factor between lean and green is the waste elimination. Lean focuses on cutting the waste from the whole supply chain (i.e. suppliers, manufacturing processes, and customers). The waste definition from the green perspective is the environmental waste. It refers to unnecessary use of resources or substances released to the air, water, or land that could harm the environment (Khan et al., 2012). The waste in the traditional view of lean are grouped into seven types. However, the innovative thinking of lean concept goes beyond these seven waste types. Lean philosophy has been evolved from manufacturing shop floor toward lean thinking (i.e. lean in non-industrial sectors). Therefore, waste classification has been broadened to include other forms. The underutilization of employee's creativity is considered as waste (Womack & Jones, 2003). Dües et al. (2013) considered green as the new lean. In this case, the waste identification expanded to cover the green waste or environmental footprints (see Figure 8). This waste includes energy, water, materials, garbage's, transportation, emissions and biodiversity (Wills, 2011).

Figure 8. Green waste representation

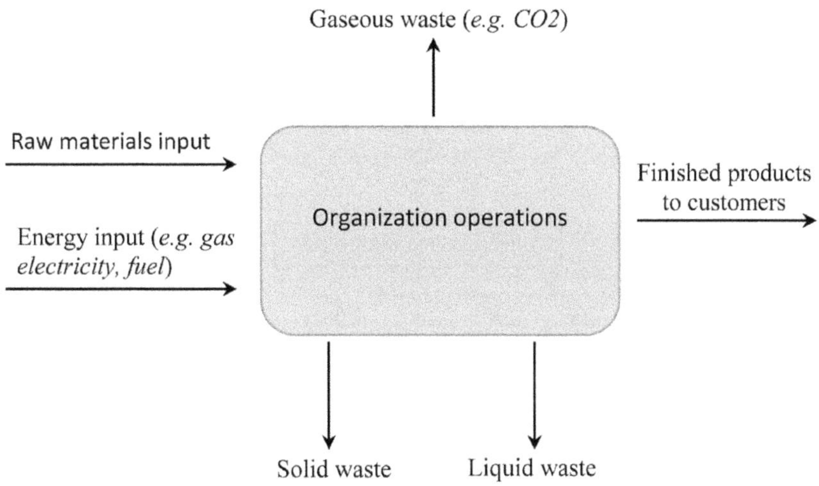

Some organizations have been achieved an improvement in their environmental results as the secondary effect from applying lean only (Rothenberg et al., 2001; Sawhney, Teparakul, Bagchi, & Li, 2007). Despite the positive results, some studies agree that lean practice might conflict with the environmental performance in the areas related to the pollution control technologies (Larson & Greenwood, 2004; Sawhney et al., 2007). Recently, some studies have simplified the understanding of the relationship between lean and green. It is stated that the connection between lean and environment performance should be mediated by the environmental practices inherited in the green concept. Lean facilitates the implementation of environmental practices. However, the resources should be managed by green practices for increasing potentials of environmental results (Hajmohammad, Vachon, Klassen, & Gavronski, 2013; Martínez-Jurado & Moyano-Fuentes, 2013; Vinodh et al., 2011).

WASTE ELIMINATION PROCESS AS A LEAN SUSTAINABLE APPROACH

The waste elimination process suggested in this chapter is considered as a lean sustainable approach. It is in line with the previous studies that highlighted the positive impacts of waste removal using lean and sustainable improvements (EPA, 2009; Moreira et al., 2010; Tice et al., 2005). The waste elimination process includes three consecutive phases as demonstrated in Figure 9. The first phase is waste documentation. This phase is branched into three stages: acknowledgment of the value stream, understanding of the nine types of waste, and locating the waste in the value

Figure 9. Proposed waste elimination process

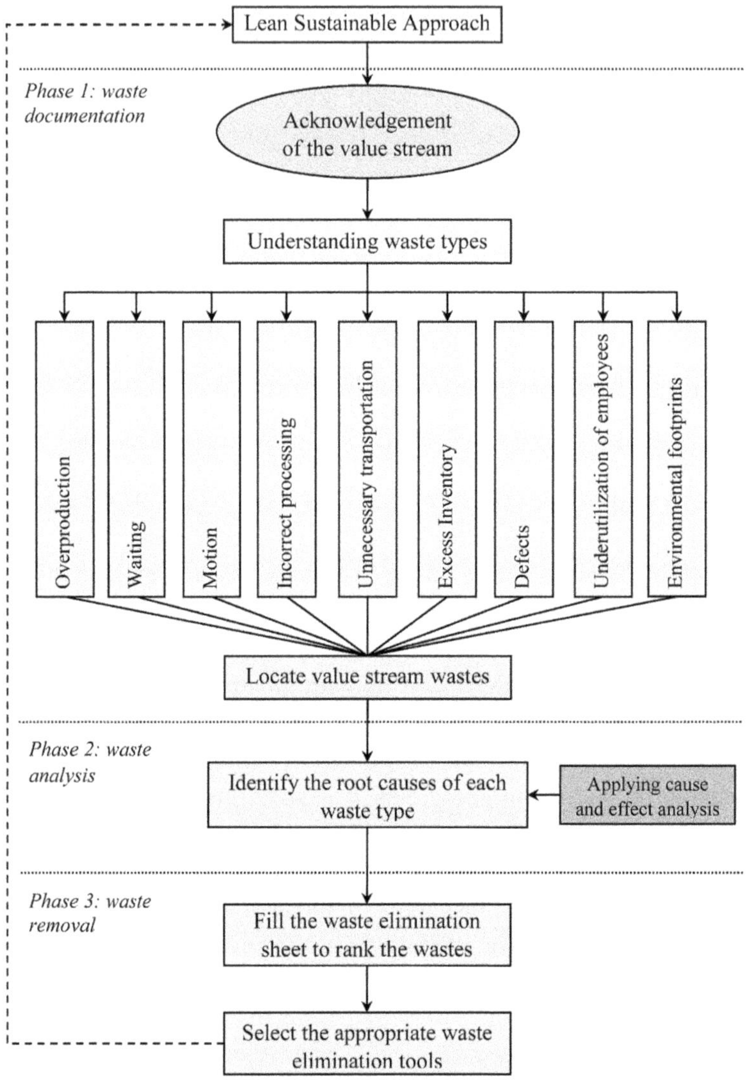

stream. The second phase provides an analysis of each waste type. In this phase, the cause-and-effect diagram (also called Ishikawa diagram) is used to identify the root causes of each waste type. The last phase concerned with waste removal. This phase processed in two stages: ranking of the root causes under each waste type and selection of waste elimination lean tools. In the first stage, the root causes of each waste are filled in the waste elimination template. The root cause then rated based on the removal cost, complexities, and associated effects of waste on other waste causes. A waste priority rank is given to each waste type. In the second stage, the Quality Function deployment (QFD) is employed to associate the appropriate lean tools for eliminating each waste root cause. The discussion of the phases and its contained stages are as follows.

Phase 1: Waste Documentation

Stage 1: Acknowledging the Value Stream

Acknowledging the value stream begins with identifying the customer value. The value is what the customer is willing to pay for (product or service) (Khalid, Hashim, & Salleh, 2014). The value stream comprises all processes of producing and delivering a product/service to customer. These processes might be carried out inside in an organization, or it could include extra external processes (e.g. suppliers) (Locher, 2011). Recognizing the value stream is significant for as it allows the organization to group products according to their manufacturing processes. Moreover, identifying the value stream brings the organization focus on processes required to deliver the customer value. Identification of the value stream requires crossing the manufacturing processes with the products. Products with similar manufacturing processes classified under the same value stream (Dolcemascolo, 2006).

Stage 2: Understanding Waste Types

In waste elimination, it is important to understand each type of waste and where it occurs in the value stream. Toyota production system (TPS) identified seven cardinal waste types namely; overproduction, waiting, unnecessary transport, incorrect processing, excess inventory, unnecessary movement and defects (Ohno, 1988). Womack and Jones (2003)_ENREF_88_ENREF_88_ENREF_51_ENREF_51 extended one more type of waste related to the underutilization of employees' potentials to improve the performance of the organization. As a result, waste has been regrouped into eight types. However, Canon production system (CPS) identified nine categories of waste (Brunt & Butterworth, 2001). The wastes are work-in-process (WIP), quality, facilities, expenses, indirect labor, talent, motion, product design, and new product run-up. The similarities between TPS and CPS categorization of waste have been clarified by Brunt and Butterworth (2001). Four types of waste show large similarities except overproduction waste which is the key difference. Ohno (1988) considered the overproduction waste as main root of other wastes. It is noted that unnecessary transportation and in appropriate processing waste types were not included within the CPS classification. The Canon's nine waste categories include aspects of design, new product run-up, expenses and talents. Nevertheless,

Brunt and Butterworth (2001) argued that Canon's waste types are less focused than Toyota's wastes as it reduces practitioner's capabilities to identify and eliminate waste. As a result of the argument proposed by Brunt and Butterworth (2001) and suggestion of Womack and Jones (2003), this chapter focuses on the eight types of waste in the lean environment. These wastes result in environmental footprints or green waste (Khan et al., 2012). This will be considered as the ninth form of waste (see Table 2).

More descriptions and grouping of the waste types are further explained and utilized in later research works. Hines and Rich (1997) discussed the seven accepted waste types in the TPS. Melton (2005) has outlined a description of the main seven waste types and their analogies in the process industry. Carreira (2005) discussed in details the general seven areas of waste. Sternberg et al. (2013) defined the seven types of waste from production and service prospective. Soltan (2009) classified the waste into obvious, less obvious, and unobvious waste. The obvious waste includes inventory, unneeded processes, excessive setup times, unreliable

Table 2. Waste types in an organization

Waste Type	Description
Overproduction	Producing items too much or too soon, resulting in poor flow of information or goods and excess inventory.
Defects	Frequent errors in paperwork or material/product quality problems resulting in scrap and/or rework, as well as poor delivery performance
Excess inventory	Excessive storage and delay of information or products, resulting in excess inventory and costs, leading to poor customer service
Inappropriate processing	Using inappropriate set of tools, procedures or systems, often when a simpler approach may be more effective. Inappropriate processing due to poor tools and product design, causing unnecessary motion and producing defects. Waste is generated when providing higher-quality products than necessary
Excessive transportation	Excessive movement of information or materials, resulting in wasted time and cost
Waiting	Long periods of inactivity for people, information or goods, resulting in poor flow and long lead times
Unnecessary motion	Poor workplace organization, resulting in poor ergonomics, e.g., excessive bending or stretching and frequently lost items
Underutilization of employee potential	Unused employee creativity and skills to improve the processes and practices. This refers to wasting the available knowledge, experience or skill of the staff/workforce by under-employing them or not using them in the proper department. Poor hiring techniques and task delegation on the part of management are the causes of underutilization of employee.
Environmental footprints (green waste)	Environmental waste is an unnecessary or excessive use of resources or hazardous substances released to air, water, or land that could harm human health or the environment.

machines, and rework. The less obvious waste occurs as a result of variability sources such as process times, delivery times, yield rates, staffing levels, and demand rates (Hopp & Spearman, 2004). Rawabdeh (2005) classified the seven wastes into three main groups related to: man, machine, and material. The man-group includes motion, waiting, and overproduction. The machine-group contains over processing. The material group contains transportation, inventory and defect waste.

The waste types are primarily allocated on the shop-floor. Brunt and Butterworth (2001) performed the waste identification and documented new seven types of waste namely wasted power and energy, wasted human potential, environmental pollution, unnecessary overhead, inappropriate design, departmental culture, and inappropriate information in addition to the existing types. Their identification process encompassed the supply chain perspective. The wasted power and energy increase in service costs of electricity, gas and water as a result of the increase in using automation on shop-floors and offices. The wasted human potential occurs wherever people are not effectively utilized for any potential improvement. The environmental pollution waste is relevant to usage of disposal materials such as packing, chemical discharge, noise levels and healthy safety issues. The waste of unnecessary overhead includes capitalising more for the required output. It comprises hiring many indirect labors, procuring of many tools and equipment, and conducting training that is not necessary or not implemented correctly. The inappropriate design waste is associated to the functionality, features, and manufacturing process of a product. The departmental culture waste and the inappropriate information waste occur whenever departmental politics and complicated IT systems slow up an organisation.

Stage 3: Locate the Value Stream Waste

Identifying the value stream and understanding each type of waste help for mapping the value stream. Mapping the value stream is considers as waste identification tool that used to locate the waste forms and spot areas for improvement within the value stream. The waste identification tools mentioned previously are used to visualizing the value stream. These tools are traditional VSM and dynamic VSM.

Phase 2: Waste Analysis

Identifying and understanding each waste type in an organization is the input for waste analysis (Furterer, 2009). Waste analysis is the first step for assessing and improving any organization performance towards lean. Waste analysis highlights the areas of waste and for improvement within the workplace environment or specific part in an organization. Waste analysis refers to find the root causes ‘sources’ of each waste type. This can be achieved using brainstorming technique for experts within an organization. In this chapter, a cause and Effect (C&E) diagram is used to explain the root causes of each waste type. The root causes can be grouped into four main groups namely; manpower, machine, method, material, and measurement. Figure 10 and Figure 11 display an example of using C&E analysis for pinpointing the possible root causes of inappropriate processing and excess inventory waste.

It is seen in Figure 10 and Figure 11 that each waste is traced back to its root cause. The inappropriate processing is due to lack of unclear standards and specifications. Therefore, employees are not aware of what actually adds value to a product/service from customer view. The inappropriate processing has been influenced by four sources manpower, method, machine and material. The manpower is contributing to the inappropriate processing through poor training, lack of com-

Figure 10. Cause and effect analysis of inappropriate processing waste

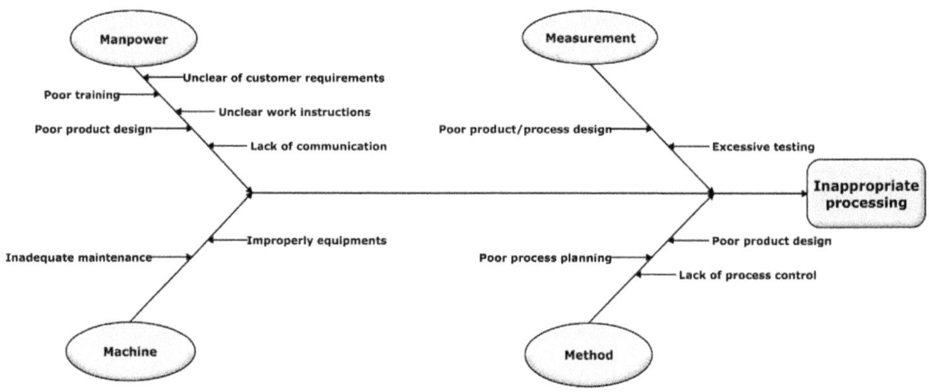

Figure 11. Cause and effect analysis of unnecessary inventory waste

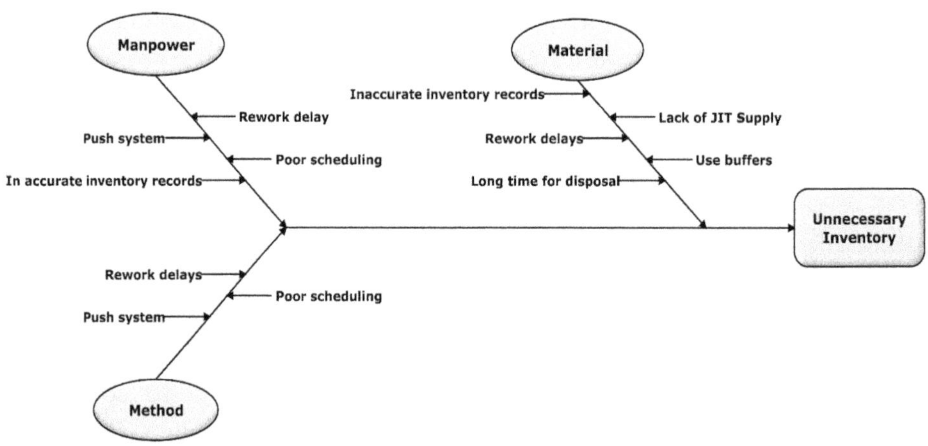

munication and unclear about the customer requirements. Inadequate maintenance and improper machines resulted in improper processing of the product. The poor product and process design, and lack or process control methods resulted in inappropriate processing. For unnecessary inventory, the three main sources of waste are manpower and method. Manpower could cause this waste through poor scheduling, using push system, delay of rework and inaccurate inventory records. The method contributes to unnecessary inventory through lack of using JIT supply system, using buffers, and long time for parts disposal.

Phase 3: Waste Removal

There are a number of techniques that have been developed to understand and eliminate waste. Kobayashi (1990) constructed a 'treasure mountain map' method to rank waste into three levels according to seriousness. The very wasteful activities are considered as gold, whereas other smaller degrees of wastefulness are considered as silver, copper, or flatlands. Once these categories of waste have been highlighted, boosting the work groups involved and developing an improvement plans can create an efficient workplace to eliminate the waste. Imai (1986) introduced the continuous improvement or Kaizen concept. Waste elimination was one of the three principles in Kaizen concept. In waste elimination, cross-functional teams play important role in analyzing the value stream activities to identify and eliminate waste. Each Kaizen event could produce around 50% reduction in waste compared to 5% annual waste reduction achieved by batch-and-queue businesses (Emiliani, 2001).

Rawabdeh (2005) investigated the waste types in a job shop environment. He proposed an assessment model to explore the relationship between the seven types of waste (i.e. wastes identified by Toyota's corporation). The proposed model would help to identify the root causes of each waste type. Seth, Seth, and Goel (2008) conducted a survey about employing the 5S programs with the waste minimization at a manufacturing organization. The survey concluded that 5S was an effective practice in improving the organization performance and safety standards. It reduced the waste at the shop floor and ergonomic level.

Mohanraj, Sakthivel, and Vinodh (2011) proposed a holistic model for continuous improvement. The model consists of three inter-connected cycles: investigation, method and implementation cycle. The three cycles were based on the 4M (man, machine, material, and method) framework. The investigation cycle have been used for examining what are the actions needed, what are the current constraints, and how to overcome these constraints to improve the performance. The method cycle describes the main activities to move an organisation towards lean performance. It includes the key system components for waste elimination such as 5S, flow manufacturing, and JIT-supply and standard operations and procedures. The implementation cycle is required to achieve the performance improvement. The plans, do, check and act (PDCA) cycle has been suggested within this cycle. They noted that the 5S framework forms the basis for a total improvement program. They acknowledged that the waste sources include the hard 'technical' and the soft drivers of any operation. The Kanban system illustrates how the supporting procedures and format of documentation need to be engineered or re-engineered to create efficient waste-less manufacturing operations.

The waste removal strategy for the nine types of waste documented in this chapter has been rarely located. Most of the previous studies emphasized on seven types identified by TPS. In this chapter, the third phase, waste removal, consists of completing the waste elimination template in order to rank the significance of each waste type and the appropriate selection of lean tools for waste removal.

Stage 1: Completing the Waste Elimination Template

The waste elimination template is employed to rank the importance of each waste type (see Table 3). The design of the proposed waste elimination template is developed from the Failure Mode and Effect Analysis (FEMA) template. Here, waste is regarded as failure to optimize the resource utilization. The modes of this failure (lean waste) are the root causes of each waste type which are placed in the second column in the template (Table 3). Then the root causes are evaluated with respect to five criteria; cost of removal (COR), ease of removal (EOR), impact (I) on other causes, occurrence (O), and detection (D). The actions to be undertaken within an organization are the key elements for the waste removal process. Therefore, organization should consider the COR and EOR for each root cause. COR represents how much cost an organization spends on removal of the waste root cause. The EOR is the complexity of actions in removing the waste root cause. The complexity challenges include number of employees involved, time required, and newness of the action.

The other three criteria (I, O, and D) are related to the impact of waste root cause removal on other causes, occurrence of the cause, and the detection mechanism of the cause. The risk priority number (RPN) is then calculated for each cause containing the impact, occurrence and detection of the cause as in Equation 1. The waste priority number (WPN) is then calculated for each waste type. The number represents a summation of the COR, EOR and RPN of each root cause under the same waste type as in Equation 2.

$$RPN_i = I_i \times O_i \times D_i \quad (1)$$

Table 3. Waste elimination template

Waste Types	Root Causes of Waste	Cost of Removal (COR)			Ease of Removal (EOR)		Impact (I)					Occurrence (O)		Detection (D)		Risk Priority Number	Waste Priority Number
		High	Medium	Low	Easy	Difficult	Strong	Moderate	No Effect	Moderate	Strong	Very infrequent	Nearly Always	High	Low		
		1	5	10	10	1	10	5	1	-5	-10	1	10	1	10		
Overproduction	Poor scheduling																
	Use of batches																
	Push system																
	Equipment breakdowns																
	Long process setup time																
	Inspection delays																
	Lack of coordination with customers																
Unnecessary Motion	Use buffers																
	Long time for disposal																
	Rework delays																
	Lack of JIT supply																
	Inaccurate inventory records																
	Rework delays																
	Push system																

continued on following page

Table 3. Continued

Waste Types	Root Causes of Waste	Cost of Removal (COR)			Ease of Removal (EOR)		Impact (I)					Occurrence (O)		Detection (D)		Risk Priority Number	Waste Priority Number
		High	Medium	Low	Easy	Difficult	Strong	Moderate	No Effect	Moderate	Strong	Very infrequent	Nearly Always	High	Low		
		1	5	10	10	1	10	5	1	-5	-10	1	10	1	10		
Inappropriate processing	No tools or wrong tools																
	Unplanned maintenance																
	Long setup times																
	Push system																
	Quality problems																
	Handling between operations																
	Poor scheduling																
	Functional process layout																
	Inadequate preventive maintenance																
	Poor work coordination																
	No employees cross-training																
	Lack of employee empowerment																
Excessive Transportation	WIP Storage																
	Misplaced materials																
	Distribution system																
	Storing away from point of use																
	Poor handling																
	Poor layout																

continued on following page

Table 3. Continued

Waste Types	Root Causes of Waste	Cost of Removal (COR)			Ease of Removal (EOR)		Impact (I)					Occurrence (O)		Detection (D)		Risk Priority Number	Waste Priority Number
		High	Medium	Low	Easy	Difficult	Strong	Moderate	No Effect	Moderate	Strong	Very infrequent	Nearly Always	High	Low		
		1	5	10	10	1	10	5	1	-5	-10	1	10	1	10		
Unnecessary Motion	Push system																
	Poor housekeeping																
	Functional process layout																
	Poor design of process flow																
	Lack of one-piece flow																
	Poor work design																
	Poor layout																
	Misunderstanding																
	Lack of cellular layout																
Defects	Improper equipment's																
	Poor quality of inputs																
	Lack of mistake proofing																
	Wrong process setup																
	No process control																
	Improper machining																
	Inspection																
	Human errors																
	Lack of understand customer needs																
	Inappropriate handling																
	Training																

continued on following page

Table 3. Continued

Waste Types	Root Causes of Waste	Cost of Removal (COR)			Ease of Removal (EOR)		Impact (I)					Occurrence (O)		Detection (D)		Risk Priority Number	Waste Priority Number
		High	Medium	Low	Easy	Difficult	Strong	Moderate	No Effect	Moderate	Strong	Very infrequent	Nearly Always	High	Low		
		1	5	10	10	1	10	5	1	-5	-10	1	10	1	10		
Inappropriate processing	Lack of process control																
	Poor product design																
	Poor process planning																
	Excessive testing																
	Poor process design																
	Lack of communication																
	Unclear work instructions																
	Inadequate maintenance																
	Improper equipment's																
Underutilization of employees potential	Organizational structure																
	Poor hiring practices																
	Poor Training																
	Poor leadership																
	High Employee turnover																
	No employee involvement																

$$WPN_j = \sum_{j=1}^{8}\sum_{i=1}^{m} COR_{i,j} + \sum_{j=1}^{8}\sum_{i=1}^{m} EOR_{i,j} + RPN_i \quad (2)$$

where,

I_i : Impact of cause i
O_i : Occurrence of cause i
D_i : Detection of cause i
RPN_i : Risk priority number of cause i
WPN_j : Priority number of waste j
n: Number of waste types
m: Number of root causes under each waste type
i: Refers to root cause *I, i=1, 2, ...,m*
j: Refers to waste type *j, j=1, 2, 3,............,8*

Stage 2: Select the Appropriate Waste Elimination Tools

Lean manufacturing is a set of tools/practices to eliminate waste in an organization. Pavnaskar, Gershenson, and Jambekar (2003) emphasized on a complete understanding of lean tools and selection of appropriate tools for effective waste elimination process. Mostafa et al. (2013)_ENREF_59_ENREF_59 highlighted that not all lean tools can eliminate the same waste, and not all waste can be eliminated with a single tool. The selection of lean tools is based on using lean toolbox guide (Bicheno, 2004). Decision making tools including analytic hierarchy process (AHP)-Data envelopment analysis (DEA) (Anvari, Zulkifli, Sorooshian, & Boyerhassani, 2014), and analytic network process (ANP) (Anand & Kodali, 2009) have been used for appropriate lean tools selection. In this chapter, the quality function deployment (QFD) technique is used to display an association between the lean tools and the eight types of waste. Employing the QFD in ranking waste is established according to the study of Hussain, Tsironis, and Ajmal (2011). It was concluded in their study that QFD is an effective technique to translate the voice of customer (VOC) into product design parameters. The study was focused on applying the QFD matrix in service sector (Telecom organizations). It was proven that QFD can be extendedly applied to other aspects associating lean tools with the eight types of waste (See Figure 12).

The QFD has two principal parts: WHAT's and HOW's (see Figure 10). The WHAT's contain information relevant to the eight types of waste while the HOW's represent the corresponding lean tools. The basic process underlying QFD exists in

Figure 12. House of waste and lean tools

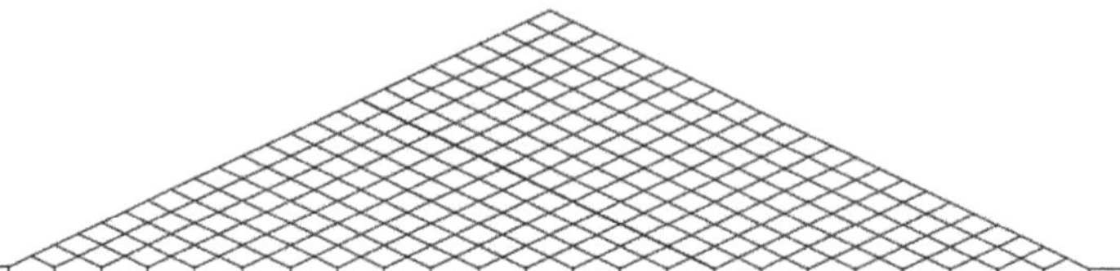

Hows (Practices) / Whats (To remove)	Priority	Pull System/Kanban	SMED	Cellular Manufacturing	Production Smoothing	Takt Time	Focused Factory	Work Standardization	Line Balancing	Kaizen/5S's	Kansei Engineering	VSM	Poka-Yoke	Benchmarking	Jidoka	Quality Circles	DFX	Training programs	Work teams	Preventive maintenance	HR Motivation	Job design	Small Lot Size Production	Importance
Overproduction																								
Defects																								
Unnecessary inventory																								
Inappropriate processing																								
Excessive transportation																								
Waiting																								
Unnecessary motion																								
Under-utilization of employees																								
How Much																								
Score																								
Rank																								

the centre of the QFD matrix where the waste types and practices intersect. The interrelationship between waste and tools follows a scaling system such as (+1= weak, +3 positive, +9 strong positive).

FUTURE RESEARCH IN GREEN AND SUSTAINABLE MANUFACTURING

Linking Lean Wastes and Green Wastes

The chapter has introduced eight types of waste under lean manufacturing. Most literature highlighted that the green waste is regarded as the outcome of lean waste. However, all individual types of lean waste may not have their own distinguishable categories under the green waste. For example, overproduction is identified as a type of lean waste but it cannot be labeled to any specific category under the green waste. It is significant for an effective waste elimination process to create a relation between individual lean waste and the environmental waste. Moreover, the establishment of a quantifying technique to measure individual lean waste against the environmental waste can help to explain waste behaviors and patterns in manufacturing systems.

Establishment of Green Manufacturing Systems

In manufacturing sector, besides lean manufacturing various manufacturing systems such as agile manufacturing, mass customization and mass production have been applied. Whilst lean has been confirmed its association to sustainability attainment, it is not evident that other manufacturing systems has pursued their path in producing in green manners. All manufacturing systems typically share some commonalities in managing the entire supply chain. Nevertheless, it seems only lean manufacturing that has built a firm connection with the sustainability concept. To take part in the development towards the green environment, the current manufacturing systems should establish coexisting environmental considerations as demonstrated in Figure 13.

CONCLUSION

This chapter has identified the types of waste in the lean manufacturing system. Among the waste types, the environmental waste has been included into the proposed waste elimination process. The aim of the process is to attain lean sustainability.

Figure 13. Role of manufacturing systems in the green concept

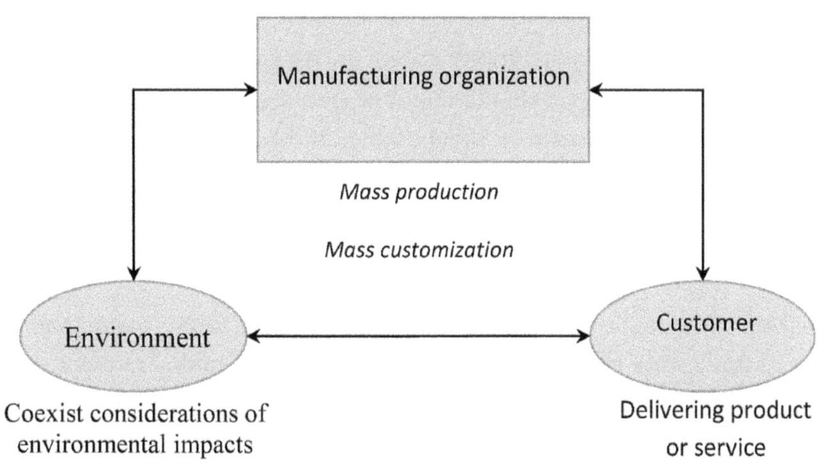

Lean is known to the manufacturing system as a concept to maximize the end-user value by cutting the non-value-added activities or waste. The connection between lean and sustainability has been established through the integration of environmental concerns. This means the lean waste types and the environmental waste types must be associated and studied in parallel. One component of the waste elimination process is to utilize the waste identification tools. The value stream mapping (VSM) is suggested to visualize the flows of materials and information as well as to locate where the waste occurs. For any complex manufacturing environment, capturing waste can be performed using more dynamic tools such as RFID-based VSM or simulation-based VSM. To enhance the waste elimination process, a lean sustainable approach and its framework have been introduced. The approach offers a holistic thinking of how waste should be acknowledged, classified, analyzed for root causes, and removed using lean tools/practices.

Additionally, the chapter provides a waste removal strategy using waste elimination equations and template. As a result, an association between waste types and root causes of waste can be established. The strategy also allows documentation of removal cost, ease of removal, impact, occurrence, and detection of waste which leads to risk and waste priority in lean and sustainable manufacturing. Nevertheless, future research in green and sustainable manufacturing is required to create an in-depth understanding on the linkage between lean and green waste. This allows individual relationships between the two categories of waste to be identified. The future manufacturing systems should be established in responding to green environment. This will provide opportunities for other manufacturing systems to be incorporated with sustainability ideology.

REFERENCES

Abdulmalek, F., & Rajgopal, J. (2007). Analyzing the benefits of lean manufacturing and value stream mapping via simulation: A process sector case study. *International Journal of Production Economics, 107*(1), 223–236. doi:10.1016/j.ijpe.2006.09.009

Ahi, P., & Searcy, C. (2013). A comparative literature analysis of definitions for green and sustainable supply chain management. *Journal of Cleaner Production, 52*(0), 329–341. doi:10.1016/j.jclepro.2013.02.018

Amini, M., & Bienstock, C. C. (2014). Corporate sustainability: An integrative definition and framework to evaluate corporate practice and guide academic research. *Journal of Cleaner Production, 76*, 12–19. doi:10.1016/j.jclepro.2014.02.016

Anand, G., & Kodali, R. (2009). Selection of lean manufacturing systems using the analytic network process – a case study. *Journal of Manufacturing Technology Management, 20*(2), 258–289. doi:10.1108/17410380910929655

Angeles, R. (2013). Using the Technology-Organization-Environment Framework and Zuboff's Concepts for Understanding Environmental Sustainability and RFID: Two Case Studies. *International Journal of Social. Human Science and Engineering, 7*(11), 1607–1617.

Anvari, A., Zulkifli, N., Sorooshian, S., & Boyerhassani, O. (2014). An integrated design methodology based on the use of group AHP-DEA approach for measuring lean tools efficiency with undesirable output. *International Journal of Advanced Manufacturing Technology, 70*(9-12), 2169–2186. doi:10.1007/s00170-013-5369-z

Ashby, A., Leat, M., & Hudson-Smith, M. (2012). Making connections: A review of supply chain management and sustainability literature. *Supply Chain Management: An International Journal, 17*(5), 497–516. doi:10.1108/13598541211258573

Bergmiller, G. G. (2006). *Lean manufacturers transcendence to green manufacturing: Correlating the diffusion of lean and green manufacturing systems. (PhD).* University of South Florida.

Bicheno, J. (2004). *The new lean toolbox: towards fast, flexible flow*. Buckingham, UK: PICSIE Books.

Brandenburg, M., Govindan, K., Sarkis, J., & Seuring, S. (2014). Quantitative models for sustainable supply chain management: Developments and directions. *European Journal of Operational Research, 233*(2), 299–312. doi:10.1016/j.ejor.2013.09.032

Brunt, D., & Butterworth, C. (2001). Waste elimination–a supply chain perspective. In D. Taylor & D. Brunt (Eds.), *Manufacturing Operations and Supply Chain Management, The LEAN Approach* (pp. 79–87). Thomson Learning.

Bruun, P., & Mefford, R. N. (2004). Lean production and the Internet. *International Journal of Production Economics, 89*(3), 247–260. doi:10.1016/j.ijpe.2003.10.007

Chen, J. C., Cheng, C.-H., Huang, P. T. B., Wang, K.-J., Huang, C.-J., & Ting, T.-C. (2013). Warehouse management with lean and RFID application: A case study. *International Journal of Advanced Manufacturing Technology, 69*(1-4), 531–542. doi:10.1007/s00170-013-5016-8

Dolcemascolo, D. (2006). *Improving the extended value stream: lean for the entire supply chain*. Productivity Press.

Dombrowski, U., Mielke, T., & Schulze, S. (2012). Lean Production Systems as a Framework for Sustainable Manufacturing. In G. Seliger (Ed.), *Sustainable Manufacturing* (pp. 17–22). Springer Berlin Heidelberg. doi:10.1007/978-3-642-27290-5_3

Dotoli, M., Fanti, M. P., Iacobellis, G., & Rotunno, G. (2012). *A lean manufacturing strategy using Value Stream Mapping, the Unified Modeling Language, and discrete event simulation.* Paper presented at the International Conference on Automation Science and Engineering (CASE), Seoul.

Drake, P. R., Lee, D. M., & Hussain, M. (2013). The lean and agile purchasing portfolio model. *Supply Chain Management: An International Journal*, *18*(1), 3–20. doi:10.1108/13598541311293140

Dües, C. M., Tan, K. H., & Lim, M. (2013). Green as the new Lean: How to use Lean practices as a catalyst to greening your supply chain. *Journal of Cleaner Production*, *40*(0), 93–100. doi:10.1016/j.jclepro.2011.12.023

El-Haik, B., & Al-Aomar, R. (2006). *Simulation-based Lean Six-sigma and Design for Six-sigma*. John Wiley & Sons. doi:10.1002/0470047720

Emiliani, M. L. (2001). Redefining the focus of investment analysts. *The TQM Magazine*, *13*(1), 34–51. doi:10.1108/09544780110359400

EPA. (2003). *Lean Manufacturing and the Environment: Research on Advanced Manufacturing Systems and the Environment and Recommendations for Leveraging Better Environmental Performance*. Washington, DC: United States Environmental Protection Agency.

EPA. (2007). *The Lean and Environment Toolkit (U. S. E. P. Agency, Trans.)*. Washington, DC: U.S. Environmental Protection Agency.

EPA. (2009). *The Environmental Professional's Guide to Lean & Six Sigma*. Washington, DC: U.S. Environmental Protection Agency.

Fliedner, G. (2008). *Sustainability: a new lean principle.* Paper presented at the 39th annual meeting of the decision sciences institute, Baltimore, MD.

Furterer, S. L. (2009). *Lean Six Sigma in service: Applications and case studies.* Boca Raton, FL: CRC Press. doi:10.1201/9781420079104

Gehin, A., Zwolinski, P., & Brissaud, D. (2008). A tool to implement sustainable end-of-life strategies in the product development phase. *Journal of Cleaner Production*, *16*(5), 566–576. doi:10.1016/j.jclepro.2007.02.012

Gold, S., Seuring, S., & Beske, P. (2010). Sustainable supply chain management and inter-organizational resources: A literature review. *Corporate Social Responsibility and Environmental Management*, *17*(4), 230–245. doi:10.1002/csr.207

Gort, R. E. (2008). *Lean And Sustainability: How Can They Reinforce Each Other? (MBA)*. Tilburg, The Netherlands: Tilburg University and Eindhoven University of Technology.

Green, K. W. Jr, Zelbst, P. J., Meacham, J., & Bhadauria, V. S. (2012). Green supply chain management practices: Impact on performance. *Supply Chain Management: An International Journal*, *17*(3), 290–305. doi:10.1108/13598541211227126

Gurumurthy, A., & Kodali, R. (2011). Design of lean manufacturing systems using value stream mapping with simulation: A case study. *Journal of Manufacturing Technology Management*, *22*(4), 444–473. doi:10.1108/17410381111126409

Hajmohammad, S., Vachon, S., Klassen, R. D., & Gavronski, I. (2013). Reprint of Lean management and supply management: Their role in green practices and performance. *Journal of Cleaner Production*, *56*, 86–93. doi:10.1016/j.jclepro.2013.06.038

Haque, B., & James-moore, M. (2004). Applying lean thinking to new product introduction. *Journal of Engineering Design*, *15*(1), 1–31. doi:10.1080/0954482031000150125

Herrmann, C., Thiede, S., Stehr, J., & Bergmann, L. (2008). An environmental perspective on Lean Production. In M. Mitsuishi, K. Ueda, & F. Kimura (Eds.), *Manufacturing Systems and Technologies for the New Frontier* (pp. 83–88). Springer London. doi:10.1007/978-1-84800-267-8_16

Hicks, B. J. (2007). Lean information management: Understanding and eliminating waste. *International Journal of Information Management*, *27*(4), 233–249. doi:10.1016/j.ijinfomgt.2006.12.001

Holweg, M. (2007). The genealogy of lean production. *Journal of Operations Management*, *25*(2), 420–437. doi:10.1016/j.jom.2006.04.001

Hopp, W. J., & Spearman, M. L. (2004). Commissioned Paper To Pull or Not to Pull: What Is the Question? *Manufacturing & Service Operations Management*, *6*(2), 133–148. doi:10.1287/msom.1030.0028

Hussain, M., Tsironis, L., & Ajmal, M. M. (2011). A QFD strategy for improving customer satisfaction: Case study of telecom companies of Pakistan. *Asian Journal on Quality*, *12*(3), 282–295. doi:10.1108/15982681111187119

Ilin, V., Simić, D., & Simić, S. (2012). Lean RFID approach enhancing the information and material flows in emergency department. *Journal of Medical Informatics & Technologies, 21*, 49–55.

Imai, M. (1986). *Kaizen: the key to Japan's competitive success* (1st ed.). New York: McGraw-Hill.

Jain, D. (2014). RFID and its impact on supply chain. *International Journal of Logistics & Supply Chain Management Perspectives, 2*(4), 741–756.

Kaur, M., Sandhu, M., Mohan, N., & Sandhu, P. S. (2011). RFID Technology Principles, Advantages, Limitations & Its Applications. *International Journal of Computer and Electrical Engineering, 3*(1), 151–157. doi:10.7763/IJCEE.2011.V3.306

Keiner, M. (2006). *The future of sustainability*. Springer Netherlands. doi:10.1007/1-4020-4908-0

Khalid, N. K. M., Hashim, A. Y. B., & Salleh, M. R. (2014). On Value Stream Mapping and Its Industrial Significance. *Journal of Industrial and Intelligent Information, 2*(2), 88–90. doi:10.12720/jiii.2.2.88-90

Khan, M., Jaber, M. Y., & Glock, C. H. (2012). *Impact of learning on the environmental performance of a two level supply chain.* Paper presented at the 17th International Symposium on Inventories, Budapest, Hungary.

King, A. A., & Lenox, M. J. (2001). Lean and Green? An empirical examination of the relationship between lean production and environmental performance. *Production and Operations Management, 10*(3), 244–256. doi:10.1111/j.1937-5956.2001.tb00373.x

Kobayashi, I. (1990). *20 keys to workplace improvement* (W. S. Warren, Trans.). Cambridge, MA: Productivity Press, Inc.

Kumar, A., & Rahman, S. (2014). RFID-enabled process reengineering of closed-loop supply chains in the healthcare industry of Singapore. *Journal of Cleaner Production*, (0). doi:10.1016/j.jclepro.2014.04.037

Langenwalter, G. (2007). *Life is Our Ultimate Customer: From Lean to Sustainability*. Retrieved 20 May 2014, from AME http://www.ame.org/search/node/Life%20is%20Our%20Ultimate%20Customer%3A%20From%20Lean%20to%20Sustainability

Larson, T., & Greenwood, R. (2004). Perfect complements: Synergies between lean production and eco-sustainability initiatives. *Environmental Quality Management, 13*(4), 27–36. doi:10.1002/tqem.20013

Leung, J., Cheung, W., & Chu, S.-C. (2014). Aligning RFID applications with supply chain strategies. *Information & Management*, *51*(2), 260–269. doi:10.1016/j.im.2013.11.010

Lian, Y. H., & Van Landeghem, H. (2002). *An application of simulation and value stream mapping in lean manufacturing*. Paper presented at the 14th European Simulation Symposium: Simulation in Industry (ESS), Dresden, Germany.

Liker, J. K. (2004). *The Toyota way: 14 management principles from the world's greatest manufacturer*. New York: McGraw-Hill.

Locher, D. A. (2011). *Value stream mapping for lean development: a how-to guide for streamlining time to market*. CRC Press.

Lu, J.-W., Chang, N.-B., & Liao, L. (2013). Environmental Informatics for Solid and Hazardous Waste Management: Advances, Challenges, and Perspectives. *Critical Reviews in Environmental Science and Technology*, *43*(15), 1557–1656. doi:10.1080/10643389.2012.671097

Martínez-Jurado, P. J., & Moyano-Fuentes, J. (2013). Lean Management, Supply Chain Management and Sustainability: A Literature Review. *Journal of Cleaner Production*, (0). doi:10.1016/j.jclepro.2013.09.042

McDonald, T., Van Aken, E. M., & Rentes, A. F. (2002). Utilising Simulation to Enhance Value Stream Mapping: A Manufacturing Case Application. *International Journal of Logistics Research and Applications*, *5*(2), 213–232. doi:10.1080/13675560210148696

Mehta, V., & Shah, H. (2005). Characteristics of a work organization from a lean perspective. *Engineering Management Journal*, *17*(2), 14–20. doi:10.1080/10429247.2005.11415282

Melton, T. (2005). The Benefits of Lean Manufacturing: What Lean Thinking has to Offer the Process Industries. *Chemical Engineering Research & Design*, *83*(6), 662–673. doi:10.1205/cherd.04351

Mohammed, J., & Sadique, A. (2011). Lean Value Stream Manufacturing for Sustainability. In G. Seliger, M. M. K. Khraisheh, & I. S. Jawahir (Eds.), *Advances in Sustainable Manufacturing* (pp. 365–370). Springer Berlin Heidelberg. doi:10.1007/978-3-642-20183-7_52

Mohanraj, R., Sakthivel, M., & Vinodh, S. (2011). QFD integrated value stream mapping: An enabler of lean manufacturing. *International Journal of Productivity and Quality Management*, *7*(4), 501–522. doi:10.1504/IJPQM.2011.040546

Molamohamadi, Z., & Ismail, N. (2013). Developing a New Scheme for Sustainable Manufacturing. *International Journal of Materials. Mechanics and Manufacturing*, *1*(1), 1–5. doi:10.7763/IJMMM.2013.V1.1

Monden, Y. (1998). *Toyota production system: an integrated approach to just-in-time*. Norcross, GA: Engineering & Management Press.

Moreira, F., Alves, A. C., & Sousa, R. M. (2010). Towards Eco-efficient Lean Production Systems. In Á. Ortiz, R. Franco, & P. Gasquet (Eds.), *Balanced Automation Systems for Future Manufacturing Networks* (Vol. 322, pp. 100–108). Springer Berlin Heidelberg. doi:10.1007/978-3-642-14341-0_12

Mostafa, S., Dumrak, J., & Soltan, H. (2013). A framework for lean manufacturing implementation. *Production & Manufacturing Research*, *1*(1), 44–64. doi:10.1080/21693277.2013.862159

Nagel, M. H., & Tomiyama, T. (2004). *Intelligent sustainable manufacturing systems, management of the linkage between sustainability and intelligence - an overview.* Paper presented at the IEEE International Conference on Systems, Man & Cybernetics.

Narasimhan, R., Swink, M., & Kim, S. W. (2006). Disentangling leanness and agility: An empirical investigation. *Journal of Operations Management*, *24*(5), 440–457. doi:10.1016/j.jom.2005.11.011

Naylor, B. J., Naim, M. M., & Berry, D. (1999). Leagility: Integrating the lean and agile manufacturing paradigms in the total supply chain. *International Journal of Production Economics*, *62*(1–2), 107–118. doi:10.1016/S0925-5273(98)00223-0

Ohno, T. (1988). *Toyota production system: beyond large-scale production*. Portland, OR: Productivity Press.

Pavnaskar, S. J., Gershenson, J. K., & Jambekar, A. B. (2003). Classification scheme for lean manufacturing tools. *International Journal of Production Research*, *41*(13), 3075–3090. doi:10.1080/0020754021000049817

Porter, M. E., & Kramer, M. R. (2006). The link between competitive advantage and corporate social responsibility. *Harvard Business Review*, *84*(12), 78–92. PMID:17183795

Powell, D., & Skjelstad, L. (2012). RFID for the extended lean enterprise. *International Journal of Lean Six Sigma*, *3*(3), 172–186. doi:10.1108/20401461211282691

Ramadan, M., Zelong, Wang, & Noche, B. (2012). *RFID- Enabled dynamic Value Stream Mapping*. Paper presented at the International Conference on Service Operations and Logistics, and Informatics (SOLI), Suzhou.

Ramesh, V., Prasad, K. V. S., & Srinivas, T. R. (2008). Implementation of a Lean Model for Carrying out Value Stream Mapping in a Manufacturing Industry. *Journal of Industrial and Systems Engineering*, *2*(3), 180–196.

Rawabdeh, I. A. (2005). A model for the assessment of waste in job shop environments. *International Journal of Operations & Production Management*, *25*(8), 800–822. doi:10.1108/01443570510608619

Rawal, A. (2009). RFID: The Next Generation Auto-ID Technology. *Microwave Journal*, *52*(3), 58–62.

Rohana, A., Nooririnah, O., Isa, H., Kamat, S., & Mehad, M. (2013). Lean Waste Analysis and Improvement Using Dynamic Value Stream Mapping. *Global Engineers & Technologists Review*, *3*(2), 1–8.

Rohrmus, D., Döricht, V., Weinert, N., Holst, J.-C., & Kiener, C. (2013). Green Cycle Factory. In A. Azevedo (Ed.), *Advances in Sustainable and Competitive Manufacturing Systems* (pp. 1653–1663). Springer International Publishing. doi:10.1007/978-3-319-00557-7_133

Rothenberg, S., Pil, F. K., & Maxwell, J. (2001). Lean, green, and the quest for superior environmental performance. *Production and Operations Management*, *10*(3), 228–243. doi:10.1111/j.1937-5956.2001.tb00372.x

Rother, M., & Shook, J. (1999). *Learning to see: value stream mapping to create value and eliminate muda*. Brookline, MA: Lean Enterprise Institute, Inc.

Rother, M., & Shook, J. (2003). *Learning to see: value stream mapping to add value and eliminate muda*. Cambridge, MA: Lean Enterprise Institute.

Saremi, S., & Taghizadeh, M. (2013). RFID Adoption by Supply Chain Organizations in Malaysia. *International Proceedings of Economics Development & Research, 59*.

Sawhney, R., Teparakul, P., Bagchi, A., & Li, X. (2007). En-Lean: A framework to align lean and green manufacturing in the metal cutting supply chain. *International Journal of Enterprise Network Management*, *1*(3), 238–260. doi:10.1504/IJENM.2007.012757

Schmidtke, D., Heiser, U., & Hinrichsen, O. (2014). A simulation-enhanced value stream mapping approach for optimisation of complex production environments. *International Journal of Production Research*, 1–15. doi:10.1080/00207543.2014.917770

Seth, D., & Gupta, V. (2005). Application of value stream mapping for lean operations and cycle time reduction: An Indian case study. *Production Planning and Control*, *16*(1), 44–59. doi:10.1080/09537280512331325281

Seth, D., Seth, N., & Goel, D. (2008). Application of value stream mapping (VSM) for minimization of wastes in the processing side of supply chain of cottonseed oil industry in Indian context. *Journal of Manufacturing Technology Management*, *19*(4), 529–550. doi:10.1108/17410380810869950

Shah, R., & Ward, P. T. (2007). Defining and developing measures of lean production. *Journal of Operations Management*, *25*(4), 785–805. doi:10.1016/j.jom.2007.01.019

Singh, B., Garg, S. K., & Sharma, S. K. (2011). Value stream mapping: Literature review and implications for Indian industry. *International Journal of Advanced Manufacturing Technology*, *53*(5-8), 799–809. doi:10.1007/s00170-010-2860-7

Singh, R. K., Murty, H. R., Gupta, S. K., & Dikshit, A. K. (2009). An overview of sustainability assessment methodologies. *Ecological Indicators*, *9*(2), 189–212. doi:10.1016/j.ecolind.2008.05.011

Solding, P., & Gullander, P. (2009). *Concepts for simulation based Value Stream Mapping*. Paper presented at the 2009 Winter Simulation Conference (WSC), Austin, TX. doi:10.1109/WSC.2009.5429185

Sullivan, W. G., McDonald, T. N., & Van Aken, E. M. (2002). Equipment replacement decisions and lean manufacturing. *Robotics and Computer-integrated Manufacturing*, *18*(3–4), 255–265. doi:10.1016/S0736-5845(02)00016-9

Tice, J., Ahouse, L., & Larson, T. (2005). Lean production and EMSs: Aligning environmental management with business priorities. *Environmental Quality Management*, *15*(2), 1–12. doi:10.1002/tqem.20075

Torielli, R. M., Abrahams, R. A., Smillie, R. W., & Voigt, R. C. (2011). Using lean methodologies for economically and environmentally sustainable foundries. *China Foundry*, *8*(1), 74–88.

Uygun, Y., & Straub, N. (2012). *Human-centred Model for Application of Lean Production Systems in Networks. In Enabling Manufacturing Competitiveness and Economic Sustainability* (pp. 660–665). Springer. doi:10.1007/978-3-642-23860-4_108

Vinodh, S., Arvind, K. R., & Somanaathan, M. (2011). Tools and techniques for enabling sustainability through lean initiatives. *Clean Technologies and Environmental Policy, 13*(3), 469–479. doi:10.1007/s10098-010-0329-x

Wills, B. (2011). *Green intentions: creating a green value stream to compete and win*. Boca Raton, FL: CRC Press.

Wolf, C., & Seuring, S. (2010). Environmental impacts as buying criteria for third party logistical services. *International Journal of Physical Distribution & Logistics Management, 40*(1), 84–102. doi:10.1108/09600031011020377

Womack, J. P., & Jones, D. T. (2003). Lean thinking: Banish waste and create wealth in your corporation (2nd ed.). New York: Free Press, Simon & Schuster, Inc.

KEY TERMS AND DEFINITIONS

Design for Environment (DfE): Is a systematic consideration of environmental component during design of new production and process development. It is used to discover product innovations that achieve customer requirements as well as reduce pollution and waste throughout the life-cycle.

Green Supply Chain Management (GSCM): Is an integration of environmental thinking into supply chain management. It includes product design, material sourcing and selection, manufacturing processes, delivery of the final product to the consumers, and end-of-life management of the product.

Green Waste: Is an unnecessary or excessive use of resources or hazardous substances released to air, water, or land that could harm human health or environment.

Lean Manufacturing: Is as a systematic approach focusing on efficient use of resources by identifying and eliminating non-value-added activities or waste.

Supply Chain Management (SCM): Is the management of a relationship network within an organization and between interdependent organizations consisting of material suppliers, purchasing, production facilities, logistics, marketing, and related systems that facilitate the forward and reverse flow of materials, services, finances and information from the original producer to final customer with the benefits of adding value, maximizing profitability through efficiencies, and achieving customer satisfaction.

Sustainable Supply Chain Management (SSCM): Is defined as a degree to which an organization strategically collaborates with its supply chain stakeholders and collaboratively manages intra- and inter-organization processes for sustainability.

Value Stream Mapping (VSM): Is a lean technique for visualizing the entire production process, representing information and material flows, and identifying wastes and its sources.

Waste Elimination Template: Is a template under the waste removal strategy. The template contains key information about types, root causes, removal cost, ease of removal, impact, occurrence, and detection of waste. The template allows risk and waste priority to be determined.

Chapter 6
Green Supply Chain Management Model for Sustainable Manufacturing Practices

Surajit Bag
Tega Industries South Africa Pty Ltd, South Africa

Neeraj Anand
University of Petroleum and Energy Studies, India

Krishan Kumar Pandey
O.P. Jindal Global University, India

ABSTRACT

The purpose of this chapter is to identify the dimensions of green supply chain and their impact on manufacturing practices. In this study, the authors used two extended strategies. First thorough review of literature was done considering articles from reputed journals. Second the factors identified from literature review was further refined through experts by forming a problem solving group consisting of seven experts from the manufacturing sector. These factors were used to develop the green supply chain management model using Interpretive structural modeling. Further MICMAC analysis was used to identify the driving and dependence power of the factors. The results of the analysis are very encouraging. Finally, the authors have presented the relationship management strategy for sustainable manufacturing practices.

DOI: 10.4018/978-1-5225-0635-5.ch006

INTRODUCTION

The paradigm shift of policies towards green economy is forcing manufacturers worldwide to consider the green initiatives seriously. There is need for the companies especially manufacturing companies to take a proactive approach rather than reactive approach. According to UNEP (2011) report, global manufacturing sector accounts for 35% of the total electricity consumed worldwide and responsible for 20% of the world's carbon dioxide emissions, which is detrimental to lives on the earth. While the above arguments indicate, that this is the high time for research on green supply chain management and implementation of GSCM practices, especially for developing countries like India which is becoming one of the global manufacturing hubs next to China.

The term "green" is now widely used interchangeably on the more established "sustainability" concept, which points to a more holistic view of environmental, social and economic dimensions. World leading firms including IBM, Dell, HP, Sony, Toyota, Nokia have adopted GSCM as a strategic initiative. A framework is developed which portray the various complex relationships involved in a green supply chain with the aid of interpretive structural modeling (ISM). Before we funnel down to the framework, let us understand the basics of sustainability and green supply chain management, how firms exploit it for gaining competitive advantage, and how to build a contextual framework using ISM methodology.

LITERATURE REVIEW

We have adopted "*Systematic Literature Review (SLR)*" in our paper as suggested by Tranfield et al. (2003). The present section is divided into two stages and its phases.

Stage 1: Planning the Review

Phase 0: Identification for the Need for a Review

The field of Manufacturing and Operations has seen radical changes through the years. It started with the "*The Industrial Revolution*" in the late 1700s, has been through several phases but real challenge emerged in front of manufacturing was its sustainability. The issues that have forced developed economies to shift their manufacturing base to India were cost and other was to put check on greenhouse gas emissions. However the approach of these developed economies was questionable. Shifting the location of manufacturing hubs is never going to solve problems but there is need for more sustainable solution which can optimize between cost and environment. The search of amicable solution has led to the growth of green supply chain management. In our present work we will discuss theoretical concept used for building a green supply chain management model. We will review latest research works related to evolution of green supply chain dimensions, green supply chain as a source of competitive advantage.

Phase 1: Preparation of a Proposal for a Review

In this section we will identify reputable journals which publishes potential works related to green supply chain and its related areas.

To begin with green supply chain management it's very important to understand "Green Manufacturing" and its dimensions. Though there are lot of definitions and arguments, but from various literatures it can be derived that during manufacturing or production, lot of wastes are generated in the form of solid, liquid and gas, which are harmful to earth and to lives supported by the earth. In order to minimize the harmful effect of these wastes, innovative manufacturing practices to be deployed which is blend of technology, culture, leadership and policies, at economical cost so that firm should not be overburdened with excessive cost of production which force them to outsource which is the present scenario in the western part of the world.

Phase 2: Development of Review Protocol

We decided initially to review all published work. In order to eliminate our biasness towards any work we reviewed all possible articles published in past fifteen years in reputed journals related to green supply chain management, sustainable manufacturing & related concepts so that we do not end up with sketchy literature or irrelevant literature.

Stage 2: Conducting a Review

Phase 3: Identification of Research

We reviewed over 100 articles related to SSCM, green manufacturing, strategy & methodologies.

Based on review, we have derived two important questions which will further drive our investigation.

Green Supply Chain Management

In the present section we have highlighted some of the respected studies carried out by past researchers in the area of green supply chain management.

We must understand the concept of supply chain management before piercing down into green supply chain management (GSCM). Within each organization, the supply chain includes all functions involved in receiving and filling a customer request. These functions include, but are not limited to new product development, marketing, operations, finance, distribution and customer service. A supply chain is dynamic and involves the constant flow of information, product and funds between different stages.

If we look back into the history, we see that before 1780 during the Agrarian era supply chain type was 'standard' and firm characteristics used to be small, organic, horizontal hierarchy and depended on quality of product to sell. Between 1780 and 1960 it was the era of Industrial revolution where a major change took place and the standard supply chain type shifted towards lean with expansion in size, variety, complexity and diversity. After 1960, the post industrial revolution phase witnessed another sea change globally and the trend shifted from lean to agile, hybrid and presently green, (Nelson et al, 2009).Thus the journey of SCM theory development has started long back and is still evolving.

Sarkis et al., (2011) has documented about various organizational theories applied to GSCM related study such as Complexity theory, Ecological modernization theory, Information theory, Institutional theory, Resource based view, Resource dependence theory, Social network theory, Stakeholder theory, and Transaction cost economics. Some of the selected studies have been presented in Table 1.

Some examples to show the impact of green practices on supply chains are evident in Table 2.

Table 1. Literature review on green supply chain management (GSCM)

Author(s), Year	Objective of Study	Key Findings	Variables Identified
Brandenburg et al. (2014)	To understand and review mathematical models focusing on environmental/ social factors in the forward supply chains.	Major publications and models were found in a limited set of six journals (JCLP, IJPR, IJPE, TRE, EJOR & DSS). AHP, ANP and LCA were commonly used tool for developing the models.	Environmental; Social and Economic dimensions
Mirhedayatian et al. (2014)	To propose a novel network DEA model for evaluating the GSCM in the presence of dual role factors, undesirable outputs and fuzzy data.	The proposed model can be easily computerized so as to serve as a decision making tool in decision making.	Cost of quality; Supplier flexibility; Carbon dioxide emission; Satisfaction; Facility technology level

continued on following page

Table 1. Continued

Author(s), Year	Objective of Study	Key Findings	Variables Identified
Suering (2013)	To review research on quantitative models for green or sustainable supply chains.	Different kinds of models are applied in this area but the social dimension is not taken into consideration. On the modeling side there are three popular methods: Equilibrium models, MCDM & AHP.	Environmental; Social and Economic dimensions
Zhu and Lai (2013)	To develop and empirically test a theoretical model.	The research contributes to the literature on institutional theory in corporate environmental practices.	Institution pressure: Coercive, Normative, Competitive
Reefke and Trocchi (2013)	To develop a framework to facilitate a balanced approach to performance measurement for SSCM.	A scorecard design customized for sustainable supply chain is proposed along with the development and implementation process.	Cost savings; Profit; customer satisfaction; quality management
Liu et al. (2012)	To integrate green marketing and SSCM.	Development of a new hub and spoke integration model. Drivers: Improve company's sustainable supply chain capabilities; reach green customers before competitors; government regulations; green customers demand; community expectation.	Products; Promotion; Planning; Process; People and Project
Hoejmose et al. (2012)	To understand the general engagement with GSCM in both B2B & B2C supply chains.	Firms in B2B market is generally less engaged with green practices compared to firms in B2C markets. Developing Trust with supply chain partners and top management commitment is a crucial GSCM driver among firms in B2B markets.	Trust; Top management commitment
Ageron et al. (2012)	To develop a Sustainable Supply Management framework.	External pressures have positive impact on the development of SSM; Waste reduction programmes have greater impact on greening supply chains; MNC & SMEs' have differential impact on SSM; Financial barriers have more impact on SSM; Top management support is a critical success factor in SSM; Key benefits such as customer satisfaction, supplier innovation, quality & capacity have greater positive impact on SSM.	Top Management Commitment; Govt. regulatory requirements; Quality; Flexibility; Waste Reduction; Clean Programs; Reducing carbon footprint; Financial costs; ROI; Green Investments; Customer Satisfaction; Innovativeness
Bose and Pal (2012)	To investigate the influence of GSCM initiatives on stock prices of firms.	Firms observe greater positive change in stock price by undertaking green initiatives. Firms with high R&D expenses show strong positive impact. Early adopters of GSCM show greater positive impact.	R&D; Size of firm; Stock prices
Dekker et al. (2012)	To review research on green logistics.	The review highlighted the contribution of operations research to green logistics	Mode choice; Intermodal transport; Equipment choice and efficiency; fuel choice and carbon intensity
Chaabane et al. (2012)	To present a generic mathematical model to assist decision makers in designing sustainable supply chains over their entire life cycle.	The model can serve as a tool that facilitates the understanding of optimal SC strategies under different environmental policies.	Economic: Cost; revenue; taxes; transfer Environmental variables: Carbon footprint; Raw material use; Energy use; Social variables: Noise; pollution

continued on following page

Table 1. Continued

Author(s), Year	Objective of Study	Key Findings	Variables Identified
Dubey and Bag (2013)	To explore sustainable manufacturing practices that improves environmental and business performance.	Green Purchasing, SRM, Green logistics and regulatory norms are positive determinants of firm business performance and Environmental performance.	Green Purchasing; SRM; Green logistics and Regulatory norms
Gunasekaran and Spalanzani (2012)	To bring important issues related to sustainable business development in both manufacturing and services sector.	Developed a framework for sustainable development along with strategies, techniques and tools.	Sustainable Product and process design; sustainability in supply operations; sustainability in production operations; sustainability in distribution operations; sustainability through reverse logistics
Gimenez et al. (2012)	To analyse the impact of environmental programmes on each dimension of the triple line.	Internal environmental programmes have a positive impact on the three components of the triple bottom line. Social initiatives have a positive impact only on two components: social and environmental performance.	Environmental; Social and Economic performance; Internal and external action programs
Hassisni et al. (2012)	To review literature on sustainable supply chains from 2000-2010.	Based on the findings authors have developed a framework for sustainable supply chain metrics.	Market Forces; Policy& Regulations; Science& Technology; Product Development; Process Capability; Sourcing& Operations; Logistics; Marketing &PR; Social issues
Barari et al. (2012)	To provide integrated and holistic conceptual framework that combines the practical aspects of green supply chain with the objective of profit maximization.	A strategy set have been designed based on the game theory approach that promises maximum economic benefit.	Green tax; green burden
Kang et al. (2012)	To establish the framework for strategy development to construct the sustainable supply chain.	Identified the factors leading to SSCM.	Leadership for knowledge sharing; Innovation of product and process
Wang et al. (2011)	To provide a multi objective mixed integer formulation for the supply chain network design with environmental concerns.	The model can be effectively used in the strategic planning for green supply chain.	Demand and Supply of Product; Carbon dioxide emission; capacity of facility; environmental investment cost; environmental protection; Transportation cost; Handling cost for products
Wu and Pagell (2011)	To understand how organizations balance short term profitability and long term environmental sustainability when making supply chain decisions under uncertainty.	Factors contributing to the uncertain decision environment are: Uncertainty about environmental outcomes & future regulation, the saliency of each environmental issue to multiple stakeholders, lack of visibility and influence in one's supply chain.	Operating principles; Technical standard

continued on following page

Table 1. Continued

Author(s), Year	Objective of Study	Key Findings	Variables Identified
Gupta and Desai (2011)	To review the current state of academic research in sustainable supply chain management.	Authors developed an integrative framework summarizing the existing literature under four broad categories: strategic considerations, decisions at functional interfaces, regulation and government policies, integrative models and decision support tools.	Product design and product life cycle; Regulation and government policies
Azevedo et al. (2011)	To investigate the relationships between green practices and supply chain performance.	A conceptual model of the relationships between green practices and SC performance was developed.	Environmental friendly practices in purchasing; Environmental collaboration with suppliers; Minimizing waste; Environmental collaboration with customers; Reverse logistics
Pereira (2009)	To understand how Information Technology can foster information management and help sort out supply chain problems.	IT must be used to develop strategies to make supply chains more robust and resilient.	Information Technology
Longo and Mirabelli (2008)	To present an advance modeling approach and a simulation model and provide a decision making tool for supply chains.	An advance model is proposed based on programming code, tables and event generators and provides the user with a simulator capable of high efficiency for executing simulation runs.	Inventory Control; Lead Time; Demand Intensity; Demand variability
Linton et al. (2007)	To prepare a background to better understand current trends in the area of sustainable supply chains.	Research on Sustainable supply chain is still at an infant stage. It has strong links with government policy.	Product life extension; Product design; Recovery process at end of life.
Zhu and Sarkis (2004)	To determine the economic and environmental relationships of GSCM practices among Chinese firms.	Significant win-win opportunities exist for Chinese firms practicing GSCM. Strong relationship exists between GSCM practice and positive economic performance.	Commitment of GSCM from senior managers; Total quality environmental management; Environmental compliance and auditing programs; Supplier relationship; green design
Croom et al. (2000)	To conduct literature review.	Lack of theoretical work in the field as compared to empirical based studies	Sourcing strategy; attitude and commitment to collaborative improvement program

It is clear that green supply chain has attracted attention of top global companies. Green practices are being used presently by these companies in some form or other. Companies are focusing mostly in reduction of natural resources, reuse of material and energy recovery program. This is really working like wonder and can potentially stop tons of green house gases in entering the atmosphere. Moreover, these companies are significantly saving costs across the supply chain. Ongoing, company-wide sustainability program has generated significant environmental and

Table 2. GSCM practices

Companies	Green Supply Chain Practices
Wal-Mart	Focused on green packaging; Reduction in packaging costs and reduction in carbon dioxide emissions
Nestle	Eco friendly product packaging and focus on source reduction, re-use, recycling, and energy recovery
Heinken	Conservation of Natural resources and energy savings
Environmental Lubricants Mfg. Inc.	Technological up gradation to improve productivity and eco friendly manufacturing process
Bank of America	Eco-friendly operations and recycling program
General Electric's	Development of green products
Dupont	Green operations and lowering emissions of airborne carcinogens and greenhouse gases
Starbucks	Green operations, use of recycling paper cups and protecting environment
Coca-Cola	Focus on three environmental goals: water stewardship, sustainable packaging, and climate & energy protection
Toyota	Development of hybrid vehicle. Saving natural resources
Honda	Developed fuel efficient cars
Dell	Recycling program
Hewlett Packard	Owning and operating "e-waste" recycling plants

financial benefits. Commitment to reduce fuel and electricity use is a good approach to protect scarce natural resources. Converting to eco friendly operations has resulted in increase in production tenfold while reducing manufacturing costs. This has also attracted more customers due to enhanced customer satisfaction.

Use of alternate source of energy such as solar panels, CNG is widely used by manufacturers. The step toward more sustainable operations has drastically lowered emissions of airborne carcinogens and greenhouse gases. Another company has also gone a step ahead by appointing an ex-Greenpeace head as an advisor to the board. And this company successfully reduced greenhouse gas emissions. Top companies has also partnered up with many environmental organizations, such as Conservation International and the Earthwatch Institute, in efforts to do right for the communities it operates in. Waste-elimination is now in the corporate philosophy to make the operations more eco-friendly than ever.

Lanxess and China Petroleum and Chemical Industry Federation (CPCIF) joined hands to present the Green Rubber Day on December 2nd, 2013 at the China World Summit in Beijing. Among other topics, Green Rubber Day dealt with the subject of high-performance tires, also known as "green tyres". The use of these tyres can help significantly to reduce rolling resistance, which is the friction that tires encounter when rolling. A tire's rolling resistance is a major factor in a vehicle's fuel economy. Using a high-performance tire that has been produced with state-of-the-art rubber can decrease the fuel consumption of a car by as much as half a liter per 100 kilometers – and consequently reduce the emission of carbon dioxide by 1.2 kilograms.

Few Definitions on Green Supply Chain Management (GSCM)

- Schrettle et al. (2014) defined GSCM as a "tool which helps to position a company from strategic viewpoint".
- Gunasekaran and Spalanzani (2012) defined GSCM as "an organizational philosophy which provides competitive edge to an organization".
- Ali and Govindan (2011) defined GSCM as "an organizational philosophy to reduce environmental risks".
- Zhu et al. (2008) defined GSCM as "adoption of eco friendly practices considering internal environmental management, green purchasing, cooperation with customers, and eco design for developing corporate and operational strategies for the firm's environmental sustainability".
- Seuring and Muller (2008) defined GSCM as "the management of material, information, and capital flows as well as cooperation among companies along the supply chain while taking goals from all three dimensions of sustainable development" which are derived from customer and stakeholder requirements".
- Carter and Rogers (2008) defined GSCM as an "integration of environmental dimensions with traditional supply chain network".
- Srivastava (2007) defined GSCM as "integrating environmental thinking into supply chain management, including product design, material sourcing and selection, manufacturing processes, delivery of the final products to the consumers, and end-of-life management of the product after its useful life".
- Vachon and Klassen (2006) defined GSCM as a "strategy, which assists to reduce wastages in supply chain network".
- Kogg (2003) defined GSCM "the set of supply chain management policies held, actions taken and relationships formed in response to concerns related to the natural environment with regard to the design, acquisition, production, distribution, use, re-use and disposal of the firm's goods and services".
- Sarkis (2003) defined GSCM as a "combination of the activities of an environmental dimension and reverse logistics".

- Gilbert (2000) defined GSCM as an "integration of environmental criteria with traditional supply chain network by redesigning purchasing policies and involving suppliers in the entire procurement process".
- Beamon (1999) defined GSCM as "utilizing the supply chain between a central company and a cooperating company, to support the company with knowhow of eco management practices and development of clean manufacturing techniques".
- Narshiman and Carter (1998) defined GSCM as "involving the procurement strategies that reduced the use of materials in addition to recycling and reuse."
- Godfrey (1998) defined GSCM as "organizational practices that continuously monitor the environmental impact of a supply chain and improve the performance."

Based on the above definitions the authors propose GSCM definition as "integrating the green dimensions into the traditional supply chain management so that the product will impose reduced environmental impact during its entire life cycle. It is the set of clean and hygienic practices which starts right from the product design stage and flows through all activities of the supply chain such as procurement, manufacturing, packaging, transportation and end of life management. Firm's internal resources mediate the relationship to external forces and its adoption. It improves the organizational performance and is the source of competitive advantage to sustain in the turbulent business environment."

Green Supply Chain Management (GSCM) Dimensions

- **Internal Environmental Management (IEM):** IEM includes Top management commitment, Cross functional cooperation for environmental improvements, Total quality environmental management, Environmental compliance and auditing programs, ISO 14001 certification, Environmental management system.
- **External Environmental Management (EEM):** EEM includes Green Purchasing, cooperation with customers for cleaner production, green packaging and eco-design.
- **Investment Recovery (IR):** IR includes sale of excess inventories, sale of excess capital equipment.
- **Eco-Design:** Eco design includes design green products for reuse, recycle and recovery of material and energy. Design eco friendly products to reduce health hazards of human beings.

Green Supply Chain Management (GSCM) as a Source of Competitive Advantage

A respected work has proved that environmental management is partially an extension of advanced manufacturing practices, continuous improvement and supplier management (Yang et al., 2010). Therefore GSCM can be regarded as a mechanism through which sustainability can be achieved leading to manufacturing competitiveness.

GSCM improves firm performance by improving Agility, Adaptability and promotes Alignment:

- **Improves Agility:** Firms adopting GSCM practices can react fast to any changes in customer demand than competitors and hence better positioned in the market.
- **Improves Adaptability:** Firms can achieve better flexibility and can lead to innovative processes and continuous improvement.
- **Promotes Alignment:** Better alignment of business process is achieved by those firms practicing GSCM.

Theoretical Discussion on Antecedents of Green Supply Chain Management (GSCM)

The key drivers of GSCM are regulatory pressures and market pressures. In developed countries government has framed environmental policy and guidelines to provide environmental solution to the manufacturers. With globalization and the increased number of countries entering the W.T.O. has led to increased GSCM practices. Several studies have mentioned how foreign firms such as Toyota and Ford have encouraged and motivated their Chinese suppliers in adopting green practices. In European Union they have a policy (WEE) for disposing waste electrical and electronic equipment after the end of their life so as to reduce the harmful effects.

Cap and trade legislation for GHG emission is applicable for manufacturers across European Union, U.S and Australia. Aberdeen group survey (2013) reports that 50% of the companies are planning to redesign their entire supply chain in order to be more sustainable and 80% of them have to comply with new environmental policies. As a result companies face new challenges and adopt best technologies to meet their legal obligations.

Companies can earn carbon credits through one of the mechanism such as Emission trading system and carbon trading or contribution to climate change technology fund.

In an effort to reduce the GHG emissions globally the carbon trading markets were first introduced under the Kyoto Protocol, known as the United Nations framework convention on climate change (UNFCCC).

Market pressures play a vital role in driving firms to adapt GSCM. Foreign customers demand environment friendly product and services from their suppliers globally including India. Foreign customers conduct strict environmental audit of their Indian suppliers before placing the orders as they demand environmental friendly goods, (Mitra, 2004). GSCM can be used as an environmental tool to improve the environmental image with in the business arena and enhancing export sales. Suppliers and community stake holders also influence GSCM adoption.

Supporting the GSCM project requires tremendous support from the top management to highlight that the project is really important and will receive proper attention from the members of the organization. Lack of resources and financial support will lead to failure of the GSCM project. It is interesting to note that top management commitment leads to success of GSCM project. Firms having proper environmental policy are always ahead and the success rate of GSCM adoption are high. Top management must motivate junior managers and ensure that every employee within the organization is aware of the environmental policy, (Zhu et al., 2005).

Several studies on GSCM were conducted in Chinese context. Without environmental awareness it is difficult to implement green practices. Regulatory, competitive, and marketing pressures led to increased environmental awareness and are the drivers of GSCM, (Zhu et al., 2005). Chinese manufacturers have taken proactive measures in closing the supply chain loop by adopting GSCM practices. GSCM can be used as an environmental tool to improve the environmental image and gain competitiveness within the international business arena, (Zhu et al., 2008). Greening any supply chain involves imbibing environmental principles in the design and operations of supply chain. A background of environmental principles for achieving eco efficiency and building of environmentally friendly organization system have been presented and emphasis is put on application of such principle in life cycle of product (Tsoulfas & Pappis., 2006).

It is very important for any practicing manager to understand the links between GCSM practices, environmental and economic performance for successful GSCM implementation. Environmental pressure act as a driver for successful GSCM practices. Market and regulatory pressures help organization to improve environmental pressures which are believed to influence eco design and green purchasing. Manufacturers facing higher regulatory pressure tend to implement green purchasing and investment recovery. Competitive pressure significantly improves the economic benefits from adoption of different GSCM practices (Zhu et al., 2007).

Computer parts' manufacturers in Thailand have adopted Environmental Management Systems (EMS) due to strict environmental regulations and export Pressure (Seksan et al., 2010).

Environmentally and socially responsive supply chains are in the early adoption stages in Indian automobile industry and actual implementation lacks a holistic approach (Shukla et al., 2009). Major Indian manufacturing firms feel that there is lack of awareness of environmental issues which is a challenge and biggest perceived barrier to adopting GSCM is that it is not cost effective (Bhateja et al., 2011).

Important factors to implement GSCM in Indian manufacturing industry are innovative green practices, awareness level of customers; supplier motivation; technology advancement and organization adoption, organization encouragement, quality of human resources, IT enablement, top management commitment, Government support policies, International Environmental agreements (Luthra et al., 2012). External factors affecting GSCM are Environmental regulations, suppliers, consumers and community stakeholders. GSCM practices can enhance the environmental, operational and financial performance of firms.

Research Questions

- What are the key GSCM factors and the nature of relationship between them in context to Indian manufacturing sector?
- Can a GSCM model be developed for the Indian manufacturing sector?

Research Objectives

Literature review has given a direction to identify the research gaps and to develop the below two specific research objectives for the present study are as follows:

- To identify the key factors influencing GSCM practices and the underlying relationships in context of Indian manufacturing sector.
- To develop a GSCM model for the Indian manufacturing sector.

METHODOLOGY

This section discusses on theoretical model development. The purpose of research methodology is to present the research design that will be used to conduct the research for this study. Based on the synthesis of literature review and experts' opinion from industry; the key factors influencing GSCM practices have been identified. GSCM model has been developed by using ISM technique which was further refined using MICMAC analysis (See Figure 1).

Figure 1. Research process flow chart

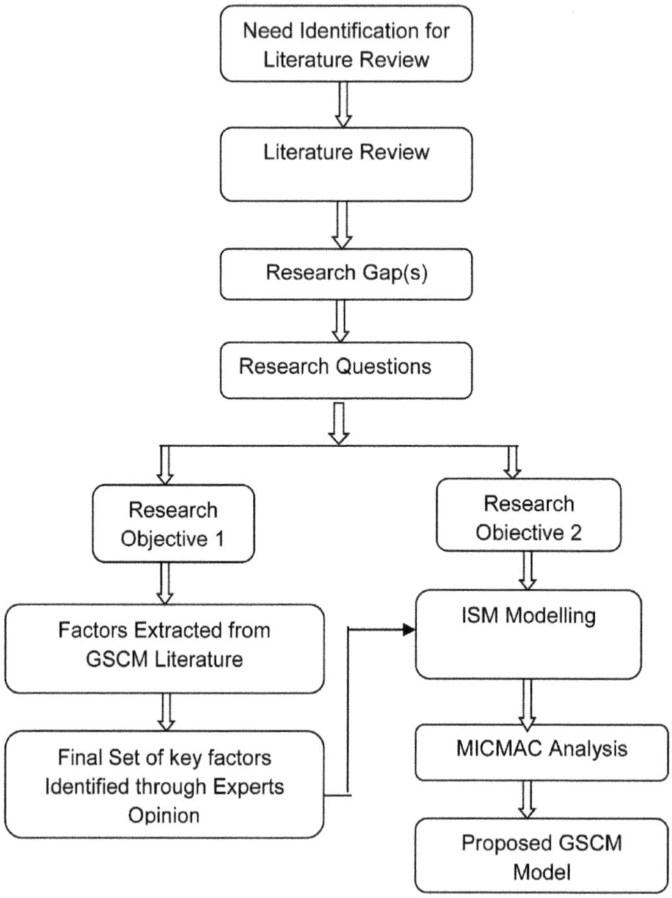

Interpretive Structural Model (ISM)-Based Methodology

ISM is a proven and popular methodology for understanding relationships among specific items that define a problem. ISM is useful to achieve the objective in presence of large number of directly and indirectly related elements and complex interactions among them which may or may not be expressed in a proper manner. ISM plays a vital role in this kind of situation and helps in understanding a structure within a system. The ISM model depicts the structure of a complex problem in a carefully designed pattern.

ISM has been used in the past by several researchers due to multiple benefits which have been mentioned in Table 3. It guides and records the results of group response on complex issues in an efficient and systematic manner, (Source: Attri et

Table 3. Review on ISM application

Author and Year	ISM Application
Hawthrone and Sage (1975)	Higher education program planning
Sage (1977)	Modeling complex situations
Jedlica and Meyer (1980)	Exploring factors involved in a cross cultural context
Saxena et al (1992)	Determining the hierarchy and class of elements in cement industry
Mandal and Desmukh (1993)	Vendor selection
Kanungo et al (1999)	Developing an IS effectiveness framework
Ravi and Shankar (2004)	Explore reverse logistics barriers
Jharkaria and Shankar (2005)	Enablers of IT implementation in SC
Ravi et al. (2005)	Indentify key reverse logistics variables
Faisal et al. (2006)	Modeling the enablers for supply chain risk mitigation
Thakkar et al. (2006)	Integrated approach with ISM and ANP to develop a balanced scorecard
Singh and Kant (2008)	Knowledge Management
Subramanian et al. (2010)	Buyer Supplier relationship
Diabat et al. (2012)	Modelling various supply chain risks
Gorane and Kant (2012)	Modelling supply chain management enablers
Khan and Haleem (2013)	Technology Management enablers
Dubey et al. (2014)	Flexible Manufacturing Systems
Dubey et al. (2014)	Modelling Antecedents of Innovation

al., 2013; Warfield 1994; Warfield 1974). ISM has been applied in different areas of the supply chain starting from purchasing to production and logistics management. Sushil (2012) have contributed in the ISM literature by providing directions to interpret the links in ISM using the tool of interpretive matrix.

The Steps Interpretative Structural Modeling

There are six basic steps under ISM and these have been presented in Figure 2.

DATA ANALYSIS AND FINDINGS

In this step the contextual relationship between elements with respect to the pairs of elements is examined. To investigate the relations between ten elements, a questionnaire was developed. The questionnaire was sent to ten referees, who are working in the position of senior management in manufacturing sector. The questionnaire

Figure 2. Flowchart for the ISM Methodology

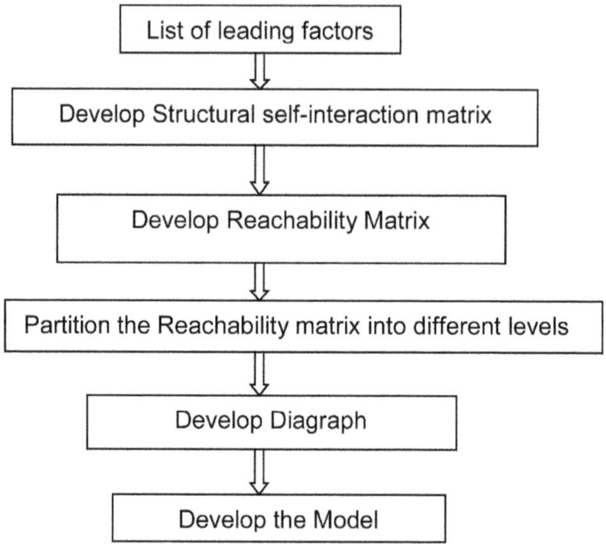

was tested for content validity. Content validity primarily depends on an appeal to the propriety of content and the way it is presented (Nunally, 1978). The instrument developed in this study demonstrates the content validity as the selection of measurement items was based on both, an exhaustive review of the literature and detailed evaluations by senior managers during pre-testing.

List of Leading Factors

The green supply chain management literature was thoroughly reviewed. As a result of the review, 20 factors were found to be influencing green supply chain management processes. A problem-solving group was formed comprising seven experts from industry who are also members of CII. The group was asked to analyze the listed factors, take out the repeated ones and rearrange the list including the ones thought to be valid in relation with Indian context. The first tour resulted with reducing each of the factors to 17. Through the second tour 12 and the third tour total 10 factors were determined.

The ten key factors shown in Table 4.

Developing the Structural Self-Interaction Matrix (SSIM)

For developing SSIM, the below symbols have been used to denote the direction of relationships between variables (i and j) (Table 5):

Table 4. Final set of key factors with items are listed

SL NO	Factors	Items
1	Top Management Commitment	Well defined environmental policy of Company
		Employee awareness about the firm's environmental policy
		Top management support to environmental programs
		Top management approval to special funds for investment in cleaner technologies
		Positive attitude of senior managers towards green practices
		Senior managers motivate and support new ideas received from junior executives
		Employees recognition for innovative ideas
2	Regulatory Pressures	Regional pollution control board pressures to adopt green practices
		Government regulations to control pollution level
		Strict monitoring by Pollution control board
		Green practices decrease incidence of penalty fee charged by pollution control board
3	Market Pressures	Company focus on Export sales
		Foreign customers sensitivity towards green practices
4	Supplier Relationship Management	Environmental criteria for supplier selection
		Environment collaboration with suppliers
		Technological integration with suppliers
		Training and educating suppliers in implementing ISO 14001
		Environmental audit for suppliers
5	Total Quality Environmental Management	Total Quality Environmental Management implementation
		Green practices promote product quality
6	Green Technology Adoption	Company focus on green design of products
		Wastage reduction from green design
		Real time information available any point of time by using Information technology infrastructure SAP/ ERP
		Usage of alternate source of energy
		Process optimization to reduce wastage
		Eco friendly packaging
		Reduction of emission of Green House Gases by use of clean technology
7	Profit	Green practices improve firm's profit
8	Market Share	Green practices improve firm's market share
9	Reduction in Carbon emission	Green practices Reduce solid waste generation
		Effluent meets CPCB norms by converting into green operations
		Green practices reduce environmental accidents & hazards
		Green practices decrease of cost of raw materials
		Green practices reduce the inventory levels
		Green practices reduce cost for energy consumption
10	Customer Relationship Management	Green practices improve customer satisfaction
		Firm recover end of life products from customers
		Customers appreciate eco friendly products

Table 5. Structural self-interaction matrix (SSIM)

	X	IX	VIII	VII	VI	V	IV	III	II	I
I	V	V	V	A	V	V	A	A	O	
II	V	V	V	A	V	V	O	A		
III	V	V	V	A	V	V	A			
IV	O	O	V	X	O	V				
V	O	V	V	A	A					
VI	V	V	V	A						
VII	O	O	V							
VIII	O	V								
IX	O									
X										

V: i lead to j but j does not lead to i
A: i do not lead to j but j lead to i
X: i lead to j and j lead to i
O: i and j are unrelated to each other

Develop Reachability Matrix

The SSIM has been converted into a binary matrix i.e., the rechability matrix (Table 6) by substituting V, A X and O by 1 and 0. The substitutions of '1' and '0' have been done as below:

1. If the (i, j) entry in the SSIM is V, then the (i.j) entry in the rechability matrix becomes '1' and (j,i) entry becomes '0'
2. If the (i, j) entry in the SSIM is A, then the (i.j) entry in the rechability matrix becomes '0' and (j,i) entry becomes '1'
3. If the (i, j) entry in the SSIM is X, then the (i.j) entry in the rechability matrix becomes '1' and (j,i) entry also becomes '1'
4. If the (i, j) entry in the SSIM is O, then the (i.j) entry in the rechability matrix becomes '0' and (j,i) entry also becomes '0'

Level Partitioning

The final reachability matrix obtained in Table 6 is now partitioned into different levels. After the first iteration, the variable classified to level 1 are discarded and the partitioning procedure is repeated on the remaining variable to determine the level 2.

Table 6. Reachability matrix

	I	II	III	IV	V	VI	VII	VIII	IX	X	Driving Power (Y)
I	1	0	0	0	1	1	0	1	1	1	6
II	0	1	0	0	1	1	0	1	1	1	6
III	1	1	1	0	1	1	0	1	1	1	8
IV	1	0	1	1	1	0	1	1	0	0	6
V	0	0	0	0	1	0	0	1	1	0	3
VI	0	0	0	0	1	1	0	1	1	1	5
VII	1	1	1	1	1	1	1	1	0	0	8
VIII	0	0	0	0	0	0	0	1	1	0	2
IX	0	0	0	0	0	0	0	0	1	0	1
X	0	0	0	0	0	0	0	0	0	1	1
Dependence Power (X)	4	3	3	2	7	5	2	8	7	5	

These iterations are continued until the level of each variable has been determined. The results for iterations 1 to 7 are shown in Tables 7-13.

Diagraph

From the final Reachability Matrix and level partitions, the structural model is generated by means of nodes and lines of edges. If there is a relationship between

Table 7. Level partitioning (iteration 1)

Variables	Reachability Set	Antecedent Set	Intersection Set	Level
1	1,5,6,8,9,10	1,3,4,7	1	
2	2,5,6,8,9,10	2,3,7	2	
3	1,2,3,5,6,8,9,10	3,4,7	3	
4	1,3,4,5,7,8	4,7	4,7	
5	5,8,9	1,2,3,4,5,6,7	5	
6	5,6,8,9,10	1,2,3,6,7	6	
7	1,2,3,4,5,6,7,8	4,7	4,7	
8	8,9	1,2,3,4,5,6,7,8	8	
9	9	9	9	*Level1*
10	10	10	10	*Level1*

Table 8. Level partitioning (iteration 2)

Variables	Reachability Set	Antecedent Set	Intersection Set	Level
1	1,5,6,8	1,3,4,7	1	
2	2,5,6,8	2,3,7	2	
3	1,2,3,5,6,8	3,4,7	3	
4	1,3,4,5,7,8	4,7	4,7	
5	5,8	1,2,3,4,5,6,7	5	
6	5,6,8	1,2,3,6,7	6	
7	1,2,3,4,5,6,7,8	4,7	4,7	
8	8	1,2,3,4,5,6,7,8	8	*Level 2*

Table 9. Level partitioning (iteration 3)

Variables	Reachability Set	Antecedent Set	Intersection Set	Level
1	1,5,6	1,3,4,7	1	
2	2,5,6	2,3,7	2	
3	1,2,3,5,6	3,4,7	3	
4	1,3,4,5,7	4,7	4,7	
5	5	1,2,3,4,5,6,7	5	*Level 3*
6	5,6	1,2,3,6,7	6	
7	1,2,3,4,5,6,7	4,7	4,7	

Table 10. Level partitioning (iteration 4)

Variables	Reachability Set	Antecedent Set	Intersection Set	Level
1	1,6	1,3,4,7	1	
2	2,6	2,3,7	2	
3	1,2,3,6	3,4,7	3	
4	1,3,4,7	4,7	4,7	
6	6	1,2,3,6,7	6	*Level 4*
7	1,2,3,4,6,7	4,7	4,7	

Table 11. Level partitioning (iteration 5)

Variables	Reachability Set	Antecedent Set	Intersection Set	Level
1	1	1,3,4,7	1	*Level 5*
2	2	2,3,7	2	*Level 5*
3	1,2,3	3,4,7	3	
4	1,3,4,7	4,7	4,7	
7	1,2,3,4,7	4,7	4,7	

Table 12. Level partitioning (iteration 6)

Variables	Reachability Set	Antecedent Set	Intersection Set	Level
3	3	3,4,7	3	*Level 6*
4	3,4,7	4,7	4,7	
7	3,4,7	4,7	4,7	

Table 13. Level partitioning (iteration 7)

Variables	Reachability Set	Antecedent Set	Intersection Set	Level
4	4,7	4,7	4,7	*Level 7*
7	4,7	4,7	4,7	*Level 7*

the factors j and i, this is shown by an arrow which points from i to j. This graph is called digraph (see Figure 3).

The analysis yields an ISM hierarchy in which Market share and Profitability is at (level 1); reduction in carbon emission (second level); Green technology adoption (third level); Total quality environmental management (fourth level); SRM & CRM (fifth level), Top management commitment (sixth level), Regulatory and market pressures (bottom level).

ISM Model

The resulting ISM model is presented in Figure 4.

Figure 3. Diagraph

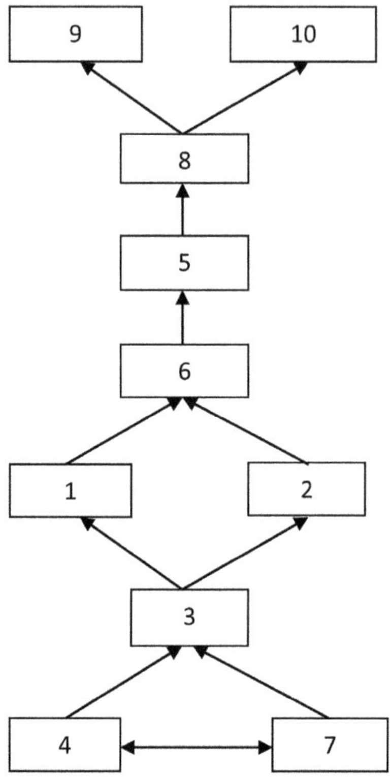

MICMAC Analysis

MICMAC analysis Matrice d' Impacts croises multiplication appliqué an classment (cross-impact matrix multiplication applied to classification) is abbreviated as MICMAC. The objective of MICMAC analysis is to analyze the drive power and dependence power of factors. Based on the drive power and dependence power the factors have been classified into four factors: autonomous factors, linkage factors, dependent and independent factors (see Figure 5 and Table 14).

The MICMAC graphical representation is shown in Figure 5.

Interpretation of MICMAC Analysis

Cluster 1: Autonomous Factors

These factors have a weak drive power and weak dependence power. In this cluster we do not have any factors.

Figure 4. ISM Model

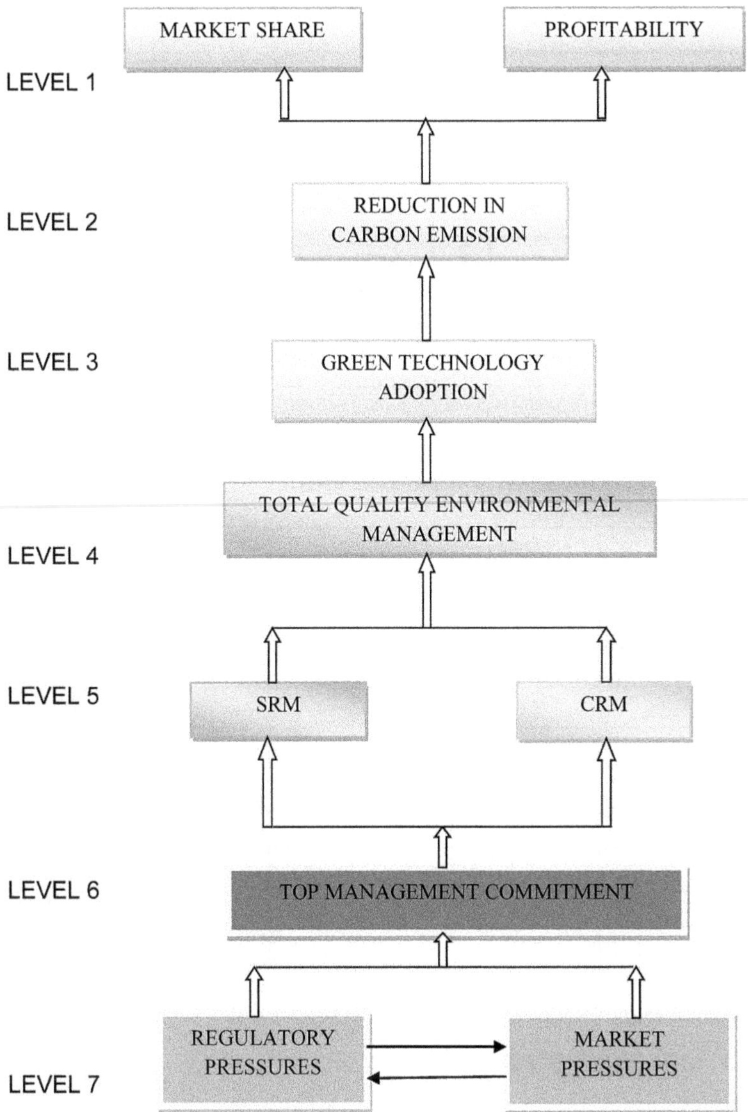

Cluster 2: Dependence Factors

These factors have a weak drive power but strong dependence power. In this cluster we have three factors, i.e., 5 (Market Pressures), 8 (Carbon Emissions Reduction), 9 (Market Share) and 10 (Profit).

Figure 5. MICMAC graphical representation

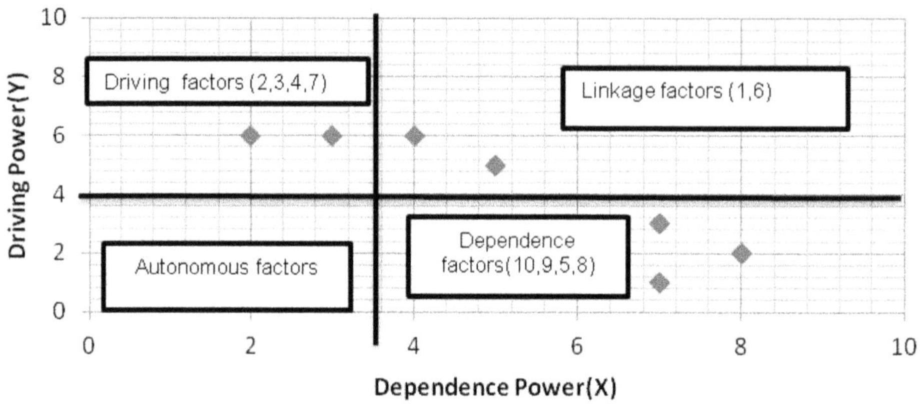

Table 14. Position coordinates of identified factors

Factors	Dependence Power(X)	Driving Power(Y)
1	4	6
2	3	6
3	3	8
4	2	6
5	7	3
6	5	5
7	2	8
8	8	2
9	7	1
10	5	1

Cluster 3: Linkage Factors

These factors have a strong drive power as well as strong dependence power. In this cluster we have two factors, i.e., 1(Supplier Relationship Management) and 6 (Green Technology Adoption).

Cluster 4: Driving Factors

These factors have a strong drive power but weak dependence power. In this cluster we have four factors, i.e., 2 (Customer Relationship Management), 3 (Top Management Commitment), 4 (Regulatory Pressures) and 7 (Total Quality Environmental Management).

We have developed appropriate research methodology to develop contextual relationship among identified factors using expert's opinion. We developed ISM model and have further refined the ISM model using MICMAC analysis in order to understand the characteristics of identified factors.

Based on ISM and MICMAC analysis we propose the below model as shown in Figure 6.

The proposed GSCM model depicts the Driving factors (Customer Relationship Management), (Top Management Commitment), (Regulatory Pressures), (Total Quality Environmental Management); Linkage factors (Supplier Relationship Management) and (Green Technology Adoption) and Dependence factors (Market Pressures), (Reduction in Carbon Emissions), (Market Share) and (Profit).

CONCLUSION

- Present study provides a systematic approach in developing a structural GSCM model for manufacturing sector. The ISM model has identified a hierarchy of factors for better decision making and building a clean system with minimum environmental impact.
- The ISM Model yields a hierarchy in which Market Share and Profitability is at level 1 (the top level) followed by Reduction in carbon emission (level 2),

Figure 6. Conclusive GSCM model

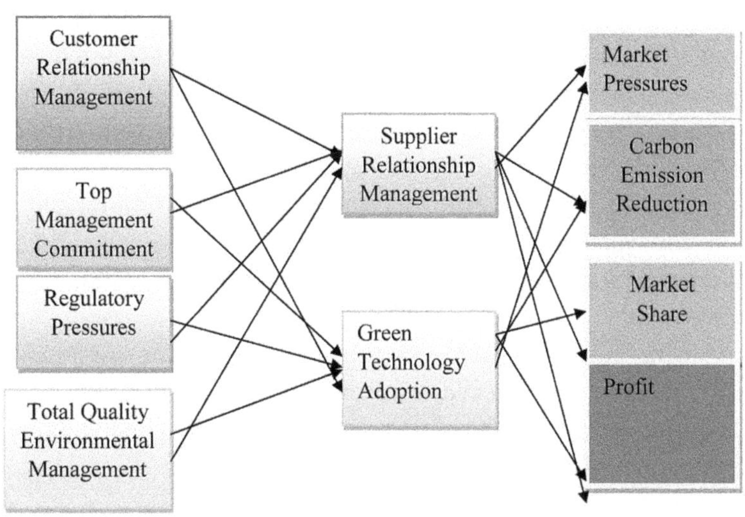

Green technology adoption (level 3), Total quality environmental management (level 4), Supplier relationship management and Customer relationship management (level 5), Top management commitment (level 6), Regulatory pressures and market pressures is at bottom level (level 7).

- Four factors are identified as driving factors and four factors as dependent factors using MICMAC analysis. Also there are two linkage factors but no autonomous factors emerged from the analysis.
- Customer relationship management, Top management commitment, Regulatory pressures and Total quality environmental management have been analyzed as driving factors. The driving factors play a vital role in successful GSCM practices. They condition the rest of the system and are called independent variables or drivers.
- The dependent factors characterize desired objectives for achieving success and identified from the analysis are Market pressures, Reduction in carbon emission, Market Share and Profit. These attributes are weak drivers but strongly dependent.
- Linkage factors are very sensitive and unstable. Any action on these factors will trigger an effect on other factors and also a feedback on themselves. Supplier relationship management and Green Technology adoption have emerged as the linkage factors.
- Based on the data analysis authors propose Relationship strategy for developing sustainable supply chain management in a manufacturing firm. Relationship strategy is based upon cooperation and collaboration with suppliers and customers.
- Buyer firms which practice supplier collaboration approach integrates procurement with other functions and achieve higher level of discipline and successful implementation of environmental certifications. Collaborative planning and information sharing helps to identify the strengths and weakness of suppliers. In this process it will be easy to educate suppliers and also pressurize for environmental certification. Collaboration and partnership develops innovation and enhances green practices.
- Cooperation leads to better understanding, Joint decision making, sourcing alternate/recyclable packaging material at economic costs, cost reduction. This creates more confidence and builds green culture within the organizational system.
- Relationship building improves company's image, gain competitive advantage, more value addition and ultimately happy customers.
- Relationship strategy leads to waste reduction, Lowers environmental costs and more profitability.

- The model presented in this study may enable management and decisions-makers involved in manufacturing companies to identify and classify the enablers that have either strong dependence or strong driving power or both strong dependence and driving power that enhances the process of GSCM practices.

SOLUTIONS AND RECOMMENDATIONS

Focus in Customer Relationship Management

There are no shortcuts to improving customer experience and driving GSCM practices. The need of the hour is to embrace GSCM practices in order to suit business goals and objectives in manufacturing sector.

CRM solutions and services that can demonstrate how the evolution of marketing, sales, customer service, customer analytics and data management processes can drive successful interaction and collaboration across customer and partner channels.

Customers are always curious about interesting organizational practices which benefit society and environment. If there is a new creative solution to a problem, that creates net positive value for customers, then certainly it will be attractive to customers. Customers will expect low risks involved in the organizational offerings.

Strong Leadership and Commitment of Top Management

Greening success is driven by dynamic leaders. In large sized manufacturing organizations greening is driven by these leaders in a more professional manner than smaller sized manufacturing organizations. Top management should prepare GSCM policy and create awareness among all employees.

Focus on Total Quality Environmental Management

Firms today are experiencing increased demands for enhanced social responsibility from investors, activists, regulators and consumers. These demands go beyond product quality, which was the focus of the total quality movement of the 1980s, and now encompass other areas, such as, responsible environmental management. Indian manufacturing firms are progressively making organizational changes to integrate the environment into their production decisions and are shifting from a regulatory driven, reactive approach to a proactive and beyond-compliance strategy towards environmental management

External market and regulatory considerations, such as the desire to improve a firm's image with customers and regulators, earning good-will with regulators and the anticipation of future regulations is associated with the adoption of TQEM.

Organizational change in the form of Total Quality Environmental Management leads firms to adopt pollution prevention practices, after controlling for the effects of various regulatory pressures and firm-specific characteristics.

There Is No Single Formula for Reducing Carbon Emissions

Every organization must chart its own greening course. Green practices, the subject of green supply chain management, and even the central objective can vary a lot from organization to organization. Organizations have different opportunities for greening, vary in resources, institutional pressures and different leaders available, each with their own innovative ideas, who do green practices distinctively. The greening strategies are unique in details across organizations. It is seen that organizations progress with green practices through standardized measurement and data capturing protocols. Reducing carbon emissions requires a sustained and focused effort. Organizations must form a cross functional team and proceed with the action plan for implementing GSCM in a holistic way. It takes long to adopt GSCM practices and therefore organizations must monitor progress on a periodic basis to understand the progress. In summary the differences in structure, size, availability of resources and market opportunities exists across organizations, which means the details of green practices are diverse.

Green organizations drive greening among their supply chain partners in the entire network. Greening organizations alter product design with emphasis on green design and waste disposal protocol to facilitate reuse and recycle in order to close the supply chain loop; this is simply to reduce environmental impact during the stages of product life cycle., i.e. from birth to the end of its useful life.

A set of ISO standards, ISO 14040 and ISO 14044, has been established and manufacturers should therefore make sure that the work in question is performed in accordance with these ISO standards.

Green Organizations Help to Increase Profitability

Green practices improve quality and lower wastages which results in significant savings and improved profitability. Green firms switch to low carbon energy sources to reduce carbon emissions. This reduces the market pressures and government pressures. The enhanced brand image of the organization helps to increase in sales volume with lower input costs. Hence GSCM is the path to achieve sustainability for those aspiring to become world class organizations in the future.

Systematically Green Necessitates an Organization to be Competent at SRM

With fierce competition, changes in market dynamics such as emergence of GSCM practices and raw material price volatility, procurement professionals in the manufacturing firms are looking for better visibility across supply chain to control costs. Supplier innovation has also become key, with procurement looking towards suppliers for new raw materials that aid in product development. Organizations that aspire to be green must emphasis on supplier relationship management. Suppliers should be provided training and education so as to implement EMS and ISO 14001. Care should be taken to ensure suppliers follow the policy strictly. Organizations that are high on SRM have a strong sense of discipline, and focus on value creation.

Greening Can Become Fully Embedded in an Organization's DNA Sequence

The promotion of Green technologies has been included in the draft strategy for the manufacturing sector prepared by the Department of Industrial Policy & Promotion (DIPP), Ministry of Commerce and Industry. Greening must be embedded within the organization system so that it flows normally through the arteries and produce desired results. Organizations with strong Information Technology (SAP/ERP) setup help to develop it more systematically. Proper MIS is generated which gives 360 degree view of the organization in terms of producing the correct production related wastage report, energy savings, and activity based costing report etc. Managers and decision makers must understand the links of the GSCM model clearly before adopting implementing the system. This leads to the development of green culture and set of behaviors, until ultimately, holistic approach is expected as a part of each activity.

Green Is Not Free: It Requires Investment in Training and Development

Organizations aspire to be green must be prepared to take risks, and a tolerance for efforts that fetches failure. Since at the initial stage it involves financial risks therefore proper training of human resources is required so as to develop a culture that undertake risks. Further investment in clean technologies and best practices is necessary which will fetch long term benefits. Focus on maximizing energy efficiency potential by replacing obsolete, inefficient equipments and adopting best available technologies is vital for organizations who aspire to develop competitive edge in the market.

Greening Helps Organizations to Attract and Retain Talents

Talented human resources at all levels are attracted to progressive organizations since these organizations do world class practices. Hence green practices support reputational advantages and help in enhancing the brand image. Greening is associated with high employee motivation, low employee turnover and ultimately, labour market success.

FURTHER RESEARCH DIRECTIONS

Like every methodology ISM also suffers from certain limitations since it is purely based on the opinion of experts and qualitative in nature. To eradicate the limitations of present research we propose to validate the model empirically in other sectors using Exploratory factor analysis and further test using regression analysis using SEM packages such as AMOS/LISREL.

REFERENCES

Ageron, B., Gunasekaran, A., & Spalanzani, A. (2012). Sustainable supply chain management: An empirical study. *International Journal of Production Economics*, *140*(1), 168–182. doi:10.1016/j.ijpe.2011.04.007

Ali, D. (2011). An analysis of the drivers affecting the implementation of green supply chain management. *Resources, Conservation and Recycling*, *55*(6), 659–667. doi:10.1016/j.resconrec.2010.12.002

Attri, R., Dev, N., & Sharma, V. (2013). Interpretive Structural Modeling (ISM) approach: An overview. *Research Journal of Management Sciences*, *2*(2), 3–8.

Azevedo, S., Carvalho, H., & Machado, C. V. (2011). The influence of green practices on supply chain performance: A case study approach. *Transportation Research Part E, Logistics and Transportation Review*, *47*(6), 850–871. doi:10.1016/j.tre.2011.05.017

Bag, S. (2013). Designing the Green Supply Chain Strategies for Indian Manufacturing Firm. *International Journal of Supply Chain Management Systems*, *2*(1), 8–18.

Bag, S., & Anand, N. (2012). Green Procurement practices and its impact on organization performance: A conceptual framework of Indian manufacturing sector. *UPES Management Review*, *1*(2), 3–11.

Barari, S., Gaurav, A., Zhang, C., Mahanty, B., & Tiwari, M.K. (2012). A decision framework for the analysis of green supply chain contracts: An evolutionary game approach. *Experts Systems with Applications*, *39*, 2965-2976.

Beamon, B. M. (1999). Designing the green supply chain. *Logistics Information Management*, *12*(4), 332-342.

Bhateja, A. K., Babbar, R., Sarbjit, S., & Sachdeva, A. (2012). Study of the critical factor finding's regarding evaluation of green supply chain performance of Indian scenario for manufacturing sector. *International Journal of Computational Engineering& Management*, *15*(1), 74–80.

Bhatja, A. K., Babbar, R., & Zhu, Q. (2011, July). Study of green supply chain management in Indian manufacturing industries: A literature review cum an analytical approach for the measurement of performance. *International Journal of Computational Engineering& Management*, *13*, 84–99.

Bose, I., & Pal, R. (2012). Do green supply chain management initiatives impact stock prices of firms? *Decision Support Systems*, *52*(3), 624–634. doi:10.1016/j.dss.2011.10.020

Brandenburg, M., Govindan, K., Sarkis, J., & Seuring, S. (2012). Quantitative models for sustainable supply chain management: Developments and directions. *European Journal of Operational Research*, *233*(2), 299–312. doi:10.1016/j.ejor.2013.09.032

Caniels, M. C. J., Gehrsitz, M. H., & Semeijn, J. (2013). Participation of suppliers in greening supply chains: An empirical analysis of German automotive suppliers. *Journal of Purchasing and Supply Management*, *19*(3), 134–143. doi:10.1016/j.pursup.2013.02.005

Carbon Trust. (2006). *Carbon Footprint in the supply chain: The next step for business*. Report No. CTC616. Carbon Trust London. Retrieved from: https://www.carbontrust.com/media/84932/ctc616-carbon-footprints-in-the-supply-chain.pdf

Carter, C. R., & Rogers, D. S. (2008). A framework of sustainable supply chain management: Moving toward new theory. *International Journal of Physical Distribution & Logistics Management*, *38*(5), 360–387. doi:10.1108/09600030810882816

Chaabane, A., Ramudhhin, A., & Paquet, M. (2012). Design of sustainable supply chains under the emission trading scheme. *International Journal of Production Economics*, *135*(1), 37–49. doi:10.1016/j.ijpe.2010.10.025

Chen, M.K., Wai, T-W. & Hung, T-Y. (2012). Component selection system for green supply chain. *Experts Systems with Applications, 39*, 5687-5701.

Cho, D. W., Lee, Y. H., Ahn, S. H., & Hwang, M. K. (2012). A framework for measuring the performance of service supply chain management. *Computers & Industrial Engineering*, *62*(3), 801–818. doi:10.1016/j.cie.2011.11.014

Croom, S., Romano, P., & Ginnakis, M. (2000). Supply chain management: An analytic framework for critical literature review. *European J. of Purchasing & Supply Management*, *6*(1), 67–83. doi:10.1016/S0969-7012(99)00030-1

Dekker, R., Bloemhof, J., & Mallidis, I. (2012). Operations research for green logistics –An overview of aspects, issues, contributions and challenges. *European Journal of Operational Research*, *219*(3), 671–679. doi:10.1016/j.ejor.2011.11.010

Diabat, A., Govindan, K., & Panicker, V. V. (2012). Supply chain risk management and its mitigation in a food industry. *International Journal of Production Research*, *50*(11), 3039–3050. doi:10.1080/00207543.2011.588619

Dubey, R., & Bag, S. (2013). Exploring the dimensions of sustainable practices: An empirical study on Indian manufacturing firms. *International Journal of Operations and Quantitative Management*, *19*(2), 101–124.

Dubey, R., Singh, T., Ali, S. S., & Tiwari, S. (2013). Contextual relationships among antecedents of truck freight using ISM and its validating using MICMAC analysis. *Int. J. Logistics and Systems Management*, *20*(1), 42–58. doi:10.1504/IJLSM.2015.065963

Dubey, R., Sonwaney, V., Aital, P., Venkatesh, V., & Ali, S. S. (2013). Antecedents of innovation and contextual relationship. *Int. J. Innovation and Research*, *9*(1), 1–14. doi:10.1504/IJBIR.2015.065950

Gimenez, C., Sierra, V., & Rodon, J. (2012). Sustainable Operations: Their impact on the triple bottom line. *International Journal of Production Economics*, *140*(1), 149–159. doi:10.1016/j.ijpe.2012.01.035

Godfrey, R. (1998). *Ethical purchasing: Developing the supply chain beyond the environment.* In T. Russel (Ed.), *Greener Purchasing: Opportunities and Innovations* (pp. 244–251). Sheffield, UK: Greenleaf Publishing. doi:10.9774/GLEAF.978-1-909493-03-2_20

Gunesekaran, A., & Spalanzani, A. (2012). Sustainability of manufacturing and services: Investigations for research and applications. *International Journal of Production Economics*, *140*(1), 35–47. doi:10.1016/j.ijpe.2011.05.011

Gupta, S., & Desai, P. O. D. (2011). Sustainable supply chain management: Review and research opportunities. *IIMB Management Review*, *23*(4), 234–245. doi:10.1016/j.iimb.2011.09.002

Hassini, E., Surti, C., & Searcy, C. (2012). A literature review and a case study of sustainable supply chains with a focus on metrics. *International Journal of Production Economics*, *140*(1), 69–82. doi:10.1016/j.ijpe.2012.01.042

Hervani, A. A., Helms, M. M., & Sarkis, J. (2005). "Performance measurement for green supply chain management" Benchmarking. *International Journal (Toronto, Ont.)*, *12*(4), 330–353.

Hojmose, S., & Brammer, S. (2012). Green supply chain management: The role of trust and top management in B2B and B2C markets. *Industrial Marketing Management*, *41*(4), 609–620. doi:10.1016/j.indmarman.2012.04.008

IPCC. (2007). *Climate change 2007: The physical science basis*. Retrieved 1 December 2013 from: <http://www.ipcc.ch/pdf/assessment report/ar4/wg1/ar4-wg1-spm.pdf>

Kang, S.-H., Kang, B., Shin, K., Kim, D., & Han, J. (2012). A theoretical framework for strategy development to introduce sustainable supply chain management. *Procedia: Social and Behavioral Sciences*, *40*, 631–635. doi:10.1016/j.sbspro.2012.03.241

Khan, J., & Haleem, A. (2013). An integrated ISM and Fuzzy MICMAC approach for modeling of the enablers of Technology Management. *Indian Journal of Applied Research*, *3*(7), 236–242. doi:10.15373/2249555X/JULY2013/74

Kogg, B. (2003). Greening a cotton-textile supply chain. *Greener Management International*, *43*(September), 53–64.

Linton, J., Klassen, R., & Jayaraman, V. (2007). Sustainable supply chains: An Introduction. *Journal of Operations Management*, *25*(6), 1075–1082. doi:10.1016/j.jom.2007.01.012

Liu, S., Kasturiratne, D., & Moizer, J. (2012). A hub and spoke model for multi dimensional integration of green marketing and sustainable supply chain management. *Industrial Marketing Management*, *41*(4), 581–588. doi:10.1016/j.indmarman.2012.04.005

Longo, F., & Mirabelli, G. (2008). An advance supply chain management tool based on modeling and simulation. *Computers & Industrial Engineering*, *54*(3), 570–588. doi:10.1016/j.cie.2007.09.008

Luthra, S., Kumar, V., Kumar, S., & Haleem, A. (2011). Barriers to implement green supply chain management in automobile industry using interpretative structural modeling technique-An Indian perspective. *Journal of Industrial Engineering and Management*, *4*(2), 231–257. doi:10.3926/jiem.2011.v4n2.p231-257

Manuj, I., & Mentzer, J. T. (2008). Global supply chain risk management strategies. *Int. J. Physical Distribution & Logistics Management*, *38*(3), 192–223. doi:10.1108/09600030810866986

Mirhedayatian, S. M., Azadi, M., & Sean, R. F. (2014). A novel network data envelopment analysis model for evaluating green supply chain management. *International Journal of Production Economics*, *147*, 544–554. doi:10.1016/j.ijpe.2013.02.009

Mitra, S. (2004). Managing Environmental Issues in Supply Chains. In B. S. Sahay (Ed.), *Emerging Issues in Supply Chain Management, IIM Calcutta working paper series* (pp. 60–68). Macmillan.

Narasimhan, R., & Carter, J. R. (1998). *Environmental supply chain management*. Arizona State University.

Nelson, D. M., Marsillac, E., & Rao, S. S. (2012). Antecedents and Evolution of the Green Supply Chain. *Journal of Operations and Supply Chain Management*, 29 – 43.

Pereira, J.V. (2009). The new supply chain's frontier: Information management. *Int. J. of Information Management, 29*, 372-379.

Reefke, H., & Trochhi, M. (2013). Balanced scorecard for sustainable supply chains: Design and development guidelines. *International Journal of Productivity and Performance Management*, *62*(8), 805–826. doi:10.1108/IJPPM-02-2013-0029

Sarkis, J. (2003). A strategic framework for Green Supply Chain Management. *Journal of Cleaner Production, 11*(4), 397–409. doi:10.1016/S0959-6526(02)00062-8

Sarkis, J., Zhu, Q., & Lai, K. (2011). An organizational theoretic review of green supply chain management literature'. *International Journal of Production Economics, 130*(1), 1–15. doi:10.1016/j.ijpe.2010.11.010

Schrettle, S., Hinz, A., MaikeScherrer-Rathje, M. & Friedli, T. (2014). Turning sustainability into action: Explaining firms' sustainability efforts and their impact on firm performance. *International Journal of Production Economics, 147*(A), 73–84

Seksan, P., Ninlawan, C., Tossapol, K., & Pilada, W. (2010). The implementation of green supply chain management practices in electronics industry. *Proceedings of the International Multi Conference of Engineers and Computer Scientists.*

Seuring, S., & Muller, M. (2008). From a literature review to a conceptual framework for supply chain management. *Journal of Cleaner Production*, *16*(15), 1699–1710. doi:10.1016/j.jclepro.2008.04.020

Shukla, A. C., Deshmukh, S. G., & Kanda, A. (2009). Environmentally responsive supply chains: Learning from the Indian auto sector. *Journal of Advances in Management Research*, *6*(2), 154–171. doi:10.1108/09727980911007181

Srivastava, S. K. (2007). Green supply chain management: A state of the art literature review. *International Journal of Management Reviews*, *9*(1), 53–80. doi:10.1111/j.1468-2370.2007.00202.x

Storey, J., Emberson, C., Godsell, J., & Harrison, A. (2006). Supply chain management: Theory, practice and future challenges. *International Journal of Operations & Production Management*, *26*(7), 754–774. doi:10.1108/01443570610672220

Suering, S. (2013). A review of modeling approaches for sustainable supply chain management. *Decision Support Systems*, *54*(4), 1513–1520. doi:10.1016/j.dss.2012.05.053

Sushil. (2012). Interpreting the Interpretive Structural Model. *Global Journal of Flexible Systems Management, 13*(2), 87-106.

Tranfield, D., Denyer, D., & Smart, P. (2003). Towards a methodology for developing evidence-Informed management knowledge by means of systematic review. *British Journal of Management*, *14*(3), 207–222. doi:10.1111/1467-8551.00375

Tsoulfas & Pappis. (2006). Environmental principles applicable to supply chain design. *Journal of Cleaner Production*, *14*(18), 1602.

UNEP Report. (2011). Retrieved from http://www.unep.org/greeneconomy/greeneconomyreport

Vachon, S., & Klassen, R. D. (2006). Extending green practices across the supply chain: The impact of upstream and downstream integration. *International Journal of Operations & Production Management*, *26*(7), 795–821. doi:10.1108/01443570610672248

Wang, F., Lai, X., & Shi, N. (2011). A multi-objective optimization for green supply chain network design. *Decision Support Systems*, *51*(2), 262–269. doi:10.1016/j.dss.2010.11.020

Warfield, J. N. (1974). Structuring complex systems. Columbus, OH: Battele Memorial Ins.

Warfield, J. N. (1994). *A science of generic design: Managing complexity through systems design*. Iowa State University press.

Warfield, J. N. (1999). Twenty laws of complexity: Science applicability in organizations. *Systems Research and Behavioral Science*, *16*(1), 3–40. doi:10.1002/(SICI)1099-1743(199901/02)16:1<3::AID-SRES241>3.0.CO;2-F

Whetten, D. A. (1989). What constitutes a theoretical contribution? *Academy of Management Review*, *14*(4), 490–495. doi:10.2307/258554

Wu, Z., & Pagell, M. (2011). Balancing priorities: Decision-making in sustainable supply chain management. *Journal of Operations Management*, *29*(6), 577–590. doi:10.1016/j.jom.2010.10.001

Yang, C. L., Lin, P. S., Chan, Y., & Sheu, C. (2009). Mediated effect of environmental management on manufacturing competitiveness: An empirical study. *International Journal of Production Economics*, *123*(1), 210–220. doi:10.1016/j.ijpe.2009.08.017

Zhu, Q., & Sarkis, J. (2004). Relationships between operational practices and performance among early adopters of green supply chain management practices in Chinese manufacturing enterprises. *Journal of Operations Management*, *22*(3), 265–289. doi:10.1016/j.jom.2004.01.005

Zhu, Q., & Sarkis, J. (2007). The moderating effects of institutional pressures on emergent green supply chain practices and performance. *International Journal of Production Research*, *45*(18-19), 4333–4355. doi:10.1080/00207540701440345

Zhu, Q., Sarkis, J., & Geng, Y. (2005). Green supply chain management in China: Pressure, practices and performance. *International Journal of Operations & Production Management*, *25*(5), 449–468. doi:10.1108/01443570510593148

Zhu, Q., Sarkis, J., & Lai, K. (2013). Institutional based antecedents and performance outcomes of internal and external of green supply chain management practices. *Journal of Purchasing and Supply Management*, *19*(2), 106–117. doi:10.1016/j.pursup.2012.12.001

Zhu, Q., Sarkis, J., Lai, K., & Geng, Y. (2008). The role of organizational size in the adoption of green supply chain management practices in China. *Corporate Social Responsibility and Environmental Management*, *15*(6), 322–337. doi:10.1002/csr.173

KEY TERMS AND DEFINITIONS

GSCM: GSCM can be defined as integrating environmental thinking into supply chain management, including product design, material sourcing and selection, manufacturing processes, delivery of the final products to the consumers, and end-of-life management of the product after its useful life.

ISM: Interpretive Structural Modelling is a proven and popular methodology for understanding relationships among specific items that define a problem.

Sustainability: SSCM is defined as an efficiency of utilizing the finite resources and achieve flexibility to changeable business environment.

Section 3
Future of Sustainable Supply Chains

Chapter 7
Recent Developments in Green Supply Chain Management:
Sourcing and Logistics

Jay R. Brown
Loyola University, USA

Andrey A. Kretinin
Kent State University, USA

Maxim A. Bushuev
Morgan State University, USA

Alfred L. Guiffrida
Kent State University, USA

ABSTRACT

In today's supply chains, green and sustainable business practices have become an integral part of long-term strategy as well as the day-to-day operations, and sourcing and logistics play fundamental roles in ensuring the success of sustainable supply chains. In this chapter, we review recent research on three classes of decision models that are used in the logistics and sourcing functions of sustainable supply chains. Specifically, we examine and summarize recent developments in sustainable decision models for supplier selection, inventory lot-sizing, and last-mile delivery. Our review of the green and sustainable features found in these three models results in a single-source consolidation of models used in sustainable sourcing and logistics. This review may prove useful to researchers who focus on sustainable supply chain management and to practitioners who seek to expand their toolkit of models for the management and control of their supply chain.

DOI: 10.4018/978-1-5225-0635-5.ch007

INTRODUCTION

Since the 1970s, there was an evolution of paradigms for improving the productivity and profitability of integrated manufacturing and distribution systems. These paradigms included the Just-In-Time (JIT) production philosophy, Lean and Agile Manufacturing, Six-Sigma, Mass Customization, and Supply Chain Management (SCM). Organizations embraced these paradigms as a means to improve their competitiveness in response to heightened levels of global competition and demands by customers for low-cost, customized products with short delivery lead times. In these paradigms, sourcing and logistics have always played an important role in ensuring that customer demands for products were met in a timely manner. As organizations recognized the strategic implications of partnering with suppliers, the traditionally functional roles of sourcing and logistics were elevated to a more strategic level. As a result, sourcing and logistics have become the cornerstones of today's widely adopted SCM philosophy. The SCM philosophy, which advocates integrating value-adding activities such as production planning and inventory control, sourcing, logistics, and customer relationship management, has been shown to improve financial performance and competitive advantage. For a detailed review of the literature on the financial impact of SCM on firm performance, the reader is directed to Shi &Yu (2013).

Today, there is heightened interest in how business practices make an impact on the environment. Concerns over resource usage, energy consumption, and waste have prompted organizations to consider green and sustainable practices in the operation of their supply chains. As a result of the adoption of green and sustainable practices into supply chain operations, the SCM philosophy has further evolved and, to reflect these adopted practices, is now commonly referred to as either the Green Supply Chain Management (GSCM) or Sustainable Supply Chain Management (SSCM). As noted by Ahi & Searcy (2013), there are numerous definitions for these terms (22 for GSCM and 12 for SSCM), and the literature suggests that SSCM is essentially an extension of GSCM. For the convenience of the reader, Table 1 contains a list of definitions of key terms that are used in this chapter.

The long-term benefits of adopting green and sustainable business practices have been investigated by several researchers. Ambec and Lanoie (2008) systematically reviewed the literature on how firms can benefit from better environmental performance and provided a summary of the relationship between environmental and economic performance. Wang and Sarkis (2013), Green, Zelbst, Meacham, and Bhadauria (2012), and Rao and Holt (2005) empirically examined how the adoption of green and sustainable practices in SCM leads to improved economic and operational performance. Recent case study-based analyses that detail the integration of GSCM in real-world supply chains were reported by Ubeda, Arcelus, &, Faulin

Table 1. Glossary of key terms

Term	Definition	Gateway References
Green Supply Chain Management (GSCM)	The integration of environmental thinking into supply chain management, including product design, material sourcing and selection, manufacturing processes, delivery of the final product to the consumers, and end-of-life management of the product after the end of its useful life.	Srivastava (2007, p. 54)
Sustainable Supply Chain Management (SSCM)	The management of material, information, and capital flows as well as cooperation among companies along the supply chain while taking goals from all three dimensions of sustainable development, i.e., economic, environmental and social, into account which are derived from customer and stakeholder requirements.	Seuring & Müller (2008, p. 1700)
The Last Mile Problem (LMP)	Optimization of the last leg of the business-to-consumer delivery service; delivery of the final product to the customer's door.	Boyer, Prud'homme, &, Chung (2009, p. 185)
Carbon Footprint	A measure of the exclusive total amount of carbon dioxide emissions that is directly and indirectly caused by an activity or is accumulated over the life stages of a product.	Wiedmann & Minx (2008, p. 4)
Greenhouse Gases (GHGs)	A "basket" of six GHGs: carbon-dioxide (CO_2), methane (CH_4), nitrous oxide (N_2O), hydrofluorocarbons (HFCs), perfluorocarbons (PFCs), and sulfur hexafluoride (SF_6).	Rao & Riahi (2006, p. 1)
Life Cycle Assessment (LCA)	A methodological framework for estimating and assessing the environmental impact attributed to the life cycle of a product.	Rebitzer, Ekvall, Frischknecht, Hunkeler, Norris, Rydberg, Schmidt, Suh, Weidema, &, Pennington (2004, p. 702)
Environmental Management Systems (EMSs)	Strategic management approaches that define how an organization will address its impacts on the natural environment.	Darnall, Jolley, &, Handfield (2008, p. 30)
ISO14001	An international standard that provides a process-oriented framework for the establishment of an environmental management system that can be audited and certified.	Delmas (2001, p. 345)
Reverse Logistics	A process in which a manufacturer systematically accepts previously shipped products or parts from the point for consumption for possible recycling, remanufacturing, or disposal.	Dowlatshahi (2000, p. 143)
Closed-Loop Supply Chain	Includes traditional forward supply chain activities and the additional activities of the reverse supply chain (acquisition of product from end-users, reverse logistics, testing and sorting, refurbishing, and remarketing).	Guide, Harrison, &, Van Wassenhove (2003, p. 3)

continued on following page

Table 1. Continued

Term	Definition	Gateway References
Environmental Sustainability	The extent to which a strategy's success is compatible with the firm's general environment over the long term. Environmental sustainability considerations include such issues as the natural environment and ecology, political-legal, and regulatory concerns as well as crisis management.	Parnell (2008, p. 38)
Environmental Purchasing	The set of purchasing policies held, actions taken, and relationships formed in response to concerns associated with the natural environment. These concerns relate to the acquisition of raw materials, including supplier selection, evaluation and development; suppliers' operations; in-bound distribution; packaging; recycling; reuse; resource reduction; and final disposal of the firm's products.	Zsidisin & Siferd (2001, p. 69)

(2011) for Eroski, a Spanish food distributor, and by Lee (2011) for Hyundai, a Korean automobile manufacturer.

The contribution of this chapter to the literature on GSCM business practices is threefold and is summarized in our research objectives. In satisfying these research objectives, our chapter will be of interest to researchers who currently work in the field of green and sustainable supply chains, researchers who may be entering the field, and practitioners who are interested in learning about new research trends in the field.

First, we summarize quantitative modeling research on supplier selection models and examine the evolution of supplier selection models to include green and sustainable sourcing criteria. Sourcing plays a fundamental role in the operation of any supply chain. Concerns over the carbon footprint and related environmental costs associated with sourcing decisions in supply chains have generated considerable interest in the literature. In response to these concerns, researchers have contributed many empirical studies that demonstrate the importance and contribution of sustainable sourcing with supply chain performance and have contributed a wide array of quantitative models for selecting suppliers. In fulfilling this first research objective, we contribute to the literature by providing a single-source consolidation of supplier selection models.

Second, we review recent research on inventory lot-sizing models that incorporate green practices and sustainability into their decision frameworks. Within the sourcing policy used by an organization, the lot-sizing decision flows naturally from the supplier selection decision. Once a supplier has been selected, the lot-size decision is the next step. The literature on environmentally conscious inventory lot-sizing models is scattered. In fulfilling this second research objective, we contribute to the literature by providing a review of inventory lot-sizing models that managers can use to reduce green costs and to better manage overall materials planning and control in their supply chains.

Third, we examine recent research on the last-mile problem (LMP) in SCM. The logistical burden of LMP is considered to be one of the most costly and highest polluting segments of the supply chain (Ülkü, 2012; Gevaers, Van de Voorde, &, Vanelslander, 2011). Last-mile delivery has proven to incur the highest transportation costs in the supply chain (Chopra, 2003). In fulfilling this third research objective, we contribute to the literature by providing a summary of the environmental impact of last-mile delivery options (customer pick-up versus carrier delivery) and strategies for managing failed home deliveries.

BACKGROUND

Firms engage in a variety of sourcing policies to support supply chain operations. These policies include single sourcing, multiple sourcing, cross-sourcing, and dual sourcing (see Coronado, 2007, for a review). In today's global marketplace, buyers have a multitude of potential external suppliers to choose from when making sourcing decisions. The Internet and web-based technologies have allowed easy access to a large portfolio of qualified suppliers. Supplier selection and the ensuing purchasing lot-size decision are critical components of a company's overall sourcing policy and have a direct impact on a company's performance (Su, 2013; Chiang, Kocabasoglu-Hillmer & Suresh, 2012; Prajogo & Olhager, 2012). However, suppliers are substantially different from each other in terms of many factors, including prices charged, quality of goods, and services offered (Araz & Ozkarahan, 2007; Weber, Current, &, Benton, 1991). The large number of suppliers and their unique characteristics contribute to making the supplier selection and lot-sizing decisions complicated (Talluri, DeCampos, &, Hult, 2013; Aissaoui, Haouari, &, Hassini, 2007).

The foundation for decision models in lot sizing and supplier selection was established in the literature many years ago. The first model to aid managers in determining how much to order appeared 100 years ago with the contribution of the economic order quantity model (EOQ) of Harris (1913). In the years following the introduction of the Harris EOQ model, numerous variants and extensions of the EOQ were developed, and the literature on inventory models has been ever expanding. For a comprehensive review of this literature, see Drake and Marley (2014). Nearly 50 years ago, Dickson (1966) identified 23 different criteria that have been shown to be important in the evaluation of suppliers. Challenges to utilizing all of these criteria when making a supplier selection decision occur when there is incomplete information available to buyers and their unique requirements (Lin, 2012). A substantial body of research has provided several types of decision models and algorithms to assist decision makers with evaluating and choosing appropriate suppliers. Despite these tools, there is still no consensus among researchers and practitioners as to which supplier selection model is the most efficient and practical (Degraeve, Labro, &, Roodhooft, 2000).

SUSTAINABLE SOURCING AND LOGISTICS

Concerns about the environment have an impact on how organizations design, coordinate, and manage their supply chains. In response to customer-driven and/ or legislative pressures over carbon emissions, greenhouse gases, water and air pollution, and the depletion of raw material stocks, many organizations have re-examined their fundamental business strategy from an environmental perspective. Central to the planning, creation, and implementation of an environmental friendly and sustainability-based competitive strategy is an overall guiding model that is supported by an operational framework and performance measurement system. The literature has seen models of strategic and operational decision making under green and sustainable strategies evolve along a modeling continuum that ranges from exploratory to conceptual to highly theoretical. Three models that illustrate this continuum are as follows: Peattie and Ratnayaka (1992) presented a three-phase model to guide companies that are under customer/legislative-driven pressures to become more sustainable. The model delineates driving forces, potential barriers to action, and strategies for being green. Linton, Klassen, &, Jayaraman (2007) noted the interdisciplinary nature of sustainability and present a model that illustrates the interrelationship between the natural and social sciences when developing sustainable supply chain operating policies. Deif (2011) adopted a systems theory modeling approach and illustrates how planning and control activities in the manufacturing phase of a supply chain evolve from a less green to a greener, more eco-efficient manufacturing paradigm.

Frameworks for integrating green and sustainable practices into supply chains can be found in Schaltegger and Burritt (2014), Sarkis (2012, 2003), Carter and Rogers (2008), and Vachon and Klassen (2006). Hervani, Helms, and, Sarkis (2005) provided an overview of green supply chain performance measurement and offer insight into the roles that inputs, metrics (financial, customer, and internal processes), organization members, and outputs play in the measurement of the performance of a green supply chain. The transition to an environmentally based supply chain strategy under a given model or framework should not be static or myopic. Green innovation is a key antecedent of the success of a GSCM strategy (Routroy, 2009), and Chang (2011) empirically showed the benefit of green product innovation and green process innovation on maintaining a competitive advantage under a sustainability driven corporate strategy.

Supplier Selection

The selection of suppliers (vendors) is fundamental to ensuring an uninterrupted supply of inputs to a supply chain. The supplier selection process consists of a series of hierarchical phases that begins with supplier pre-selection and concludes with final supplier selection (De Boer, Labro, &, Morlacchi, 2001). Within this process, multiple criteria must be evaluated for each supplier. These criteria are both quantitative and qualitative and address a supplier's performance across multiple dimensions, including (1) antecedents of performance (e.g., training, labor relations), (2) operations process (e.g., technical capabilities, operating control), and (3) final outcomes (e.g., quality, delivery, price). In many instances, interactions among the criteria themselves further complicate the selection process (Kannan & Haq, 2007). Hence, no single factor dominates, and a comprehensive evaluation should address the "totality of features" (Carter, 1995). Given the complexity of evaluating a portfolio of competing suppliers over a wide set of criteria, researchers have formulated a multitude of models to assist in making the final selection decision.

The broad scope of research on supplier evaluation models is evident by the 359,000 hits that result from a search using the phrase "supplier evaluation model" in Google Scholar. The vast number of published supplier evaluation models eliminates any possibility of providing a complete review of each model in the form of a journal article, dissertation, textbook, or even an edited volume. Thus, we adopt a methodology to examine and classify published literature reviews on supplier evaluation models. By adopting this methodology, we can contribute a comprehensive review of the literature that serves the following objectives: (1) provide in one book chapter a consolidation of research that can serve as a single source to keep researchers up to date with the research developments in supplier selection models, (2) identify gaps in the field that could stimulate new research agendas, and (3) identify areas in current research activities that need further development.

The papers included in this review are narrative reviews of published supplier selection models. Given the narrative feature of the papers reviewed, a formal structured literature review methodology or a meta-analysis-based literature review methodology is not applicable. Hence, we adopted a narrative literature review methodology in this chapter. Research papers considered for inclusion in this review were found by searching the open literature using the keyword searches "supplier selection models" and "vendor selection models" in EBSCOhost, Google Scholar, ProQuest, and Science Direct. Only papers written in English were considered.

Guiffrida (1999) reviewed 50 supplier evaluation models that used at least one of the evaluation criteria noted by Dickson (1966). The evaluation models were grouped into four types: Cost-Based, Weighted Point, Mathematical Programming, and Analytic Hierarchy Process (AHP). Cost-based models use either a cost index or total cost measure, which is based on a set of cost drivers, which measure the impact of supplier performance on the buyer. Weighted-point models attach a weighting scheme to the evaluation criteria and an overall performance score that can be compared across suppliers, as determined based on the sum of the product of the performance ratings and the criteria weights. For the mathematical-programming model, linear programming is used for models with a single selection objective. Goal programming/multiple criteria programming is used for models with multiple selection objectives. Of the 50 suppler selection models reviewed, weighted-point models were most frequently used (19), followed by mathematical programming (15), AHP (10), and cost-based models (6).

Aissaoui, Haouari, and Hassini (2007) examined the supplier selection problem from a systematic perspective and illustrated how the supplier selection decision is integrated into the overall purchasing decision process. The authors emphasize that, under this systematic approach, the supplier selection decision must be connected to inventory management. Models are classified according to a sourcing framework (single versus multiple), with further sub classifications that identify whether the model is integrated with inventory management decisions and whether quantity discounts apply.

Jain, Wadhwa, and Deshmukh (2009) provided a review of the literature on supplier-related issues in SCM that addresses the topics of supplier selection, supplier-buyer relationships, and supplier-buyer flexibility. They identified 20 different methodologies for modeling the supplier-selection process and 39 applications of these methodologies. Mathematical-programming applications (linear programming, integer programming, mixed-integer programming, and goal programming) have the highest frequency of model citations (10), followed by AHP (5). An examination of Table 1 of Jain et al. (2009, p. 3016) shows the use of advanced statistical methodologies, such as structural equation modeling and factor analysis, data envelopment analysis (DEA), fuzzy logic, and neural networks in modeling the supplier selection problem.

Agarwal, Sahai, Mishra, Bag, and Singh, (2011) reviewed 60 multi-criteria supplier selection models published from 2000 to 2011 and argue that a multiple-criteria decision making (MCDM) approach is needed across the wide span of evaluation criteria because the emphasis on one measure often requires sacrificing another measure. The percentages of citations of MCDM and its subcategories include DEA

(30%), mathematical programing (17%), case-based reasoning (11%), criteria-based models (7%), fuzzy-set theory (6%), analytical network process (ANP) (3%), simple multi-attribute rating technique (3%), and genetic algorithms (2%). The authors attribute the high adoption rate of DEA to its robustness, whereby suppliers and their processes are modeled from a systems-efficiency perspective.

Wu and Barnes (2011) reviewed 140 papers on partner selection decision-making published between 2001 and 2011. The focus of the review is research that is relevant for supply partner selection in agile supply chains. The authors presented a four-phase partner selection framework that summarized the decision models for each phase. The overwhelming majority of decision models is found in the final selection phase of the partner selection framework and includes the following model types: linear weighing, mathematical programming, fuzzy-set theory, and AHP/ANP. Each model type is further delineated in terms of its integration with inventory management (no integration or integration) and number of products (single or multiple). In the sub classification of single product with no inventory integration, the partner selection models and their frequency are as follows: linear weighting (5), mathematical programing (21), and fuzzy-set theory (29). For multiple products with inventory integration, the models and their frequencies are AHP/ANP (13) and mathematical programing (12).

Ware, Sing, and Banwet (2012) reviewed the literature on supplier selection models over a 20-year period, from 1991 to 2011. To highlight trends in model usage, the authors partitioned a total of 153 supplier selection models into eight categories, with each category subdivided into five-year periods. The distinct modeling methodologies and the frequency of papers cited using each methodology were as follows: AHP (16), ANP (7), fuzzy techniques (24), DEA (3), linear/nonlinear programming (10), multi-objective processes (16), optimization techniques (11), and other approaches (66). Over the last five-year period (2006 to 2011), fuzzy and multi-objective supplier selection models were the most prevalent. A practical feature of their review paper is the step-by-step and detailed illustration of five selected decision methodologies: fuzzy multi-objective, nonlinear multi-objective, fuzzy AHP, AHP, and the fuzzy technique for order of preference by similarity to ideal solution (TOPSIS). This tutorial approach to these selected methodologies may be useful to researchers who seek to adopt these methodologies into their research toolkits.

In a review of 123 supplier selection models published from 2008 to 2012, Chai, Liu, and Ngai (2013) identified 26 decision-making techniques that are used in the supplier-selection models and classify them from three perspectives: MCDM, mathematical programming, and artificial intelligence. The specific decision-making techniques with the highest frequency of use within each classification group were AHP in MCDM (24.4%), linear programming within mathematical programming (15.4%), and genetic algorithms within artificial intelligence (6.5%). The models were further analyzed along the dimensions of level of decision integration present in the model and level of uncertainty found in the decision-model environment.

Pal, Gupta, and Garg (2013) classified 17 supplier selection methods into six categories, which include methods for prequalification of suppliers (categorical methods, DEA, and cluster analysis) plus five categories that include MCDM, mathematical programming, artificial intelligence, fuzzy logic, and combined approaches. The MCDM category includes AHP, ANP, TOPSIS, multi-attribute theory (MAUT), and outranking methods. Mathematical programming models include linear programming, multiple objective programming, and goal programming. The authors summarized case-based reasoning and neural networks within the artificial intelligence classification. Fuzzy logic represents a fifth classification category. The sixth category consists of combined approaches, such as AHP and TOPSIS, and mathematical programming combined with AHP and MAUT.

Govindan, Rajendran, Sarkis, and Murugesan (2013) reviewed the literature on green supplier evaluation criteria and green multi-criteria supplier-selection models. Their review spans 1996 to 2011 and they identified 120 distinct green evaluation criteria and 33 green multi-criteria supplier-selection models. They grouped the green suppler selection models by frequency into the following decision methodology categories: AHP (6), ANP (4), mathematical programing (1), and other approaches, such as DEA, statistical methods, and fuzzy inference (14). A highly useful feature of this review paper is the authors' mapping of the decision methodology used in the supplier-selection model to the environmental criteria used in the model. The authors observed that the implementation of green issues within the supplier selection process is limited, as demonstrated by the limited number of green supplier-selection models. The authors noted that a greater emphasis on sensitivity analysis of model criteria weights and data is needed.

Green/Sustainable Inventory Lot-Sizing

Lot-sizing models in inventory management have been traditionally used by inventory managers in addressing two fundamental questions: (1) How large should an inventory replenishment order be? and (2) When should an inventory replenishment order be placed? Addressing these two questions is fundamental to the efficient operation of an inventory management system that is designed to provide an acceptable level of customer service. There is a direct connection between inventory management and environmental concerns. For example, inventory decisions can be linked to the level of carbon emissions, as lot-sizing decisions have an impact on the amount of freight that requires transport and warehousing in the supply chain. Despite this direct connection, the integration of inventory lot-sizing models with green and sustainable practices is in its infancy, with no seminal review of this important literature's having been conducted. Research on green and sustainable aspects of lot sizing has been scattered (see Bouchery, Ghaffari, Jemai, &, Dallery, 2012; Bonney & Jaber, 2012; and Jaber, 2009, for an introduction to this literature).

The following four lot-size models are commonly found in the inventory management literature: economic order quantity (EOQ) model, economic production quantity (EPQ) model, newsvendor (NV) model, and joint economic lot size (JELS) model. In Table 2, we present a review of 22 inventory lot-sizing papers that integrate green/sustainable criteria using these four model types.

Table 2. Summary of inventory lot-size models with sustainability criteria

Authors	Type of Inventory Model				Sustainability Criteria
	EOQ	EPQ	NV	JELS	
Arıkan & Jammernegg (2014)			X		Carbon emissions
Bozorgi, Pazour, &, Nazzal (2014)	X				Carbon emissions
Mukhopadhyay & Goswami (2014)		X			Volume (mass) of pollutant
Toptal, Özlü, &, Konur (2014)	X				Carbon emissions
van der Veen & Venugopal (2014)	X				Energy consumption
Zanoni, Mazzoldi, &, Jaber (2014)				X	Carbon emissions
Zhiyong, Lingyu, &, Guicheng (2014)	X				Carbon emissions
Arslan &Turkay (2013)	X				Carbon emissions
Chen, Benjaafar, &, Elomri (2013)	X				Carbon emissions
Jaber, Glock, &, El Saadany (2013)				X	Carbon emissions
Rosič & Jammernegg (2013)			X		Carbon emissions
Zanoni, Bettoni, &, Glock (2013)				X	Energy consumption
Bouchery, Ghaffari, Jemai, &, Dallery (2012)	X				Carbon emissions
Zhang & Xu (2013)			X		Carbon emissions
Choi & Chiu (2012)			X		Unsold product
Glock, Jaber, &, Searcy (2012)				X	Carbon emissions
Song & Leng (2012)			X		Carbon emissions
Hua, Cheng, &, Wang (2011)	X				Carbon emissions
El Saadany, Jaber, &, Bonney (2011)				X	Greenes-quality measure
Wahab, Mamun, &, Ongkunaruk (2011)	X			X	Carbon emissions
Manikas & Godfrey (2010)			X		Emission permits
Tao, Guiffrida, &, Troutt (2010)	X	X			Fixed carbon cost

As seen in Table 2, the EOQ is the most frequently used model for sustainable inventory lot sizing, followed by the NV, JELS and EPQ models. Carbon emissions (typically measured in units of carbon dioxide gas) are the most commonly used sustainability criteria. Fixed carbon cost, unsold product, pollutant volume, and the number of emission permits also have been used as sustainability criteria in inventory lot-sizing models. The widespread use of carbon emissions may be the result of the relative ease by which carbon emission can be measured; however, carbon dioxide gas represents only one component of overall greenhouse gas emissions.

The JELS model is an attempt to advance inventory-based sustainability modeling to the scope of a supply chain, as these models integrate the inventory lot-sizing decision between the buyer and supplier. With the exception of the work of Bouchery, Ghaffari, Jemai, and Dallery (2012) and Zhang and Xu (2013), the number of models that concern sustainable inventory lot sizing across multiple products is very limited.

The Last Mile Delivery Problem

The importance that the logistical component plays in the performance of a supply chain is well documented (Srisorn, 2013; Chopra, 2003; Stank, Goldsby, Vickery, &, Savitskie, 2003). In a climate of enhanced awareness of environmentally sustainable business practices, the issue of carbon emissions as a result of freight transportation in supply chains is rapidly becoming a key managerial concern (Demir, Bektaş, &, Laporte, 2014; Van Woensel, Creten, &, Vandaele, 2001). Golicic, Boerstler, and Ellram (2010) noted that developing a sustainable supply chain transportation strategy is a key concern of organizations. Nevertheless, as reported in Supply Chain and Logistics (2009, p. 42), only 10% of companies are actively modeling their supply chain carbon footprints and have implemented successful sustainability initiatives. Of particular concern to supply chain managers is last mile problem (LMP), which is considered to be one of the most costly and highest polluting segments of the supply chain (Ülkü, 2012; Gevaers, Van de Voorde, &, Vanelslander, 2011). According to Goodman (2005), up to 28% of all transportation costs are incurred in last-mile delivery. Thus, more efficient management of the LMP can lead to cost savings in supply chain operations.

The majority of research on LMP has investigated how to achieve higher levels of customer service and cost savings based on improvements in the delivery process (Esper, Jensen, &, Turnipseed, 2003), delivery routing (McLeod, Cherrett, &, Song, 2006), completing the final delivery (Punakivi, Yrjölä, &, Holmström, 2001), delivery security (McKinnon & Tallam, 2003), delivery scheduling (Boyer et al. 2009), and coordinating the delivery fulfillment process (Kull, Boyer, &, Calantone, 2007). The last mile framework also has been adopted in models of the delivery process of relief supplies from aid distribution centers to natural disaster sites (Huang, Smilowitz, &, Balcik, 2011; Stapleton, Pedraza Martinez, &, Van Wassenhove, 2009; Balcik, Beamon, &, Smilowitz, 2008). The research cited above highlights the importance of the LMP in improving customer service and reducing costs in supply chain operations. However, as noted by McIntyre, Smith, Henham, and Pretlove (1998), traditional time- and cost-based measures support only short-term performance objectives and fail to support the longer-term logistical decisions that are critical to a sustainable and environmentally compatible green business strategy.

The environmental aspects of LMP have been investigated through two research streams. The first stream concerns the environmental tradeoffs between home deliveries and customer pick-up strategies and reflects research by Siikavirta, Punakivi, Karkkainen, and Linnanen (2003), Edwards, McKinnon, Cherrett, McLeod, and Song (2010), and Brown and Guiffrida (2014). The second research stream involves the environmental impact of designating alternative outlets as local collection and delivery points (CDPs) and comprises research by McLeod et al. (2006), Edwards et al. (2010), and Song, Guan, Cherrett, and Baowen (2013).

Siikavirta et al. (2003) studied greenhouse gas (GHG) emissions that results from home delivery versus customer pick-up in the e-grocery industry in Helsinki, Finland. GHG emissions were measured as a function of the travel distance incurred under the competing delivery strategy methods. A simulation methodology was used, and the results demonstrated that, in comparison to customer pick-up, home delivery can lead to reductions in GHG emissions in the range of 18% to 87%.

Edwards et al. (2010) compared the level of CO_2 emissions that result from home delivery versus customer pick-up in retail online shopping for non-food products. The emission factors for carbon-fuel based vehicles used under each delivery strategy were estimated from secondary data sources. The study considered delivery failure rates for home delivery that ranged from 2% to 25% as well as a subjective estimate of the degree of trip chaining (a customer's picking up multiple items from multiple locations during one trip) that occurred under the customer pick-up strategy. Edwards et al. concluded that it is likely (given a set of defined assumptions) that, when customers buy fewer than 24 items per shopping trip, the CO_2 per item purchased will be lower under home delivery.

Brown and Guiffrida (2014) compared the level of carbon emissions that resulted from conventional shopping (customer pick-up with trip chaining) versus home delivery of retail products purchased online. A break-even methodology was used to construct a decision framework for assessing the degree of carbon emission impact of home delivery versus customer pick-up. The decision framework determines the break-even number of customers for carbon emissions equivalence under the two delivery strategies to support a desired delivery service level for a given demand density and delivery capacity. For example, if home delivery is used when the number of customers is lower than for the break-even number of customers, the level of carbon emissions will be higher than if the customers had picked up their purchases themselves (conventional shopping). The robustness of the decision framework and break-even analysis methodology is demonstrated using empirical data collected by the authors and secondary data found in the literature.

A second LPM research stream concerns the environmental impact of failed home deliveries of packages. McLeod et al. (2006) analyzed the tradeoff between two options for rectifying failed home deliveries. The vehicle mileage incurred by carriers that redeliver a package after a failed delivery was compared to the aggregate travel mileage incurred by a carrier that took the failed delivery to a local CDP plus the mileage incurred by the customer who retrieved the item from the CDP. Post offices, grocery stores, and petrol stations commonly serve as CDPs in the United Kingdom (UK). A case study conducted in Winchester, UK, using CDPs to solve the failed delivery problem, showed no significant reduction in the carriers' mileage but the potential for reductions of customers' mileage by up to 80%. The authors noted that the results achieved in their case study were heavy influenced by the parameter estimates used for the delivery failure rate, the number of available CDPs, and the extent of customer trip chaining.

To evaluate the carbon dioxide generated by customers' retrieving their packages after a failed home delivery, Edwards et al. (2010) developed a spreadsheet-based carbon-auditing model. The analysis was based on 120 deliveries to be made over a 50-mile distance in a typical UK city. First-time delivery failure rates varied from 10% to 30%. The study found that 85% to 95% of the carbon dioxide emissions due to first-time failed home delivery resulted from the travel of the customer to retrieve the failed delivery item from the carrier's depot and not the carrier's repeated home delivery attempt following a delivery failure. Using CDPs was found to reduce carbon dioxide emission, and, thus, CDPs represent a more environmentally friendly approach to mitigating failed deliveries than redelivery by the carrier or customer retrieval from the carrier's depot.

Song et al. (2013) quantified GHG emissions generated from redelivery attempts for failed home delivery by the carrier and customer pick-up at CDPs or the home depot of the carrier. In this study, delivery data for households in Winchester, UK, and West Sussex, UK, were used. The first-time delivery failure rate varied from 10% to 50% with the four different CPDs options (railway station, petro station, supermarket, and post office). The authors conclude that a CPD network would reduce GHG emissions when: (i) at least 30% of customers travel to the carrier depot to retrieve a first-time failed home delivery, (ii) the proportion of first-time home deliveries is significant, and (iii) post offices are used as CDPs.

CONCLUSION

Models for supplier selection, inventory lot sizing, and last mile delivery all play important roles in the logistics and sourcing operations of a sustainable supply chain. Nine literature reviews on supplier selection models, spanning the time period from 1999 to 2013, were summarized. The methodologies used in the supplier selection models reviewed was diverse and consisted of single-and multi-criteria optimization methods, advanced statistical methods, and heuristics. Only one paper (Govindan et al. 2013) was found that provided a review of supplier-selection models that used sustainability criteria within the decision model. A total of 22 sustainable inventory lot-size models were found in the literature, and carbon-emission models based on the EOQ model were most common. Decision models for LMP took into account the carbon emission tradeoffs associated with strategies for home delivery of items by carrier versus customer pick-up at the carrier's depot or at a designated CPD.

In summary, the literature reviewed in this chapter indicates that attempts have been made to introduce sustainability into decision models that are useful for sourcing and logistics under the GSCM strategy. Models for sustainable supplier selection could be expanded along the following dimensions: First, greater emphasis could be given to the role that supplier development plays in the evaluation of suppliers. Buyers may find it necessary to share expertise and even to financially invest in suppliers to achieve a portfolio of suppliers that meet their long-term goals of sustainable contribution. Second, existing sustainable supplier selection models need to place more emphasis on the sensitivity analysis of key model inputs. Models for sustainable inventory lot-sizing are also somewhat limited in their reliance on the low-hanging fruit of carbon emissions as a means to introduce sustainability in inventory decisions. Alternative measures of sustainability are needed that are greater in scope than the carbon footprint. In addition, the inventory lot-sizing modeling environment needs to expand from the restrictive single-product EOQ model to the multiproduct JELS model that can integrate the decision-making process across multiple echelons of the supply chain. Decision models for LMP rely heavily on case-based studies of the level of carbon emission to evaluate competing last mile delivery strategies. A more robust investigation of decision models for solving LMP must address multiple environmental performance measures used in a formally designed experiment that controls for population density and delivery infrastructure.

REFERENCES

Agarwal, P., Sahai, M., Mishra, V., Bag, M., & Singh, V. (2011). A review of multi-criteria decision making techniques for supplier evaluation and selection. *International Journal of Industrial Engineering Computations*, *2*(4), 801–810. doi:10.5267/j.ijiec.2011.06.004

Ahi, P., & Searcy, C. (2013). A comparative literature analysis of definitions for green and sustainable supply chain management. *Journal of Cleaner Production*, *52*, 329–341. doi:10.1016/j.jclepro.2013.02.018

Aissaoui, N., Haouari, M., & Hassini, E. (2007). Supplier selection and order lot sizing modeling: A review. *Computers & Operations Research*, *34*(12), 3516–3540. doi:10.1016/j.cor.2006.01.016

Ambec, S., & Lanoie, P. (2008). Does it pay to be green? A systematic overview. *The Academy of Management Perspectives*, *22*(4), 45–62. doi:10.5465/AMP.2008.35590353

Araz, C., & Ozkarahan, I. (2007). Supplier evaluation and management system for strategic sourcing based on a new multicriteria sorting procedure. *International Journal of Production Economics*, *106*(2), 585–606. doi:10.1016/j.ijpe.2006.08.008

Arıkan, E., & Jammernegg, W. (2014). The single period inventory model under dual sourcing and product carbon footprint constraint. *International Journal of Production Economics*. DOI: 10.106/j.ijpe.2014.04.027

Arslan, M., & Turkay, M. (2013). EOQ revisited with sustainability considerations. *Foundations of Computing and Decision Sciences*, *38*(4), 223–249. doi:10.2478/fcds-2013-0011

Balcik, B., Beamon, B. M., & Smilowitz, K. (2008). Last mile distribution in humanitarian relief. *Journal of Intelligent Transportation Systems*, *12*(2), 51–63. doi:10.1080/15472450802023329

Bonney, M., & Jaber, M. Y. (2012). Environmentally responsible inventory models: Non-classical models for a non-classical era. *International Journal of Production Economics*, *133*(1), 43–53. doi:10.1016/j.ijpe.2009.10.033

Bouchery, Y., Ghaffari, A., Jemai, Z., & Dallery, Y. (2012). Including sustainability criteria in inventory models. *European Journal of Operational Research*, *222*(2), 229–240. doi:10.1016/j.ejor.2012.05.004

Boyer, K. K., Prud'homme, A. M., & Chung, W. (2009). The last mile challenge: Evaluating the effects of customer density and delivery window patterns. *Journal of Business Logistics, 30*(1), 185–201. doi:10.1002/j.2158-1592.2009.tb00104.x

Bozorgi, A., Pazour, J., & Nazzal, D. (2014). A new inventory model for cold items that considers costs and emissions. *International Journal of Production Economics.* DOI:10.1016/j/ijpe.2014.01.006

Brown, J. R., & Guiffrida, A. L. (2014). Carbon emissions comparison of last mile delivery versus customer pick up. *International Journal of Logistics and Applications, 17*(6), 503–521. doi:10.1080/13675567.2014.907397

Carter, C. R., & Rogers, D. S. (2008). A framework of sustainable supply chain management: Moving toward new theory. *International Journal of Physical Distribution & Logistics Management, 38*(5), 360–387. doi:10.1108/09600030810882816

Carter, R. (1995, April). The seven Cs of effective suppler evaluation. *Purchasing and Supply Management*, 44-45.

Chai, J., Liu, J. N. K., & Ngai, E. W. T. (2013). Application of decision making-techniques in supplier selection: A systematic review of literature. *Expert Systems with Applications, 40*(10), 3872–3885. doi:10.1016/j.eswa.2012.12.040

Chang, C. H. (2011). The influence of corporate environmental ethics on competitive advantage: The mediation role of green innovation. *Journal of Business Ethics, 104*(3), 361–370. doi:10.1007/s10551-011-0914-x

Chen, X., Benjaafar, S., & Elomri, A. (2013). The carbon-constrained EOQ. *Operations Research Letters, 41*(2), 172–179. doi:10.1016/j.orl.2012.12.003

Chiang, C. Y., Kocabasoglu-Hillmer, C., & Suresh, N. (2012). An empirical investigation of the impact of strategic sourcing and flexibility on firm's supply chain agility. *International Journal of Operations & Production Management, 32*(1), 49–78. doi:10.1108/01443571211195736

Choi, T.-M., & Chiu, C.-H. (2012). Mean-downside-risk and mean-variance newsvendor models: Implications for sustainable fashion retailing. *International Journal of Production Economics, 135*(2), 552–560. doi:10.1016/j.ijpe.2010.10.004

Chopra, S. (2003). Designing the distribution network in a supply chain. *Transportation Research, 39*(2), 123–140. doi:10.1016/S1366-5545(02)00044-3

Coronado, J. L. (2007). *An Optimization Model for Strategic Supply Chain Design under Stochastic Capacity Disruptions*. (Dissertation). Department of Industrial and Systems Engineering, Texas A&M University, College Station, TX.

Darnall, N., Jolley, G. J., & Handfield, R. (2008). Environmental management systems and green supply chain management: Compliments for Sustainability? *Business Strategy and the Environment*, *17*(1), 30–45. doi:10.1002/bse.557

De Boer, L., Labro, E., & Morlacchi, P. (2001). A review of methods supporting supplier selection. *European Journal of Purchasing & Supply Management*, *7*(2), 75–89. doi:10.1016/S0969-7012(00)00028-9

Degraeve, Z., Labro, E., & Roodhooft, F. (2000). An evaluation of vendor selection models from a total cost of ownership perspective. *European Journal of Operational Research*, *125*(1), 34–58. doi:10.1016/S0377-2217(99)00199-X

Deif, A. M. (2011). A system model for green manufacturing. *Journal of Cleaner Production*, *19*(14), 1553–1559. doi:10.1016/j.jclepro.2011.05.022

Delmas, M. (2001). Stakeholders and competitive advantage: The case of ISO 14001. *Production and Operations Management*, *10*(3), 343–358. doi:10.1111/j.1937-5956.2001.tb00379.x

Demir, E., Bektaş, T., & Laporte, G. (2014). A review of recent research on green road freight transportation. *European Journal of Operational Research*, *237*(3), 775–793. doi:10.1016/j.ejor.2013.12.033

Dickson, G. W. (1966). An analysis of vendor selection systems and decisions. *Journal of Purchasing*, *2*(1), 5–17.

Dowlatshahi, S. (2000). Developing a theory of reverse logistics. *Interfaces*, *30*(3), 143–155. doi:10.1287/inte.30.3.143.11670

Drake, M. J., & Marley, K. A. (2014). A century of the EOQ. In T.-M. Choi (Ed.), *Handbook of EOQ Inventory Problems* (pp. 3–22). Springer. doi:10.1007/978-1-4614-7639-9_1

Edwards, J., McKinnon, C., Cherrett, T., McLeod, F., & Song, L. (2010). Carbon dioxide benefits of using collection-delivery points for failed home deliveries in the United Kingdom. *Transportation Research Record: Journal of the Transportation Research Board*, (2191), 136-143.

Edwards, J. B., McKinnon, A. C., & Cullinane, S. L. (2010). Comparative analysis of the carbon footprints of conventional and online retailing: A "last mile" perspective. *International Journal of Physical Distribution & Logistics Management*, *40*(1/2), 103–123. doi:10.1108/09600031011018055

El Saadany, A. M. A., Jaber, M. Y., & Bonney, M. (2011). Environmental performance measures for supply chains. *Management Research Review*, *34*(1), 1202–1221.

Esper, T. L., Jensen, T. D., Turnipseed, F. L., & Burton, S. (2003). The last mile: An examination of effects of online retail delivery strategies on consumers. *Journal of Business Logistics*, *24*(2), 177–203. doi:10.1002/j.2158-1592.2003.tb00051.x

Gevaers, R., Van de Voorde, E., & Vanelslander, T. (2011). *Characteristics and typology of last-mile logistics from an innovative perspective in an urban context*. Paper presented at the Transportation Research Board Annual Meeting, Washington, DC.

Glock, C. H., Jaber, M. Y., & Searcy, C. (2012). Sustainability strategies in an EPQ model with price and quality sensitive demand. *The International Journal of Logistics Management*, *23*(3), 340–359. doi:10.1108/09574091211289219

Golicic, S. L., Boerstler, C. N., & Ellram, L. M. (2010). Greening transportation in the supply chain. *Sloan Management Review*, *51*(2), 47–55.

Goodman, R. W. (2005). Whatever you call it, just don't think of last-mile logistics, last. *Global Logistics and Supply Chain Strategies*, *9*(12), 46–51.

Govindan, K., Rajendran, S., Sarkis, J., & Murugesan, P. (2013). Multi criteria decision making approaches for green supplier evaluation and selection: A literature review. *Journal of Cleaner Production*. doi:10.1016/j.jclepro.2013.06.046

Green, K. W. Jr, Zelbst, P. J., Meacham, J., & Bhadauria, V. S. (2012). Green supply chain management practices: Impact on performance. *Supply Chain Management: An International Journal*, *17*(3), 290–305. doi:10.1108/13598541211227126

Guide, V. D. R., Harrison, T. P., & Van Wassenhove, L. N. (2003). The challenge of closed-loop supply chains. *Interfaces*, *33*(6), 3–6. doi:10.1287/inte.33.6.3.25182

Guiffrida, A. L. (1999). *A Cost-Based Model for Evaluating Vendor Delivery Performance*. (Thesis). Department of Industrial Engineering, State University of New York at Buffalo, Buffalo, NY.

Harris, F. W. (1913). How many parts to make at once? *Factory, the Magazine of Management*, *135-136*, 152.

Hervani, A. A., Helms, M. M., & Sarkis, J. (2005). Performance measurement for green supply chain management. *Benchmarking: An International Journal, 12*(4), 330–353. doi:10.1108/14635770510609015

Hua, G., Cheng, T. C. E., & Wang, S. (2011). Managing carbon footprints in inventory management. *International Journal of Production Economics, 132*(2), 178–185. doi:10.1016/j.ijpe.2011.03.024

Huang, M., Smilowitz, K., & Balcik, B. (2011). Models for relief routing: Equity, efficiency and efficacy. *Transportation Research Part E, Logistics and Transportation Review, 48*(1), 2–18. doi:10.1016/j.tre.2011.05.004

Jaber, M. Y. (2009). *Inventory Management: Non-Classical Views*. Boca Raton, Florida: CRC Press. doi:10.1201/9781420079982

Jaber, M. Y., Glock, C. H., & El Saadany, A. M. A. (2013). Supply chain coordination with emission reduction incentives. *International Journal of Production Research, 51*(1), 69–82. doi:10.1080/00207543.2011.651656

Jain, V., Wadhwa, S., & Deshmukh, S. G. (2009). Select supplier-related issues in modelling a dynamic supply chain: Potential, challenges and direction for future research. *International Journal of Production Research, 47*(11), 3013–3039. doi:10.1080/00207540701769958

Kannan, G., & Haq, A. N. (2007). Analysis of interactions of criteria and subcriteria for the selection of supplier in the built-in-order supply chain environment. *International Journal of Production Research, 45*(17), 3831–3852. doi:10.1080/00207540600676676

Kull, T. J., Boyer, K., & Calantone, R. (2007). Last-mile supply chain efficiency: An analysis of learning curves in online ordering. *International Journal of Operations & Production Management, 27*(4), 409–434. doi:10.1108/01443570710736985

Lee, K.-H. (2011). Integrating carbon footprint into supply chain management: The case of Hyundai Motor Company (HMC) in the automobile industry. *Journal of Cleaner Production, 19*(11), 1216–1223. doi:10.1016/j.jclepro.2011.03.010

Lin, R. (2012). An integrated model for supplier selection under a fuzzy situation. *International Journal of Production Economics, 138*(1), 55–61. doi:10.1016/j.ijpe.2012.02.024

Linton, J. D., Klassen, R., & Jayaraman, V. (2007). Sustainable supply chains: An introduction. *Journal of Operations Management*, *25*(6), 1075–1082. doi:10.1016/j.jom.2007.01.012

Manikas, A., & Godfrey, M. (2010). Inducing green behavior in a manufacturer. *Global Journal of Business Research*, *4*(2), 27–38.

McIntyre, K., Smith, H. A., Henham, A., & Pretlove, J. (1998). Logistical performance measurement and greening supply chains: Diverging mindsets. *The International Journal of Logistics Management*, *9*(1), 57–68. doi:10.1108/09574099810805744

McKinnon, A. C., & Tallam, D. (2003). Unattended delivery to the home: An assessment of the security implications. *International Journal of Retail and Distribution Applications*, *31*(1), 30–41. doi:10.1108/09590550310457827

McLeod, F., Cherrett, T., & Song, L. (2006). Transport impacts of local collection/delivery points. *International Journal of Logistics Research and Applications*, *9*(3), 307–317. doi:10.1080/13675560600859565

Mukhopadhyay, A., & Goswami, A. (2014). Economic production quantity models for imperfect items with pollution costs. *Systems Science & Control Engineering: An Open Access Journal*, *2*(1), 368–378. doi:10.1080/21642583.2014.912571

Pal, O., Gupta, A. K., & Garg, R. K. (2013). Supplier selection criteria and methods in supply chains: A review. *International Journal of Management Sciences*, *7*(10), 137–143.

Parnell, J. A. (2008). Sustainable strategic management: Construct, parameters, research directions. *International Journal of Sustainable Strategic Management*, *1*(1), 35–45. doi:10.1504/IJSSM.2008.018125

Peattie, K., & Ratnayaka, M. (1992). Responding to the green movement. *Industrial Marketing Management*, *21*(2), 103–110. doi:10.1016/0019-8501(92)90004-D

Prajogo, D., & Olhager, J. (2012). Supply chain integration and performance: The effects of long-term relationships, information technology and sharing, and logistics integration. *International Journal of Production Economics*, *135*(1), 514–522. doi:10.1016/j.ijpe.2011.09.001

Punakivi, P., Yrjölä, H., & Holmström, J. (2001). Solving the last mile issue: Reception box or delivery box? *International Journal of Physical Distribution & Logistics Management*, *31*(6), 427–439. doi:10.1108/09600030110399423

Rao, P., & Holt, D. (2005). Do green supply chains lead to competitiveness and economic performance? *International Journal of Operations & Production Management, 25*(9), 898–916. doi:10.1108/01443570510613956

Rao, S., & Riahi, K. (2006). The role of non-CO 2 greenhouse gases in climate change mitigation: Long-term scenarios for the 21st century. *The Energy Journal*, 177-200.

Rebitzer, G., Ekvall, T., Frischknecht, R., Hunkeler, D., Norris, G., Rydberg, T., & Pennington, D. W. et al. (2004). Life cycle assessment: Part 1: Framework, goal and scope definition, inventory analysis, and applications. *Environment International, 30*(5), 701–720. doi:10.1016/j.envint.2003.11.005 PMID:15051246

Rosič, H., & Jammernegg, W. (2013). The economic and environmental performance of dual sourcing: A newsvendor approach. *International Journal of Production Economics, 143*(1), 109–119. doi:10.1016/j.ijpe.2012.12.007

Routroy, S. (2009). Antecedents and drivers for green supply chain management implementation in manufacturing environment. *ICFAI Journal of Supply Chain Management, 6*(1), 20–35.

Sarkis, J. (2003). A strategic decision making framework for green supply chain management. *Journal of Cleaner Production, 11*(4), 397–409. doi:10.1016/S0959-6526(02)00062-8

Sarkis, J. (2012). A boundaries and flows perspective of green supply chain management. *Supply Chain Management: An International Journal, 17*(2), 202–216. doi:10.1108/13598541211212924

Schaltegger, S., & Burritt, R. L. (2014). Measuring and managing sustainability performance of supply chains. Review and sustainability supply chain management framework. *Supply Chain Management: An International Journal, 19*(3), 2–2. doi:10.1108/SCM-02-2014-0083

Seuring, S., & Müller, M. (2008). From a literature review to a conceptual framework for sustainable supply chain management. *Journal of Cleaner Production, 16*(15), 1699–1710. doi:10.1016/j.jclepro.2008.04.020

Shi, M., & Yu, W. (2013). Supply chain management and financial performance: Literature review and future directions. *International Journal of Operations & Production Management, 33*(10), 123–1317. doi:10.1108/IJOPM-03-2012-0112

Siikavirta, H., Punakivi, M., Karkkainen, M., & Linnanen, L. (2003). Effects of e-commerce on greenhouse gas emissions. *Journal of Industrial Ecology*, *6*(2), 83–97. doi:10.1162/108819802763471807

Song, J., & Leng, M. (2012). Analysis of single-period problem under carbon emissions policies. In T.-M. Choi (Ed.), *Handbook of Newsvendor Problems Models, Extensions and Applications* (pp. 297–313). New York: Springer. doi:10.1007/978-1-4614-3600-3_13

Song, L., Guan, W., Cherrett, T., & Baowen, L. (2013). Quantifying the greenhouse gas emissions of local collection-and-delivery points for last-mile deliveries. *Transportation Research Record: Journal of the Transportation Research Board*, (2340), 66-73.

Srisorn, W. (2013). The benefit of green logistics to organization. *International Journal of Social. Human Science and Engineering*, *7*(8), 1182–1185.

Srivastava, S. K. (2007). Green supply-chain management: A state-of-the-art literature review. *International Journal of Management Reviews*, *9*(1), 53–80. doi:10.1111/j.1468-2370.2007.00202.x

Stank, T. P., Goldsby, T. J., Vickery, S. K., & Savitskie, K. (2003). Logistical service performance: Estimating its influence on market share. *Journal of Business Logistics*, *24*(1), 27–55. doi:10.1002/j.2158-1592.2003.tb00031.x

Stapleton, O., Pedraza Martinez, A., & Van Wassenhove, L. N. (2009). *Last mile vehicle supply chain in the international federation of Red Cross and Red Crescent societies*. Working Paper. INSEAD, Fontainebleau, France.

Su, J. (2013). Strategic sourcing in the textile and apparel industry. *Industrial Management & Data Systems*, *113*(1), 23–38. doi:10.1108/02635571311289647

Supply Chain and Logistics. (2009). *"Green" supply chain movement hits a speed bump*. Retrieved from: www.industryweek.com

Talluri, S., DeCampos, H. A., & Hult, G. T. M. (2013). Supplier rationalization: A sourcing decision model. *Decision Sciences*, *44*(1), 57–86. doi:10.1111/j.1540-5915.2012.00390.x

Tao, Z., Guiffrida, A. L., & Troutt, M. D. (2010). A green cost based economic production/order quantity model.*Proceedings of the First Annual Kent State International Symposium on Green Supply Chains*.

Toptal, A., Özlü, H., & Konur, D. (2014). Joint decisions on inventory replenishment and emission reduction investment under different emission regulations. *International Journal of Production Research*, *52*(1), 243–269. doi:10.1080/00207543.2013.836615

Ubeda, S., Arcelus, F. J., & Faulin, J. (2011). Green logistics at Eroski: A case study. *International Journal of Production Economics*, *131*(1), 44–51. doi:10.1016/j.ijpe.2010.04.041

Ülkü, M. A. (2012). Dare to care: Shipment consolidation reduces not only cost, but also environmental damage. *International Journal of Production Economics*, *139*(2), 438–446. doi:10.1016/j.ijpe.2011.09.015

Vachon, S., & Klassen, R. D. (2006). Extending green practices across the supply chain: The impact of upstream and downstream integration. *International Journal of Operations & Production Management*, *26*(7), 795–821. doi:10.1108/01443570610672248

van der Veen, J. A., & Venugopal, V. (2014). Economic and environmental performance of the firm: Synergy or trade-off? Insights from the EOQ model. In T.-M. Choi (Ed.), *Handbook of EOQ Inventory Problems* (pp. 121–137). Springer. doi:10.1007/978-1-4614-7639-9_6

Van Woensel, T., Creten, R., & Vandaele, N. (2001). Managing the environmental externalities of traffic logistics: The issue of emissions. *Production and Operations Management*, *10*(2), 207–223. doi:10.1111/j.1937-5956.2001.tb00079.x

Wahab, M. I. M., Mamun, S. M. H., & Ongkunaruk, P. (2011). EOQ models for a coordinated two-level international supply chain considering imperfect items and environmental impact. *International Journal of Production Economics*, *134*(1), 151–158. doi:10.1016/j.ijpe.2011.06.008

Wang, Z., & Sarkis, J. (2013). Investigating the relationship of sustainable supply chain management with corporate financial performance. *International Journal of Productivity and Performance Management*, *62*(8), 871–888. doi:10.1108/IJPPM-03-2013-0033

Ware, N., Sing, S., & Banwet, D. (2012). Supplier selection problem: A state-of-the-art review. *Management Science Letters*, *2*(5), 1465–1490. doi:10.5267/j.msl.2012.05.007

Weber, C. A., Current, J. R., & Benton, W. C. (1991). Vendor selection criteria and methods. *European Journal of Operational Research*, *50*(1), 2–18. doi:10.1016/0377-2217(91)90033-R

Wiedmann, T., & Minx, J. (2008). A definition of 'carbon footprint. In C. C. Pertsova (Ed.), *Ecological Economics Research Trends* (pp. 1–11). Hauppauge, NY: Nova Science Publishers.

Wu, C., & Barnes, D. (2011). A literature review of decision-making models and approaches for partner selection in agile supply chains. *Journal of Purchasing and Supply Management, 17*(4), 256–274. doi:10.1016/j.pursup.2011.09.002

Zanoni, S., Bettoni, L., & Glock, C. H. (2013). Energy implications in the single-vendor single-buyer integrated production inventory model. In C. Emmanoulisdis, M. Taisch, & D. Kiritsis (Eds.), *Advances in Production Management Systems. Competitive Manufacturing for Innovative Products and Services* (pp. 57–64). Berlin: Springer. doi:10.1007/978-3-642-40352-1_8

Zanoni, S., Mazzoldi, L., & Jaber, M. Y. (2014). Vendor-managed inventory with consignment stock agreement for single vendor–single buyer under the emission-trading scheme. *International Journal of Production Research, 52*(1), 20–31. doi: 10.1080/00207543.2013.812812

Zhang, B., & Xu, L. (2013). Multi-item production planning with carbon cap and trade mechanism. *International Journal of Production Economics, 144*(1), 118–127. doi:10.1016/j.ijpe.2013.01.024

Zhiyong, T., Lingyu, H., & Guicheng, S. (2014). Carbon footprint and order quantity in logistics. *Journal of Industrial Engineering and Management, 7*(2), 475–490. doi:10.3926/jiem.1031

Zsidisin, G. A., & Siferd, S. P. (2001). Environmental purchasing: A framework for theory development. *European Journal of Purchasing & Supply Management, 7*(1), 61–73. doi:10.1016/S0969-7012(00)00007-1

ADDITIONAL READING

Abdallah, T., Farhat, A., Diabat, A., & Kennedy, S. (2012). Green supply chains with carbon trading and environmental sourcing: Formulation and life cycle assessment. *Applied Mathematical Modelling, 36*(9), 4271–4285. doi:10.1016/j.apm.2011.11.056

Amindoust, A., Ahmed, S., Saghafinia, A., & Bahreininejad, A. (2012). Sustainable supplier selection: A ranking model based on fuzzy inference system. *Applied Soft Computing, 12*(6), 1668–1677. doi:10.1016/j.asoc.2012.01.023

Bhattacharyya, K., & Guiffrida, A. L. (2014). An optimization framework for improving supplier delivery performance. *Applied Mathematical Modelling* (ahead of print, doi:10.1016/j.apm.2014.12.004).

Cleophas, C., & Ehmke, J. F. (2014). When are deliveries profitable? *Business and Information Systems Engineering*, *6*(3), 153–163. doi:10.1007/s12599-014-0321-9

Esposito, E., & Passaro, R. (2013). Purchasing and supply management in a changing world. *Journal of Purchasing and Supply Management*, *19*(3), 119–121. doi:10.1016/j.pursup.2013.07.002

Feng, B., Fan, Z. P., & Li, Y. (2011). A decision method for supplier selection in multi-service outsourcing. *International Journal of Production Economics*, *132*(2), 240–250. doi:10.1016/j.ijpe.2011.04.014

Kar, A. K., & Pani, A. K. (2014). Exploring the importance of different supplier selection criteria. *Management Research Review*, *37*(1), 89–105. doi:10.1108/MRR-10-2012-0230

Kilic, H. S. (2013). An integrated approach for supplier selection in multi-item/multi-supplier environment. *Applied Mathematical Modelling*, *37*(14), 7752–7763. doi:10.1016/j.apm.2013.03.010

North, M. (2006). Supplier performance management can make a difference. *Supply Chain Management Review*, October, 13-15.

Parthiban, P., Zubar, H. A., & Katakar, P. (2013). Vendor selection problem: A multi-criteria approach based on strategic decisions. *International Journal of Production Research*, *51*(5), 1535–1548. doi:10.1080/00207543.2012.709644

Pikousova, K., & Prusa, P. (2013). Supplier evaluation: the first step in effective sourcing: original research. *Journal of Transport and Supply Chain Management*, *7*(1), 1–4. doi:10.4102/jtscm.v7i1.87

Ramudhin, A., Chaabane, A., & Paquet, M. (2010). Carbon market sensitive sustainable supply chain network design. *International Journal of Management Science and Engineering Management*, *5*(1), 30–38.

Reed, B. D., Smas, M. J., Rzepka, R. A., & Guiffrida, A. L. (2010). Introducing green transportation costs in supply chain modeling, *Proceedings of the First Annual Kent State International Symposium on Green Supply Chains*, July 29-30, 2010, Canton, Ohio, 189-197.

Rezaei, J., & Davoodi, M. (2012). A joint pricing, lot-sizing, and supplier selection model. *International Journal of Production Research*, *50*(16), 4524–4542. doi:10.1080/00207543.2011.613866

Seuring, S. (2013). A review of modeling approaches for sustainable supply chain management. *Decision Support Systems*, *54*(4), 1513–1520. doi:10.1016/j.dss.2012.05.053

Shaw, K., Shankar, R., Yadav, S. S., & Thakur, L. S. (2013). Global supplier selection considering sustainability and carbon footprint issue: AHP multi–objective fuzzy linear programming approach. *International Journal of Operational Research*, *17*(2), 215–247. doi:10.1504/IJOR.2013.053624

Spina, G., Caniato, F., Luzzini, D., & Ronchi, S. (2013). Past, present and future trends of purchasing and supply management: An extensive literature review. *Industrial Marketing Management*, *42*(8), 1202–1212. doi:10.1016/j.indmarman.2013.04.001

Chapter 8
Integrated Sustainable Supply Chain Management:
Current Practices and Future Direction

Geevaneswary Saththasivam
Universiti Sains Malaysia, Malaysia

Yudi Fernando
Universiti Sains Malaysia, Malaysia

ABSTRACT

In today's sustainable world the thinking should be what is the life cycle costs, resources or supply chain process. The rapid industrialization had caused negative impact on the environment because of the increase in the pollution, waste, and rapid consumption of natural resources. This chapter is reviews the previous literature on current practices and future direction of Sustainable Supply Chain Management (SSCM), thus, it can increase practitioner's understanding of the importance and value of sustainable supply chain practices. A wide literature review is conducted to identify the relative important factors constitute of Sustainable Supply Chain Integration framework. Content analysis is introduced and applied for reviewing literature reviews of supply chain management, published in English-speaking peer-reviewed journals between 2000 and 2014. A descriptive evaluation of the

DOI: 10.4018/978-1-5225-0635-5.ch008

literature body is followed by a content analysis on the basis of a specific pattern of analytic categories derived from a typical research process. With best understanding the concept of SSCM and gap identification from literature review, this chapter can contribute to extended model of SSCM and enhance firm competitiveness. This understanding is vital due to the increase of economic importance of sustainable supply chains in addition to their role in enhancing competitive power of companies in international markets.

INTRODUCTION

Sustainable Supply Chain Management (SSCM) emerges as a new systematic environmental approach in supply chain management. SSCM has been increasingly accepted and practices by forward-thinking organization. The current changing in environmental requirements that influenced manufacturing activities had increased attention of scholars and practitioners in developing environmental management strategies for the supply chain. The growing importance of sustainable supply chain management is driven largely by the escalating deterioration of the environment, expanded legislation, intensified global competition, and value-seeking initiatives. Generally, adding the 'green' component to supply chain management involves addressing the influence and relationships between traditional supply chain management and the natural environment. As a result, sustainable supply chain management assists to lower the environmental impact of a firm, enhance its business efficiency, and possibly create major competitive advantages through operations and innovations.

Most organizations are paying attention to go greener their business strategy and practices as concern to environmental sustainability. They have realized the greater benefit of the green technology adoption in business operation, which also affected company eco-system included suppliers, distributors, retailers and other supporting companies such as third party logistics (3PL). As a result, sustainable supply chain management emerges as a new systematic environmental approach in supply chain management and has been increasingly accepted and practices by forward-thinking organization (Zhu & Sarkis, 2004).

The current changing in environmental requirements that influenced manufacturing activities had increased attention in developing environmental management strategies for the supply chain (Beamon, 1999). Thus, the concept of sustainable supply chain integration arises as a new systematic approach and becoming an important factor for business activities today. Zhu (2010) claimed that sustainable supply chain can be regarded as an environmental innovation. By integrating the 'green concept and sustainability to the supply chain' concept, it has created a new research agenda where the supply chain will have a direct relation to the environment and economic.

Sustainable supply chain management involves integrating environmentally and financially viable practices into the complete supply chain lifecycle, from product design and development, to material selection, (including raw material extraction or agricultural production), manufacturing, packaging, transportation, warehousing, distribution, consumption, return and disposal. Environmentally sustainable supply chain management and practices can assist organization in not only reducing their total carbon footprint, but also in optimizing their end-to-end operations to achieve greater cost savings and profitability. All supply chains can be optimized using sustainable practices. Sustainability in the supply chain encapsulates a number of different priorities such as environmental stewardship, conservation of resource, reduction of carbon footprint and financial savings and viability and social responsibility.

A concept of sustainable supply chain requires a change in how supply chains are designed and managed through explicit consideration of economic, environmental, and social performance measures in strategic, tactical, and operational decision making. An integrated sustainable supply chain, involves life-cycle management. The concept of a sustainable supply chain covers every stage in manufacturing, from the first to the final stage of the lifecycle. The definition of sustainable supply chain has ranged from green purchasing to an integrated green closed-loop supply chain. The adaptation of supply chain can be at any of the following stages, product design, material sourcing and selection, manufacturing processes, delivery of the final product to the consumer, and end-of-life management of the product after its useful life. The literature contains various definitions on sustainable or green supply chain. However, few of them consider an end-to-end supply chain with green adoption. Hervani et al. (2005) says that the green supply chain comprises of green purchasing, green manufacturing, green distribution or marketing andreverse logistics. Sustainable supply chain integration involves incorporating environmental and financial viable practices into the complete supply chain lifecycle, from product design and development, to material selection, (including raw material extraction or agricultural production), manufacturing, packaging, transportation, warehousing, distribution, consumption, return and disposal. Environmentally sustainable supply chain management and practices can assist organizations in not only reducing their total carbon footprint, but also in optimizing their end-to-end operations to achieve greater cost savings and profitability.

Whereas, green supply chain is defined as ‘integrating environmental thinking into supply-chain management, including product design, material sourcing and selection, manufacturing processes, delivery of the final product to the consumers as well as end-of-life management of the product after its useful life’ (Srivastava, 2007). To conceptualize the sustainable supply chain integration, the scope of the study will be focused on electric and electronics industry. The area of focus is within this industry because Electrical & Electronics (E&E) industry is one of the leading industries, contributing 24.5 per cent to the manufacturing sector in the Malaysia's Gross Domestic Product (GDP). E&E products have been the largest traded items for Malaysia for several decades since the industry inception in the 1960s. The industry evolution until today has turned Malaysia as one of the leading points in the global E&E value chain. In 2012, Malaysia's exports of E&E products was valued at RM231.23 billion, with 49.2 per cent share of manufactured goods exports and 32.9 per cent share of Malaysia's total exports.

Nevertheless, the growing importance of sustainable supply chain management is driven largely by the escalating deterioration of the environment, expanded legislation, intensified global competition, and value-seeking initiatives. Generally, adding the 'green' component to supply chain management involves addressing the influence and relationships between traditional supply chain management and the natural environment. The underlying issue of sustainable development is that companies must move beyond mere "green" that indicates relentless pursuit of short-term profitability towards long-term sustainability. According to Werbach (2009), society increasingly holds global businesses accountable as the only institution powerful enough to respond at the scale of the challenges that our planet faces. As a result, sustainable supply chain integration assists to lower the environmental impact of a firm, enhance its business efficiency, and possibly create major competitive advantages through operations and innovations.

BACKGROUND OF THE STUDY

Sustainability in business is a dynamic state that occurs when a company creates ongoing value for its shareholders and stakeholders. Most of the E&E manufacturing companies especially electric and electronics industries compete as participants in integrated sustainable supply chains management. This initiative, which is changing the ways products are designed, produced, and delivered, has the potential to alter the manufacturing landscape as dramatically as the industrial revolution or the advent of mass production. The integration of the supply chain was described by Vachon and Klassen (2008) as an attempt to elevate the linkages within each component of the chain, by facilitating a better decision making and to get all the pieces of the chain to interact in a more efficient way by creating a visibility of supply chain and to identify bottlenecks. Therefore, the integration within a supply chain can be expected to positively influence cooperative activities related to environmental and financial issues. Sustainable supply chain integration should be enforced across all business operations.

Concept of Integrated Sustainable Supply Chain Management

Supply chain management (SCM) is a holistic approach to demand, sourcing and procurement, production and logistics process management (Chow et al. 2008; Chopra & Meindl 2010). It is a network consisting of all parties involved (e.g. manufacturer, supplier, retailer, customer, etc), both downstream and upstream, directly or indirectly, for manufacturing and delivering a product or service to the end customers (Mentzer et al. 2001). These days the enormous challenge is to bring sustainable procurement into a central decision making role in line with organizational business goals. However it is not enough for a sustainable supply chain strategy to simply attempt to mitigate effects on the environmental. To be accepted and successful, it has to be seen to drive value creation through the organization. Key to the success of any sustainable supply chain initiative is how an organisation changes the way it works with upstream and downstream trading partners. Sustainable supply chain initiatives plays a significant role in achieving the "triple bottom line" of social, environmental, and economic benefits and, therefore, contributing to sustainable development of the society. As we know sustainability has become a huge buzzword, both in today's business world and within the broader facets of society (Carter & Easton, 2011). It is necessary to implement practices that not only promote company and overall supply chain efficiency, but also that focus on social, economic, and environmental concerns. Sustainable supply chain management requires consideration of economic performance, social performance, and environmental performance.

The wider goal may be carbon reduction but for the organisation the real benefits are measured in terms of better utilisation of assets, less waste production, faster times to market, greater product innovation and increased profitability. Taking into consideration of the lifecycle approach from initial raw material extraction to eventual product disposal requires a much greater degree of collaboration, transparency and integration of supply chain processes and systems between trading partners than there has previously been. Underpinning these new working practices is intelligent supply chain management and B2B ecommerce that enables the low cost creation of global trading communities that use automation to improve their business performance while achieving significant environmental results.

An integrated supply chain framework is needed to tie the whole network together in order to reduce constant supply chain challenges such as functional issues, poor transparency of knowledge and information and the inadequate formation of appropriate customer and supplier relationships (Storey et al. 2006). As such, supply chain integration (SCI) plays an essential role in improving organisational performance (Richey et al. 2009; Boon-itt & Wong 2011). Various SCM practices are used to effectively integrate supply and demand to improve the management of a supply chain (Li et al. 2005; Wong et al. 2005; Koh et al. 2007). SCI requires a collaborative effort among suppliers, cross-functional departments and customers that are linked and coordinated by the flow of processes and information (Boon-itt & Wong 2011)

Sustainable Supply Chain Management refers to all phases of supply chain management which must comply with environmental protection requirements (Zhu & Sarkis, 2007) and can be broadly categorized into intra- and inter-organizational environmental practices (Shi et al., 2012). In addition, supply chain integration stands for the co-operation of supply chain members in both intra- and inter-organizational processes (Flynn et al., 2010). From the earlier rationalization, sustainable supply chain integration could be further defined as "the collaboration of a firm with its supply chain partners to manage both intra and inter organizational environmental practices" and could be separated into internal and external integration. Internal integration focuses on removing cross-functional barriers, enhancing firms' environmental capabilities and encouraging employees to learn about environmental initiatives (Wu et al., 2012). On the other hand, external integration involves co-operation between suppliers and customers to mutually manage cross-firm environmental issues (Vachon & Klassen, 2008).

Malaysia is moving forward to be an industrialized economy. At this era, globalization has expanded markets and accelerated technological development. Many developing countries have established competitive advantage by opening their economies to the global economy. Companies in developed countries have benefited lower production and service costs by moving their production facilities and business departments offshore. Rapid economic growth due to globalization, which has improved the quality of our lives, has also caused environmental issues. Recent environmental incidents occurring in the upstream of the supply chains of global firms caused not only financial damage to the firms but also damage to the firms' reputation. These anecdotes raised the alarm to the large buying companies how important supplier selection and support are. Supplier firms that are capable of environmental management could create potential business opportunities because large global firms are seeking such suppliers with environmental management capabilities world widely. These suppliers with the adoption of sustainable supply chain management practices would be more sustainable to keep and extend their current contracts.

Today, almost all companies operating in the market face the challenges to get the stability between their activities and the environment. Research on "green" supply chains have focused mainly on environmental aspects, the relationship occurring between the principles of supply chain operations, and activities (cooperation and assessment) undertaken by managers, suppliers and consumers. Namely, the "green" supply chains relates to activities such as reducing sources of waste and resources of consumption, but they are not typical processes for reverse logistics processes

(Zhong, 2006). The creation of "green or sustainable" supply chain integration is a process in which the environmental aspects are taking into account in every supply chain activities such as decisions regarding sourcing of raw materials and creating long-term relationships with suppliers (Gilbert, 2001). In addition, there are three main approaches that are involved in the creation of "green" supply chain: environment, strategy and logistics (Hwa, 2001).

The most common tool used in life-cycle analysis is called life cycle assessment (LCA), which is used to assist designers at the beginning of product design to identify techniques that ensure that the product components and designs are attuned to environmental standards. Standards must be set when conducting an LCA of the amount of energy and natural resources utilized, the amount of emissions caused by air and water pollution, and the waste products resulting from consumption of the product output. Life-cycle analysis evaluates various aspects of the product system, including technology and its potential impact to the environment, throughout its life cycle. It involves examining all stages of the development of a product or service, from the selection of raw materials, through product outputs and usage, to product processing after use.

In order to achieve the goal of sustainable supply chain, industries must gather all existing material systems, information management systems, and process operation systems, and integrate all the green requirements into tasks and procedures to achieve greening. For this purpose, industries strive through the sustainable supply chain system to evaluate the green performance of suppliers, assess the quality of green products, and direct overall management of suppliers. In addition, industry leaders use information technology (IT) to assist other industries to solve the problems and challenges of applying sustainable supply chain management. This is to ensure that the overall supply chain conform to the standards of sustainable supply chain integration through management systems, production processes, technique standards, verification of work, and analysis of product components.

Green Supply Chain Management and Sustainable Supply Chain Management

The difference between green supply chain management and sustainable supply chain management can be identified from the definition given by various scholars. Wee, Lee, Yu, and Wang (2011) define green supply chain management as an integration of environment considerations into supply chain management, including product design, material sourcing and selection, manufacturing processes, delivery of the final product to the consumers, and end-of-life management of the greening products. According to Carter and Rogers (2007) sustainable supply chain is the strategic, transparent integration and achievement of an organization's social, environmental, and economic goals in the systemic coordination of key inter-organizational business processes for improving the long-term economic performance of the individual company and its supply chains. Generally, green innovations can help to reduce the environmental burden of the logistics industry, particularly in terms of pollution and greenhouse gas emissions. Their collaborations with other industries are, however, critical in providing the right environment for firms to innovate. As one of the big challenges facing the industry at the present moment is concerning with creating a long-term sustainable society with the least possible negative environmental impact (Lin & Ho, 2008), it is evidently clear that there is a need to integrate green innovations in logistics.

Challenges of Supply Chain Sustainability in Malaysia

In today's "sustainable" world the thinking should be what is the life cycle costs, resources or supply chain process. As a matter of fact, Malaysia is moving forward to be an industrialized economy. Sustainability is a hot topic in the electric and electronics industry. Sustainability has become a huge buzzword, both in today's business world and within the broader facets of society (Carter and Easton, 2011). It is necessary to implement practices that not only promote company and overall supply chain efficiency, but also that focus on social, economic, and environmental concerns. Sustainable supply chain management requires consideration of economic performance, social performance, and environmental performance.

According to Carter, there are four facilitators of sustainable supply chain management:

1. **Strategy:** Holistically and purposefully identifying individual SSCM initiatives which align with and support the organizations' overall sustainability,
2. **Risk Management:** Including contingency planning for both the upstream and downstream supply chain,
3. **Organizational Culture:** Which is deeply engrained and encompasses organizational citizenship, and which includes high ethical standards and expectations, and
4. **Transparency:** In terms of proactively engaging and communicating with key stakeholders and having traceability and visibility in supply chain operations.

These facilitators make up the four corners of sustainability, and sustainability comes from the intersection of economic performance, environmental performance and social performance, for example, the triple bottom line of sustainability (Carter & Easton, 2011). Sustainable development helps to meet the needs of the present without compromising the ability of future generations to meet their own needs. An effective sustainable supply chain enables companies to change their design process to create products that are more streamlined, needless packaging, contain fewer harmful chemicals, and allow for increased reusability and recycling. Besides that, sustainable development in supply chain also creates best practices in lean manufacturing and greening of facilities which leads to companies' successful strategies and towards their contribution to the communities.

As manufacturers seek to reduce costs and maintain customer loyalty, implementing sustainable practices can help to achieve these aims while at the same time ensuring that companies are playing a responsible role in their industry and society in general. The manufacturing industry remained to be an important contributor to Malaysian's economy in recent years. But, the rapid industrialization had caused negative impact on the environment because of the increase in the pollution, waste, and rapid consumption of natural resources (Eltayeb & Zailani 2009). It is clearly showed that manufacturing industry is the most contributors to the environmental problem. Environmental issues have become very important issues of concern for the Malaysian government and the public (Eltayeb, Zailani, & Ramayah 2011). With regards to environmental concern into manufacturing industry, it is necessary for the study to investigate the level of adoption and implementation sustainable supply chain integration among electric and electronics industries in Malaysia.

For many companies, the largest opportunity for improving sustainability performances such as reducing carbon emissions, water use, toxic chemicals and addressing social and human rights concerns is in its global supply chain. For example, up to 60 percent of a manufacturing company's carbon footprint is in its supply chain. The potential benefits of improved supply chain performance are every bit as compelling as those achieved through direct action on the companies' own operations. By bringing sustainability improvements deep into supply chains, companies can better protect their reputations from human rights and environmental violations, increase productivity and save on costs related to energy, water use and reductions in waste and toxic chemical use. Therefore, implementing sustainable supply chain integration is not just about responding to concerns about climate change or product safety. As we know, sustainability is viewed as possessing environmental, economic and social dimensions.

The sustainability approaches adopted by organization and the supply chains to which they belong must include these facets. These three components of sustainability are collectively termed the "triple bottom line" or 3BL. The idea of a triple bottom line, which originated in the 1990s, implies attending to the traditional financial aspects of an organization as well as to the social and environmental criteria. The economic aspect of the 3BL refers to profit making and attaining and sustaining competitive advantage through sustainability.A greener supply chain is often smarter, leaner and more efficient than one that doesn't consider environmental, social or sustainability issues. Therefore, this study will focus on how Sustainable Supply Chain Integration contributes towards firm financial performance among E&E industries. In the electrical and electronics industries, suppliers use green procurement procedures and documentation to specifically control the use of various hazardous materials. However, Sustainable Supply Chain Integration aims to fulfill the "green requirement" for global electric and electronic companies throughout the entire supply chain management. To make customer impress, the company needs to make themselves standing out from others. Being environmental friendly is one way to differentiate them from the competitors. Furthermore, when competitors already adopted supply chain management, the company gets a pressure instead. Therefore, it is good ideas to implement Sustainable Supply Chain Integration to a large extend no matter if the competitors have adopted it or not. In many cases, customers were the one who require special treatment or special products. Therefore, the company needs to make changes to make them satisfy and stay with them. The whole idea of a sustainable supply chain is to reduce costs while helping the environment. In Malaysia, electrical and electronics industry plays a decisive role in the global information and communications technology (ICT) industry. Consequently, the validated instrument enables decision makers at ICT manufacturers to evaluate the perceptions of Sustainable Supply Chain Integration in their organizations. In addition, the critical factors of implementing sustainable supply chain practices can help companies identify those areas of supply chain integration where acceptance and improvements will be made, and in prioritizing supply chain integration efforts.

In Malaysia today companies do not see the benefits of themselves review their organizations environmental impact or developing in this area. Environmental awareness among the public is low and the lack of demand for green products result in few drivers for companies to become green. Environmental education in today's society is important due to the escalating environmental problems that require immediate attention. The need to continuously educate the public on environmental issues and promote action to minimize damage on the environment is vital. The task is monumental but efforts must be stepped up and multiplied to ensure the environment receives the protection it deserves. Zarrintaj and Sharifah (2012) highlighted that environmental education components were existed in various sources within the existing educational system, however, there are still huge gaps on public awareness in Malaysia.

Also the lack of proper legislation and compliance audits are part of the problem. Lacking proper environmental concern in professional training and development and knowledge are extensive and improvements needed. The concept of sustainable supply chain management is not wide spread but popularity is increasing steadily.

Some environmental initiatives are though performed in the industry but not under the name of the concept. Customer demands, legislation and education should be the main focus areas for developing the industry.

METHODS: LITERATURE REVIEW

Sources of literature were mainly selected from secondary sources (e.g. books, theses and the internet) and documents (mainly public documents, company documents and mass media items). The purpose of exploring the existing literature was to be familiar with the following aspects, what is already known about the research area (sustainability, supply chain integration, non-financial performance), main concept and theories. Consequently, the initial literature review formed the basis for the research at hand. The research was then focused on the research questions with content analysis as main method used.

Content analysis is a set of research tools for the scientific study of written communications with the objective of determining key ideas and themes contained within them (Cullinane & Toy, 2000). Content analysis can be both qualitative and quantitative, where the latter seeks "to quantify content in terms of predetermined categories and in a systematic and replicable manner" (Bryman & Bell, 2007, p. 302). With qualitative content analysis there is much more movement back and forth between conceptualisation, data collection, analysis and interpretation than is the case with quantitative content analysis (Bryman & Bell, 2007). Based on the initial literature review, concepts related to the research area were used to identify suitable articles. Articles selected from journals were organized and recorded in a database. In order to identify the gap in the literature of this emerging research area, a literature map, as depicted in Figure 1, was developed with the support of 30 relevant and most highly cited articles in sustainable supply chain management. Two areas that require further research includes: (1) framework to help structure the decisions taken along the supply chain integration and (2) implementation of practices that help obtain firms' initiatives of adopting sustainability as an on-going practice.

SUGGESTION FOR FUTURE STUDY

In the near future, the researcher intends to carry out an extensive review of more literatures on social aspects in order to assess how are used by companies to achieve sustainable development and improvement in their performance, to identify which sustainable supply chain practices were implemented by organizations to improve

Figure 1. Literature map of the research in sustainable supply chain integration

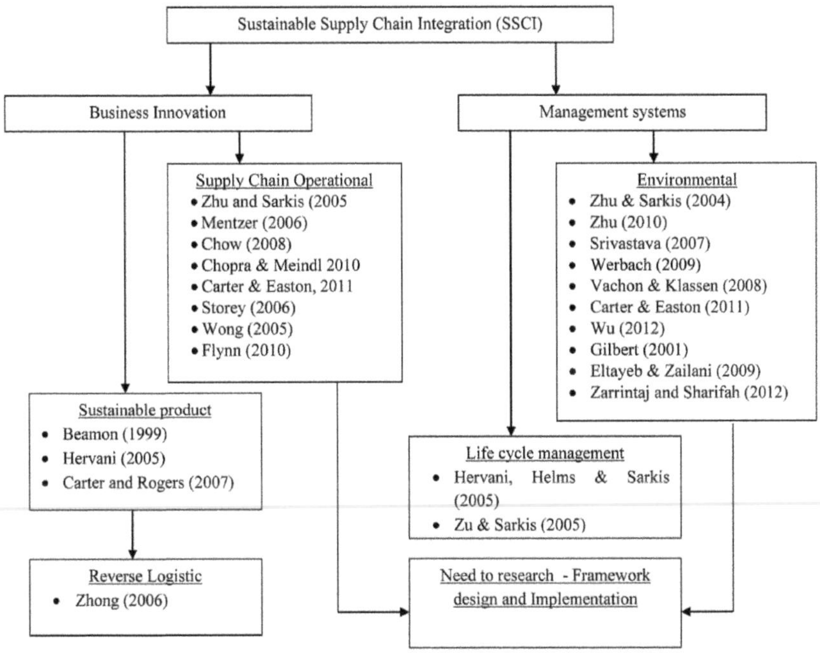

sustainable development of their businesses in a balanced way, to examine how sustainable supply chain practices affect other key performance indicators mentioned.

The research intends to address the following challenges:

- Integrating sustainability performance with traditional performance objectives.
- Measuring the operational performance of supply chain.
- Finding the relationship amongst the sustainable supply chain and performance measurements.

CONCLUSION

This paper set out to explore the role of integrated sustainable supply chain and firm non-financial performance. Based on a systematic review and content analysis of the articles we are able to outline the relationship between sustainable supply chain and firm performance in the literatures reviewed. The Sustainable supply chain integration processes can provide an effective means by which costs can be reduced and customer service levels improved. Organizations that aim to become part of an extended, integrated supply network can also expect that this will require an infrastructure enabling effective information flows and streamlined manufacturing and logistics activities. A key component of this infrastructure will be based on robust and durable collaborative arrangements with trading partners. The most effective of these networks will be those that are able to get the mix of information requirements, physical logistics and collaboration right, providing shared benefits to a majority of partner organizations. A successful integrated green supply chain organization would be able to develop a competency and competitive business environment which will lead to firm sustainability.

REFERENCES

Abu Seman, N. A., Zakuan, N., Jusoh, A., & Mohd, S. M. A. (2012). Green Supply Chain Management: A Review and Research Direction. *International Journal of Managing Value and Supply Chains*, *3*(1).

Beamon, B. M. (1999). Designing the green supply chain. *Logistics Information Management*, *12*(4), 332–342.

Boon-itt, S., & Wong, C. W. (2011). The moderating effects of technological and demand uncertainties on the relationship between supply chain integration and customer delivery performance. *International Journal of Physical Distribution & Logistics Management*, *41*(3), 253–276.

Carter, C. R., & Easton, P. L. (2011). Sustainable supply chain management: Evolution and future directions. *International Journal of Physical Distribution & Logistics Management*, *41*(1), 46–62.

Chiou, T. Y., Chan, H. K., Lettice, F., & Chung, S. H. (2011). The Influence of Greening the Suppliers and Green Innovation on Environmental Performance and Competitive Advantage in Taiwan. *Transportation Research Part E, Logistics and Transportation Review*, *47*, 822–836.

Chopra, S., & Meindl, P. (2010). *Supply chain management: strategy, planning, and operation, 4th*. Upper Saddle River, NJ: Pearson Education, Inc.

Chow, W. S., Madu, C. N., Kuei, C.-H., Lu, M. H., Lin, C., & Tseng, H. (2008). 'Supply Chain Management in the US and Taiwan: An empirical study', Omega. *International Journal of Management Sciences*, *36*, 665–679.

Cullinane, K., & Toy, N. (2000). Identifying influential attributes in freight route/mode choice decisions: A content analysis. *Transportation Research Part E, Logistics and Transportation Review*, *36*, 41–53.

Eltayeb, T. K., & Zailani, S. (2009). Going green through green supply chain initiatives towards environmental sustainability. *Operations and Supply Chain Management*, *2*(2), 93–110.

Eltayeb, T. K., Zailani, S., & Ramayah, T. (2011). Green supply chain initiatives among certified companies in Malaysia and environmental sustainability: Investigating the outcomes. *Resources, Conservation and Recycling*, *55*, 495–506.

Flynn, B. B., Huo, B., & Zhao, X. (2010). The impact of supply chain integration on performance: A contingency and configuration approach. *Journal of Operations Management*, *28*(1), 58–71.

Gilbert, S. (2001). *Greening Supply Chain: Enhancing Competitiveness Through Green Productivity*. Taipei: Academic Press.

Hervani, A.A., Helms, M.M., & Sarkis. J. (2005). Performance measurement for green supply chain management. *Benchmarking: An International Journal, 12*(4), 330–353.

Hwa T.J. (2001). *Green Productivity and Supply Chain Management, Greening Supply Chain: Enhancing Competitiveness Through Green Productivity*. Taipei: Academic Press.

Large, R. O., & Thomsen, C. G. (2011). Drivers of Green Supply Chain Management Performance: Evidence from Germany. *Journal of Purchasing and Supply Management*, *17*, 176–184.

Lin, C. Y., & Ho, Y. H. (2008). An Empirical Study On Logistics Service Providers' Intention To Adopt Green Innovations. *Journal of Technology Management & Innovation*, *3*(1), 17–26.

Liu, X., Yang, J., Qu, S., Wang, L., Shishime, T., & Bao, C. (2011). Sustainable Production:Practices and Determinant Factors of Green Supply Chain Management of Chinese Companies. *Business Strategy and the Environment.*

Mentzer, J. T., Dewitt, W., & Keebler, J. S. (2001). Defining supply chain management. *Journal of Business Logistics*, *22*(2), 1–25.

Richey, R. G. Jr, Chen, H., Upreti, R., Fawcett, S. E., & Adams, F. G. (2009). The moderating role of Barriers on the relationship between drivers to supply chain integration and firm performance. *International Journal of Physical Distribution & Logistics Management*, *39*(10), 826–840.

Shang, K. C., Lu, C. S., & Li, S. (2010). A taxonomy of green supply chain management capability among electronics-related manufacturing firms in Taiwan. *Journal of Environmental Management*, *91*, 1218–1226. PMID:20181423

Shi, V. G., Koh, S. C. L., Baldwin, J., & Cucchiella, F. (2012). Natural resource based green supply chain management. *Supply Chain Management: An International Journal*, *17*(1), 54–67.

Srivastava, S. K. (2007). 'Green supply-chain management: A state-of-the-art literature review'. *International Journal of Management Reviews*, *9*(1), 53–80.

Storey, J., Emberson, C., Godsell, J., & Harrison, A. (2006). Supply chain management: Theory, Practice and future challenges. *International Journal of Operations & Production Management*, *26*(7), 754–774.

Vachon, S., & Klassen, R. D. (2008). Environmental management and manufacturing performance: The role of collaboration in the supply chain. *International Journal of Production Economics*, *111*(2), 229–315.

Wee, H., Lee, M., Yu, J., & Wang, C. (2011). 'Optimal replenishment policy for a deteriorating green product: Life cycle costing analysis'. *International Journal of Production Economics*, *133*(2), 608–611.

Werbach, A. (2009). *Strategy for sustainability*. Harvard Business Press.

Wu, G. C., Ding, J. H., & Chen, P. S. (2012). The effects of GSCM drivers and institutional pressures on GSCM practices in Taiwan's textile and apparel industry. *International Journal of Production Economics*, *135*(2), 618–636.

Zhong, H. (2006). *Model and algorithms for supply chain Network with Bi-directional floks*. The State University of New Jersey.

Zhu, Q., Geng, Y., Fujita, T., & Hashimoto, S. (2010). Green Supply Chain Management in Leading Manufacturers: Case Studies in Japanese Large Companies. *Management Research Review*, *33*(4), 380–392.

Zhu, Q., & Sarkis, J. (2004). 'Relationships between operational practices and performance among early adopters of green supply chain management practices in Chinese manufacturing enterprises'. *Journal of Operations Management*, *22*(3), 265–289.

Zhu, Q., & Sarkis, J. (2004). Relationships between operational practices and performance among early adopters of green supply chain management practices in Chinese manufacturing enterprises. *Journal of Operations Management*, *22*, 265–289.

Zhu, Q., & Sarkis, J. (2007). The moderating effects of institutional pressures on emergent green supply chain practices and performance. *International Journal of Production Research*, *45*(18-19), 4333–4355.

Chapter 9
Advocating Sustainable Supply Chain Management and Sustainability in Global Supply Chain

Kijpokin Kasemsap
Suan Sunandha Rajabhat University, Thailand

ABSTRACT

This chapter aims to advocate Sustainable Supply Chain Management (SSCM) and sustainability in global supply chain, thus describing the overview of Supply Chain Management (SCM); the overview of SSCM; the overview of Triple Bottom Line (TBL); TBL measures in terms of social, environmental, and economic perspectives; the multifaceted applications of SSCM; the barriers to the implementation of SSCM; the overview of sustainability; the utilization of sustainability in global supply chain; and the sustainability aspects in global supply chain concerning risk management, transparency, strategy, and culture. This chapter performs a detailed literature review and uses the conceptual theory to introduce the concept of TBL to implement SSCM practices and to promote sustainability in global supply chain. The chapter argues that advocating SSCM and sustainability has the potential to enhance organizational performance and achieve business goals in global supply chain.

DOI: 10.4018/978-1-5225-0635-5.ch009

INTRODUCTION

Sustainable supply chain management (SSCM) has received much consideration from corporate and academic over the past decade (Azadi, Jafarian, Saen, & Mirhedayatian, 2015). SSCM is a concept which ensures the environmentally friendly practices in traditional supply chains (Diabat, Kannan, & Mathiyazhagan, 2014). SSCM investigates management strategies, risks, and opportunities in the market (Beske, 2012). Many business approaches have emerged that attempt to measure the contribution of firms to corporate sustainability (Schneider & Meins, 2012). As competition becomes global in nature, the importance of global supply chain rapidly grows (Connelly, Ketchen, & Hult, 2013). SSCM has become a strategic issue for any company looking to meet targets in terms of economic competitiveness, time, and quality of service characterized by the globalization of trade and the acceleration of industrial cycles (Eskandarpour, Dejax, Miemczyk, & Péton, 2015).

Consumers, governments, and other stakeholders are pressuring companies to be more responsible for their business operations and to strike a balance between profitability and sustainability (Hsueh, 2015). One key research stream, which explores why and how supply chains incorporate and manage sustainability, has generated significant attention within industry, academia, and policy circles (Silvestre, 2015). Considering the dynamic nature of business environments (e.g., the downturn and upturn of the global economy), competitive firms are flexible enough to make constant strategic adjustments and policy changes (Hong, Roh, & Rawski, 2012). The innovation processes toward sustainable development have received increasing attention in academic literature (Bos-Brouwers, 2010). The topic of sustainability has moved from the fringes of SSCM research to the mainstream and is an area of significant research activity (Pagell & Shevchenko, 2014).

Global concerns about sustainability have increased substantially (Govindan, Jafarian, & Nourbakhsh, 2015). Sustainability practices are one of the most important factors influencing the customers' perceptions and the decision-making process (Berezan, Raab, Yoo, & Love, 2013). The integration of sustainability concepts into SSCM involves the various business processes and cooperation between parties across supply chain networks (Winter & Knemeyer, 2013). The strength of this chapter is on the thorough literature consolidation of advocating SSCM and sustainability in global supply chain. The existing literature of SSCM and sustainability provides a contribution to practitioners and researchers by describing the multifaceted applications of SSCM and sustainability to appeal to the different segments of SSCM and sustainability in order to maximize the business impact of SSCM and sustainability in global supply chain.

BACKGROUND

For modern organizations competing in the greatly potential markets, the search for new sources of competitive advantage is necessary in order to sustain in the global economy (Kasemsap, 2015a). Industries currently face pressure on environmental initiatives from both government regulations and global competition in addition to customer pressure (Diabat et al., 2014). The concept of SCM was introduced in the early 1980s (Harland, 1996). The fast growing branch of research is SSCM (Seuring & Muller, 2008). The extant literature on SSCM contains studies on a diverse set of topics such as green purchasing, purchasing ethics, remanufacturing, safety management, supplier certification, carbon footprint, and reverse logistics (Winter & Knemeyer, 2013).

Globalization forces SCM to focus not only on the economic criteria but also on the good labor conditions and the environmentally friendly production (Azadi et al., 2015). Current researchers and practitioners give special attention to environmental issues to achieve the environmental and business needs (Caniato, Caridi, Crippa, & Moretto, 2011). Businesses face the new challenges and opportunities for adopting environmental practices (Sezen & Turkkantos, 2013). The definitions of sustainability have been more encompassing, and have explicitly incorporated the social, environmental, and economic dimensions of the major viewpoint by defining organizational sustainability as a balance among economic development, environmental stewardship, and social equity (Sikdar, 2003).

In recent years, increased concerns over pollution and greenhouse gas emissions have initiated a wave of policy change in both governmental, industry, and non-governmental organizations in order to reduce the overall environmental impact and ensure a sustainable future (Dadhich, Genovese, Kumar, & Acquaye, 2015). Sustainability performance can be defined as the performance of a company in all dimensions and for all drivers of corporate sustainability (Schaltegger & Wagner, 2006). The mainstream of the literature on corporate sustainability follows the win-win paradigm, according to which economic, environmental and social sustainability aspects can be simultaneously achieved (Hahn, Figge, Pinkse, & Preuss, 2010). Sustainability factors play a critical role for the long-term achievement of the SCM and the purchasing process becomes more complicated with the social and environmental pressures (Bai & Sarkis, 2010).

ADVOCATING SUSTAINABLE SUPPLY CHAIN MANAGEMENT AND SUSTAINABILITY IN GLOBAL SUPPLY CHAIN

This section describes the overview of SCM; the overview of SSCM; the overview of TBL; TBL measures in terms of social, environmental, and economic perspectives; the multifaceted applications of SSCM; the barriers to the implementation of SSCM; the overview of sustainability; the utilization of sustainability in global supply chain; and the sustainability aspects in global supply chain concerning risk management, transparency, strategy, and culture.

Overview of Supply Chain Management

Supply chain management (SCM) is considered as the integration of an organization's business processes with the objective of replacing a single isolated operational unit with a whole supply chain spanning raw material suppliers to the customer (Frohlich & Westbrook, 2001). Gibson et al. (2005) defined SCM as the planning and management of all activities involved in sourcing and procurement, conversion, demand creation and fulfillment, and all logistics management activities. The objectives of SCM are to produce the business value in terms of products and services (Winter & Knemeyer, 2013) and to optimize the whole entity of supply chain (Cooper, Lambert, & Pagh, 1997).

SCM is defined as an interorganizational arrangement, involving cooperative relationships (Das, 2006; Grant & Baden-Fuller, 2004). SCM is viewed as an organizational response by organizations to business pressures in their supply chain environment (Cravens, Piercy, & Shipp, 1996). Economic theory defines supply chain as a multidirectional system of autonomous decision makers (Kotzab, Teller, Grant, & Sparks, 2011). The management of supply chain is the result of a social negotiation process among the interested business parties regarding specific norms and standards (Halldorsson, Kotzab, Mikkola, & Skjøtt-Larsen, 2007). As an emerging technology, cloud computing changes the form and function of information technology (IT) infrastructures in global supply chain (Kasemsap, 2015b).

SCM exhibits typical characteristics of a subject at early stages of evolution including diversity and lack of conceptual clarity (Gibson et al., 2005). SCM is a business process integration that allows organizations to exploit the competitive advantages of SCM and achieve better organizational performance by lowering costs and increasing profits and customer satisfaction (Mitra & Singhal, 2008). The performance of supply chain is increased through suppliers in the supply chain (Fynes, de Burca, & Mangan, 2008). The effective execution of SCM can result in organizational performance improvement (Kotzab et al., 2011). Li et al. (2006)

stated that higher levels of SCM practice lead to an increased competitive advantage and improved organizational performance. Frohlich and Westbrook (2001) stated that the organizations with the widest supplier and customer integration have the highest rate of organizational performance improvement.

Overview of Sustainable Supply Chain Management

Sustainable supply chain management (SSCM) is defined as the pursuit of sustainability objectives through the purchasing and supply process, incorporating the social, economic, and environmental elements (Walker & Jones, 2012). Dyllick and Hockerts (2002) defined SSCM as the combination of sustainable development and SCM whereby sustainable development is defined as combining social, environmental, and economic factors. Regarding SSCM, closed-loop SCM is a major perspective for sustainable operations (Sahamie, Stindt, & Nuss, 2013).

SSCM makes the organizations available to gain sustainable competitive advantage over their competitors (Pagell & Wu, 2009). Cetinkaya et al. (2011) explained that sustainable supply chains are not limited to green supply chains, but they should consider financial issues and help to the society. In spite of the increasing research on sustainable supply chains, there is a paucity of empirical evidence and theoretical reflection on sustainable supply chains in developing and emerging economies (Silvestre & Silva Neto, 2014). To stay competitive, companies are starting a broad range of sustainability measures from increasing internal operational efficiency to making holistic process improvements toward business strategies (Greenberg & Quillian, 2012). Sustainable supply chains can occur through learning and innovative solutions (Pagell & Wu, 2009).

Sustainable supply chain effectively performs on both traditional measures of profit and loss, as well as on an expanded conceptualization of performance that includes social and natural dimensions (Pagell & Wu, 2009). Supply chain performs on the financial performance measures as the social and environmental indicators in modern organizations (Beske, 2012). Supply chains that focus only on financial performance are likely to remain irrelevant, and consequently unsuccessful (Hall, Matos, & Silvestre, 2012). Because complexity is part of the supply chain reality (Ageron, Gunasekaran, & Spalanzani, 2012), the supply chain evolves toward more sustainable practices through the continuous, evolutionary, and dynamic process of change.

The literature on environmental supply chains (Zhu, Sarkis, & Lai, 2008) argues that the impact of supply chains on the natural environment must be taken into account when making management decisions. While the environmental implications of supply chains are clearly important, the concept of sustainability must include not only the economic and environmental dimensions, but also the social impact of supply chain operations (Silvestre, 2015). Organizations move toward sustainability and engage in SSCM practices to react to organizational incentives from their environment, namely governments, non-governmental organizations (NGOs), and other various stakeholders, or to influence their environment (Seuring & Muller, 2008). SSCM relates to the long-term improvement of organizations and has implications for companies' economic bottom lines (Carter & Easton, 2011). Development issues in sustainability involve the companies and the whole supply chain (Fabbe-Costes, Roussat, & Colin, 2011).

Concerning SSCM, researchers and practitioners have examined several topics of environmental and social issues, including the development of environmental logistics strategies (Murphy, Poist, & Braunschwieg, 1996); environmental purchasing (Min & Galle, 1997); carrier selection for hazardous materials, and the transportation of hazardous materials (Sharp, Novack, & Anderson, 1991); the improvement of fuel efficiency and emission reduction from transportation equipment (McKinnon, Stirling, & Kirkhope, 1993); safety in motor carrier, rail, and airline industries (Cantor, Corsi, & Grimm, 2006; Crum, Dooley, & Morrow, 1995); diversity hiring and promotion issues concerning logistics personnel (Lynagh, Murphy, & Poist, 1996); the variety of motor carriers (Corsi, Tuck, & Gardner, 1982); and the range of industrial suppliers (Carter, Auskalnis, & Ketchum, 1999).

Focusing on how SSCM can be successfully implemented, SSCM theory points to the necessity for top management support (Gold, Seuring, & Beske, 2010), the installation of cross functional teams (Chen & Paulraj, 2004), enhanced communication, and the pursuit of win-win situations for supply chain partners (Seuring & Muller, 2008). The link between the environmental supply chain practices and the market-based, operational-based, and accounting-based forms of organizational performance is positive, providing support for the business case that SSCM results in the increased organizational performance (Golicic & Smith, 2013). Organizational practices that lead to sustainable supply chain are a mixture of best practices in traditional SSCM and new proactive behaviors toward sustainable practices. Being proactive and committed can be effective if the business model and the environmental and social elements of sustainability are suitably aligned (Pagell & Wu, 2009).

Enhancing SSCM performance is a critical approach for achieving competitive advantages for companies (Cai, Xiao, & Liu, 2009). SCM and SSCM effectively deal with the dynamics put onto organizations with the characteristics of the highly vulnerable logistics networks, the globalized competition, and the information sharing capabilities (Carter & Rogers, 2008). Efficient coordination of the flow of goods and information can increase an organization's ability to respond to customer needs for green innovative products (Dobrzykowski, Tran, & Hong, 2011). Wal-Mart and

IBM have communicated their green initiatives of SSCM across the areas of their commercial suppliers (Hong et al., 2012). Srivastava (2007) focused on reverse logistics, while Carter and Rogers (2008) focused on the link between supply chain and organizational performance.

Overview of Triple Bottom Line

The term triple bottom line was coined by John Elkington in his 1997 book Cannibals with Forks: The Triple Bottom Line of 21st Century Business (Slaper & Tanya, 2011). Elkington (1997) developed the concept of the TBL, describing that business goals are united in the societies and environments within which they operate in organizations. TBL is an accounting framework with three dimensions (i.e., social, environmental, and economic dimensions). TBL is essential for sustainability movement in global supply chain (Edgeman & Eskildsen, 2014). TBL suggests that at the intersection of social, environmental, and economic performances, there are activities that organizations can engage in which positively affect the natural environment and society in the long-term economic benefits in organizations (Carter & Rogers, 2008). TBL idea (Elkington, 2002) captures the three dimensions of sustainability (i.e., social, environmental, and economic perspectives) that have been identified as the crucial performance dimensions for sustainable supply chains (Zailani, Jeyaraman, Vengadasan, & Premkumar, 2012).

As environmental awareness grows, firms are expected to expand the scope of their environmental strategies beyond organizational boundaries and to address more comprehensively environmental issues in their supply chains and product life cycles (Comas Marti & Seifert, 2013). Sustainable development is an issue of increasing importance, in which modern organizations must practically engage multiple stakeholders and consider their business objectives when deciding on organizational actions (Epstein & Widener, 2011). Environmental and social issues have played a prominent role in industrial countries during the last few decades (Bask, Halme, Kallio, & Kuula, 2013). TBL dimensions are people, planet, and profits, referring to three pillars of sustainability. Many organizations have adopted the TBL framework to evaluate their performance in a broader context (Slaper & Hall, 2011). In traditional business accounting and common usage, the bottom line refers to either profit or loss, which is usually recorded on a statement of revenue and expenses.

Over the past decades, due to the rapid reduction of natural resources and perspectives over corporate social responsibility (CSR), sustainability has become important for researchers and scholars (Govindan, Khodaverdi, & Jafarian, 2013). An example of an organization seeking TBL will be a social enterprise run as a non-profit, but earning income by offering opportunities for handicapped people who have been labeled unemployable, to earn a living recycling. The organization earns a profit, which is controlled by a volunteer board, and ploughed back into the community. The social benefit is the meaningful employment of disadvantaged citizens, and the reduction in the society's welfare or disability costs.

Environmental impacts need to be managed in the same way as the financial and commercial performance (Lee & Kim, 2009). The environmental benefit comes from the recycling accomplished. In the private sector, a commitment to CSR implies a commitment to some form of TBL reporting. CSR encompasses the entire supply chain rather than the individual company domains (Andersen & Skjoett-Larsen, 2009; Spence & Bourlakis, 2009). The challenges of putting TBL into practice relate to the measurement of social and ecological categories through finding applicable data and determining how a policy contributes to sustainability in modern business (Slaper & Hall, 2011).

The concept of TBL demands that an organization's responsibility lies with stakeholders rather than shareholders. According to the stakeholder theory, the business entity should be used as a vehicle for coordinating stakeholder interests, instead of maximizing shareholder profit. A growing number of financial institutions incorporate a TBL approach in their work. There are challenges to putting the TBL into practice. These challenges include measuring each of the three categories (i.e., social, environmental, and economic dimensions), finding applicable data, and calculating a policy's contribution to sustainability. Many businesses and nonprofit organizations have adopted the TBL sustainability framework to evaluate their performance, and a similar approach has gained currency with governments at the federal, state, and local levels.

Triple Bottom Line Measures

TBL measures in terms of social, environmental, and economic perspectives are utilized by stakeholders, practitioners, and researchers toward improving sustainable supply chain performance.

Social Measures

The social dimension refers to both individuals and organizational levels (Winter & Knemeyer, 2013). While concrete material circumstances lie at the basis of the social dimension, the social phenomena themselves are immaterial and difficult to analyze (Lehtonen, 2004). Hall and Matos (2010) stated that the social dimension of sustainable development is the main challenge in sustainable supply chain. von Geibler et al. (2006) indicated eight perspectives for social assessment (i.e., health and safety, quality of working conditions, impact on employment, education and training, knowledge management, innovation potential, customer acceptance and societal product benefit, and social dialogue).

Companies should involve a wide range of stakeholders with different goals, demands, and opinions that may differently interpret the same condition (Winter & Knemeyer, 2013). Social variables refer to social dimensions of a community or region and could include measurements of education, equity and access to social resources, health and well-being, quality of life, and social capital. The specific examples involve unemployment rate, female labor force participation rate, median household income, the percentage of population with a post-secondary degree, the average commute time, the violent crimes per capita, and the health-adjusted life expectancy.

Environmental Measures

Environmental management is an important topic in SSCM (Cantor, Morrow, & Montabon, 2012). The environmental dimension includes the set of objectives, plans and mechanisms that promote greater environmental responsibility and encourage the development of environmentally friendly technologies (Klassen, 2001). A large proportion of sustainability research has focused on this dimension (Lehtonen, 2004). Environmental variables should represent the measurements of natural resources and reflect the potential influences to its viability.

Environmental sustainability has emerged as an important organizational capability to protect the environment and sustain businesses (Wong, 2013). Environmental management represents an important challenge for small businesses, and prior research has shown that they are less engaged with environmental issues than their larger counterparts (Brammer, Hoejmose, & Marchant, 2012). Plaza-Úbeda et al. (2009) stated that many advocates of environmental management system adoption promote the idea of win-win gains, where improvements in environmental performance are accompanied by financial rewards.

Having long-range trends available for each of the environmental variables would help organizations identify the impacts a policy would have on the area. The life cycle approach is widely used in the analysis of sustainability (Rota, Reynolds, & Zanasi, 2013). The life cycle thinking is the common theme of popular footprint analyses, such as carbon, ecological, and water footprinting (Wiedmann, Lenzen, & Barrett, 2009). The specific examples include sulfur dioxide concentration, concentration of nitrogen oxides, selected priority pollutants, excessive nutrients, electricity consumption, fossil fuel consumption, solid waste management, hazardous waste management, and change in land use.

Economic Measures

Financial analysis, considered as a suitable tool for assessing a company's financial and economic situation and guiding the decision-making processes of companies and financial markets, should embrace sustainability issues within its logic, under some kind of scheme or framework that permits the evaluation of a company's sustainable management system and the impact of sustainability issues on financial performance (Castro & Chousa, 2006).

The success and competitiveness of a company is the basis of the economic dimension (Winter & Knemeyer, 2013). The economic dimension is quantitative in nature, and is focused toward the effective use of resources and achieving a return on investment (Rumelt, 1974). Economic variables ought to be variables that deal with the bottom line and the flow of money. Specific examples include personal income, cost of underemployment, establishment churn, establishment sizes, job growth, employment distribution by sector, percentage of organizations in each sector, and revenue by sector.

Multifaceted Applications of Sustainable Supply Chain Management

Companies that utilize SSCM as a strategic tool in business management are likely to have a competitive edge over others (Wamalwa, 2014). SSCM considers the entire life cycle of the product using the TBL approach as a measure of success (Linton, Klassen, & Jayaraman, 2007; Pagell, Wu, & Wasserman, 2010). Carter and Rogers (2008) defined SSCM as the strategic integration and the achievement of the organization's social, environmental, and economic goals in the coordination of key business processes for improving the economic performance of the individual organizations and their supply chain.

There are the social and environmental supply chain activities that lie at the intersection with the economic essence. Potential economic advantages (i.e., the intersections of economic, social and environmental performances) include the following items.

1. Cost savings concerning reduced packaging waste (Mollenkopf, Closs, Twede, Lee, & Burgess, 2005; Rosenau, Twede, Mazzeo, & Singh, 1996), and the ability to design for reuse and disassembly (Christmann, 2000; Hart, 1995; Shrivastava, 1995). One key aspect might be greater awareness and dissemination of best practice that quantifies costs and benefits (Holt & Ghobadian, 2009).

2. Reduced health and safety costs, and lower recruitment and labor turnover costs resulting from safer warehousing and transportation and better working conditions (Brown, 1996; Carter, Ellram, & Tate, 2007).
3. Better working conditions can increase motivation and productivity, and reduce the absenteeism of supply chain personnel (Holmes, Power, & Walter, 1996; McElroy, Rodriguez, Griffin, Morrow, & Wilson, 1993).
4. Organizations that proactively address environmental and social concerns can influence government regulation when this regulation is modeled after an organization's existing production and supply chain processes, leading to a competitive advantage for organizations and their suppliers (Carter & Dresner, 2001).
5. Reduced costs, shorter lead times, and better product quality associated with the implementation of ISO 14001 standards, which provide a framework for environmental management systems (Hanson, Melnyk, & Calantone, 2004; Montabon, Melnyk, Sroufe, & Calantone, 2000; Tibor & Feldman, 1996). The registration of ISO 14001 can be leveraged across the supply chain into a competitive advantage (Curkovic & Sroufe, 2011).
6. Engaging in sustainable behavior can make an organization more attractive to suppliers and customers (Ellen, Webb, & Mohr, 2006), to potential employees (Capaldi, 2005), and to shareholders (Klassen & McLaughlin, 1996).

Barriers to the Implementation of Sustainable Supply Chain Management

Literature review methodology is considered as a suitable approach (Luthra, Garg, & Haleem, 2014) for indicating barriers to implementing SSCM. Luthra et al. (2015) indicated that there are various barriers to implementing SSCM (e.g., the lack of legislative framework, political instability, unawareness among society about social practices, the lack of customers support, top management commitment inadequacy, financial checks, the lack of knowledge among supply chain members, the lack of trust among supply chain members, technical obstructions, and organizational culture resistance to change). Legislative framework is important for establishing a modern business environment toward adopting SSCM and creating demand for sustainable products (Gouldson & Murphy, 2013).

Regulatory bodies should expand the appropriate support to sustain the SSCM due to political instability (Govindan, Kaliyan, Kannan, & Haq, 2014). Environmental advocacy groups force companies to think about their environmental management programs (Walker, Di Sisto, & McBain, 2008). Lack of customer support is recognized as one of the major barriers to implementing SSCM practices (Zhu & Geng, 2013). Top management commitment inadequacy of an organization often lacks the internal mechanisms to effectively value the benefits of managing environmental sustainability (Govindan et al., 2014). Implementation of SSCM requires the massive amounts of funds (Luthra, Garg, & Haleem, 2013).

Many companies struggle to implement SSCM because of the inadequacy of financial resources (Luthra et al., 2015). Lack of knowledge of supply chain members' sustainability is the major barrier to implementing SSCM (Al Zaabi, Al Dhaheri, & Diabat, 2013). The issue of trust among supply chain members plays a significant role, since lack of trust is a hurdle for collaboration (Beske & Seuring, 2014). Trust among supply chain members to implement SSCM provides the benefits to supply chain members toward implementing SSCM practices (Ageron et al., 2012). Technical complexity and the non-availability of suitable equipment within companies are the major barriers to adopting SSCM (Govindan et al., 2014). Companies fail to implement SSCM due to the cultural resistance to change (e.g., ineffective communication among supply chain members, institutional rules, and internal politics) (Carter & Rogers, 2008).

Overview of Sustainability

There is a growing realization that corporations must address the issue of sustainability (Searcy, 2012). Sustainability concepts have dramatically widened the scope of organizational performance measurement options and leading organizations are grappling with sustainability reporting (Hubbard, 2009). The term sustainability has been defined in various disciplines such as engineering science, operations management, and social sciences (Linton et al., 2007). It is important to examine the effects of the trade-offs companies have to make in the sustainability practices among three dimensions of social, environmental, and economic perspectives because of the resource limitations and the effects of virtues cycle that can be generated by sustainability practices in all three dimensions because of the positive interaction effects in the long run (Xu & Gursoy, 2015).

Companies which manage global supply chains face a high level of complexity with a large number of suppliers in the diverse socioeconomic contexts and the growing expectations of customers and standardization schemes to control the social and environmental aspects (Harms, Hansen, & Schaltegger, 2013). Teng et al. (2012) indicated that sustainability practices have become the significant determinants of customer loyalty and their willingness to pay a premium. Sustainability on a macro level embraces the integration of three basic principles (i.e., environmental integrity, social equity, and economic prosperity) (Bansal, 2005). Shrivastava (1995) defined sustainability as the potential for reducing the long-term risks associated with resource depletion, fluctuations in energy costs, product liabilities, and waste management. The area of sustainability is examined with a particular focus on using the TBL approach to evaluate sustainability efforts in a supply chain context (Winter & Knemeyer, 2013). The new stakeholder view of the firm goes beyond previous work on the TBL and balanced scorecard (Perrini & Tencati, 2006).

Sustainability is recognized as a source of competitive advantage (Nidumolu, Prahalad, & Rangaswami, 2009). Sustainability refers to an integration of social, environmental, and economic responsibilities (Carter & Rogers, 2008). There is growing awareness of the environmental and social impacts of the business activities of organizations (Yongvanich & Guthrie, 2006). Companies with sustainability integrated in their orientation and innovation processes effectively promote the value creation perspectives (e.g., the development of products to the marketplace and cooperation with stakeholders) (Bos-Brouwers, 2010).

At the business level, sustainability is equated with eco-efficiency (Dyllick & Hockerts, 2002). Eco-efficiency can be recognized either as an indicator of environmental performance or as a business strategy for sustainable development (Koskela & Vehmas, 2012). The dialogues with internal and external stakeholders enable the creation of a key performance indicator (KPI) set to account for social sustainability (von Geibler et al., 2006). The sustainability performance tends to become a strategic goal (Wu & Haasis, 2013). The impact of sustainability innovations can be improved by focusing on sustainability as an explicit goal as well as processes and projects instead of concentrating on content and structure (Arnold & Hockerts, 2011).

Utilization of Sustainability in Global Supply Chain

Sustainability is viewed as a continuous process toward sustainable continuance (Elkington, 2006; Mauerhofer, 2008). Sustainable development is an issue of increasing importance, in which organizations must engage multiple stakeholders and consider their objectives when deciding on organizational actions, but research provides little guidance regarding how managers can measure stakeholder reactions and use information to facilitate the decision-making process (Epstein & Widener, 2011). The concepts of sustainable development are increasingly linked with product design, sourcing and materials management, and with the reduction of resources and waste in the supply chain (Dangelio & Pontrandolfo, 2010; Shang, Lu, & Li, 2010; Srivastava, 2008). Procurement has a key role in sustainability as policies need to extend beyond organizational boundaries incorporating their whole supply chains (Meehan & Bryde, 2011). Procurement groups remain the key drivers of many sustainability-related strategies (Goldschmidt, Harrison, Holtry, & Reeh, 2013).

Corporate sustainability minimizes the downside risk of the business model by creating implicit contracts that reach beyond traditional shareholder value, transferring risk to suppliers, improving leadership by motivating employees, and by directing their attention to critical issues. Corporate sustainability, that is the capacity of a firm to continue operating over a long period of time, depends on the sustainability of its stakeholder relationships (Perrini & Tencati, 2006). Enterprise sustainability requires trans-functional, trans-disciplinary, trans-stakeholder, trans-aesthetic and trans-human knowledge that is possible through transdisciplinarity (Shrivastava, Ivanaj, & Persson, 2013). Organizational sustainability efforts practically reach deep into supply chains (Lee & Kashmanian, 2013). Organizations need appropriate systems to measure and control their own behavior in order to assess whether they are responding to stakeholder concerns in an effective way and to communicate the results achieved (Perrini & Tencati, 2006). The private sector takes significant responsibility for resolving some of the world's most intractable problems from global climate change to community impacts of business operations, and the health of employees.

Modern business is increasingly working with stakeholders to understand their views and incorporate them into the strategic decision-making processes. Regarding sustainability perspectives, stakeholders have the ability to influence the success or failure of the business at various levels. A primary objective of corporate stakeholder engagement is to generate a better understanding of stakeholder perspectives on key issues and build relationships with key individuals. Organizations with experience in engaging stakeholders tend to recognize four main areas of value (i.e., enabling the informed decision making by generating business intelligence; expanding markets and opportunities; building brand equity and reputation; and bringing diverse perspectives for creativity and innovation). Organizations have faced with sustainability practices at all levels (i.e., strategic, operational, and outcome measures) to gain sustainable competitive advantage (Zhu & Sarkis, 2004).

Multi-stakeholder contexts in development involve various actors from national or governmental organizations, international organizations, civil society, and the private sector encompassing various disciplines such as urban development and rural development (Thabrew, Wiek, & Ries, 2009). Stakeholder studies of environmental management have concentrated on stakeholder influence or the creation of stakeholder management models (Onkila, 2011). Plaza-Úbeda et al. (2009) suggested that the enhanced environmental performance following environmental management system adoption is linked to managers' belief in the win-win paradigm, and that managers who adhere to the win-win story also tend to integrate other stakeholder demands into their business strategies.

To get better corporate sustainability, organizations should take the experiences from organizational innovation management stressing the importance of interorganizational relationships, develop a small number of sustainable development indicators for procurement, take advantage of the relatively more-advanced environmental practices to show how these elements have socioeconomic impacts, and emphasize the triggers that lead to changes in behavior among procurement staff in global supply chain (Meehan & Bryde, 2011). Since sustainability requirements cover all the component parts that suppliers provide, external network configurations that involve suppliers and customers are important for sustainability implementation (Liao, Hong, & Rao, 2010; Roh, Min, & Hong, 2011). Internal response practices and external network configurations result in desirable sustainability outcomes (Hong, Tran, & Park, 2010; Soni & Kodali, 2010). For example, retailers incorporate matters of sustainability in general management and performance-oriented management (Hampl & Loock, 2013).

Sustainability Aspects in Global Supply Chain

Four aspects of sustainability include risk management, transparency, strategy, and culture (Henriques & Richardson, 2004; Sarkis, 2001; Savitz & Weber, 2006). Organizations can engage in environmental and social activities that positively affect the natural environment and society, and provide long-term economic benefits and a competitive advantage in organizations (Carter & Rogers, 2008; Markley & Davis, 2007). The operations management literature has considered sustainability from the ecological perspective without the explicit incorporation of social sustainability (Daily & Huang, 2001; Sarkis, 2001).

Risk Management

The concept of risk management is considered as a reoccurring theme in the sustainability literature (Carter & Rogers, 2008). Shrivastava (1995) stated that within the context of sustainability, an organization must manage the short-term financial results and risk factors resulting from its products, environmental waste, and public safety. Supplier selection adopts a risk-oriented strategic perspective whereas supplier development represents a business opportunity approach to managing supplier chains for sustainable products (Harms et al., 2013). Gladwin et al. (1995) stated that sustainable development must encompass the concept of security which demands safety from chronic threats and protection from harmful disruption including biodiversity loss, climate change, freshwater scarcity, food insecurity, and population growth.

Sustainability risk management (SRM) is a business strategy that aligns profit goals with an organization's environmental policies. The goal of SRM is to make this alignment efficient enough to sustain and grow a business while preserving the environment. One of the main drivers for SRM adoption is increasing demand for compliance with the national and global regulations. Organizations implementing SRM practically focus on the environmental effects of each business process individually and look for the ways to minimize them.

Transparency

As the focus of environmental policy and management shifts from cleaner production at the process level toward greener products as a whole, stakeholders ask for transparency throughout the entire value chain (Kaenzig, Friot, Saadé, Margni, & Jolliet, 2011). Transparency deals with the idea that by having an engaging and open environment in the company as well as the community will improve performance and increase profits. It is an open culture that promotes employee involvement in the innovation and creative processes.

Reaching out to the community creates a much bigger team is extremely cheap and provides evaluation from various perspectives. Companies are looking inward and realizing changes must be made to fulfill environment needs such as energy efficiency, limiting product waste and toxicity, and designing innovative products. Transparency can be improved through vertical coordination across supply chain as well as horizontal coordination across networks (Carter & Rogers, 2008).

Transparency is partially driven by the quick speed of communication via the Internet and satellite television (Elkington, 1997), as well as other factors such as interoperable software and globalization of supply chain (Friedman, 2005). Transparency involves reporting to stakeholders, engaging stakeholders, and using their feedback and input to secure buy-in and improve supply chain processes (Carter & Rogers, 2008). Transparency encompasses green marketing activities within a stakeholder perspective (Rivera-Camino, 2007) as well as more traditional cause-related marketing (Drumwright, 1994).

Strategy

Strategy is the direction and the scope of an organization over the long term which achieves advantage for the organization through its configuration of resources within a changing environment and to fulfill the stakeholder expectations (Lysons & Farrington, 2012). Strategic initiatives influence modern organizations to adopt internal response practices reflecting the features of sustainability practices (Hong, Kwon, & Roh, 2009; Yang, Hong, & Modi, 2011).

Sustainability as part of the development of business strategies is broadly recognized in modern business (Oberhofer & Dieplinger, 2014). Business strategy provides an effective challenge in a global supply chain (Muller, Vermeulen, & Glasbergen, 2012). IBM (2005) described the integration of its TBL strategy with its core business strategy. An organization's sustainability initiatives and its corporate strategy must be closely interwoven, rather than separate programs that are independently managed in modern organizations (Shrivastava, 1995).

Chiarini (2014) investigated the strategies for developing SSCM such as ISO 14001, the eco-management and audit scheme (EMAS), life cycle assessment (LCA), auditing, waste management systems, reverse logistics, environmental indicators, and reuse. Dynamic scanning, identification, and reconfiguration capabilities can facilitate firms' strategic change toward sustainability and higher competitive advantage in an evolving market environment (Wu, He, Duan, & O'Regan, 2012).

Corporate strategies to address sustainability concerns can manage different stakeholders and enable the corporation to sustain the economic growth strategy (Banerjee & Bonnefous, 2011). The signaling, information provision and the adoption of standards are the major requirements for strategy commitment, mutual learning, and the establishment of ecological cycles toward the success of SSCM (Wittstruck & Teuteberg, 2012).

Culture

The task of achieving sustainability is a cultural issue (Dos Santos, Svensson, & Padin, 2013). Organizational culture and corporate environmental sustainability are jointly related (Klassen, 2000). Organizations that support employee involvement in environmental efforts have demonstrated greater success in reducing the organization's environmental impact (Ramus & Steger, 2000). By recognizing and creating organizational culture as part of corporate sustainability efforts, organizations are able to better integrate the sustainability initiatives into daily business activities (Baumgartner, 2009). Merging sustainability strategies and activities to the organizational culture reduces the risk of organization (Baumgartner, 2009).

FUTURE RESEARCH DIRECTIONS

The strength of this chapter is on the thorough literature consolidation of advocating SSCM and sustainability in global supply chain. The existing literature of SSCM and sustainability provides a contribution to practitioners and researchers by describing the multifaceted applications of SSCM and sustainability to appeal to the different segments of SSCM and sustainability in order to maximize the business impact of SSCM and sustainability in global supply chain. The classification of the extant literature in the domains of SSCM and sustainability will provide the potential opportunities for future research.

Practitioners and researchers should recognize the advantage of a more multidisciplinary approach toward research activities in implementing SSCM and sustainability to be utilized in the knowledge-based organizations. It will be useful to bring additional disciplines together (e.g., strategic management, marketing, finance, and human resources) to support a more holistic examination of SSCM and sustainability in order to combine or transfer existing theories and approaches to inquiry in this area. An examination of linkages between SSCM activities and Lean initiatives would seem to be viable for future research efforts.

There appears to be opportunities to look at the connection between managerial components and sustainability efforts, in an effort to better understand how managerial practices can influence the success or failure of sustainability initiatives. The development and validation of appropriate metrics and scorecards in support of SSCM and sustainability offers an opportunity for highly applicable research. There may be potential for researchers to conduct research related to how suppliers can engage their customers on sustainability initiatives or to better understand how sustainable supply chain initiatives can be used to promote an organization's brand and marketing efforts.

CONCLUSION

This chapter aimed to advocate SSCM and sustainability in global supply chain, thus describing the overview of SCM; the overview of SSCM; the overview of TBL; TBL measures in terms of social, environmental, and economic perspectives; the multifaceted applications of SSCM; the barriers to the implementation of SSCM; the overview of sustainability; the utilization of sustainability in global supply chain; and the sustainability aspects in global supply chain concerning risk management, transparency, strategy, and culture. SSCM involves integrating the environmentally and financially viable practices into the complete supply chain lifecycle, from product design and development, to material selection, manufacturing, packaging, transportation, warehousing, distribution, consumption, return, and disposal. SSCM can assist organizations in optimizing their end-to-end operations to achieve greater cost savings and profitability.

By managing and improving social, environmental, and economic performances throughout supply chains regarding TBL, companies can conserve resources, optimize processes, reveal product innovations, save costs, increase productivity, and promote corporate values. Companies with best practices around sustainability will continuously screen customers for unmet needs generated by sustainability developments in the marketplace. The preventive shift to a sustainable supply chain has positive business impacts including risk management, realizing gains through more efficient operations, creating sustainable products, and improving sustainability in global supply chain. SSCM allows for companies to better protect themselves from supply chain interruptions that arise due to environmental risk and unforeseen governance challenges, particularly in developing markets.

Sustainability activities can create the important efficiencies in global supply chain. Companies should design sustainability management processes that include not only their own operations, but also address the impacts of business partners in global supply chain. It is important for modern organizations to investigate their SSCM and sustainability applications, organize a business plan to regularly examine their improvements, and effectively respond to the SSCM and sustainability needs of customers in modern business. For practitioners and researchers, this chapter provides managerial guidelines for advocating SSCM and sustainability applications utilized in global supply chain. Advocating SSCM and sustainability has the potential to improve supply chain performance and reach business goals in global supply chain.

REFERENCES

Ageron, B., Gunasekaran, A., & Spalanzani, A. (2012). Sustainable supply management: An empirical study. *International Journal of Production Economics*, *140*(1), 168–182. doi:10.1016/j.ijpe.2011.04.007

Al Zaabi, S., Al Dhaheri, N., & Diabat, A. (2013). Analysis of interaction between the barriers for the implementation of sustainable supply chain management. *International Journal of Advanced Manufacturing Technology*, *68*(1/4), 895–905. doi:10.1007/s00170-013-4951-8

Andersen, M., & Skjoett-Larsen, T. (2009). Corporate social responsibility in global supply chains. *Supply Chain Management: An International Journal*, *14*(2), 75–86. doi:10.1108/13598540910941948

Arnold, M. G., & Hockerts, K. (2011). The greening Dutchman: Philips' process of green flagging to drive sustainable innovations. *Business Strategy and the Environment*, *20*(6), 394–407. doi: 10.1002/bse.700

Azadi, M., Jafarian, M., Saen, R. F., & Mirhedayatian, S. M. (2015). A new fuzzy DEA model for evaluation of efficiency and effectiveness of suppliers in sustainable supply chain management context. *Computers & Operations Research*, *54*, 274–285. doi:10.1016/j.cor.2014.03.002

Bai, C., & Sarkis, J. (2010). Integrating sustainability into supplier selection with grey system and rough set methodologies. *International Journal of Production Economics*, *124*(1), 252–264. doi:10.1016/j.ijpe.2009.11.023

Banerjee, S. B., & Bonnefous, A. M. (2011). Stakeholder management and sustainability strategies in the French nuclear industry. *Business Strategy and the Environment, 20*(2), 124–140. doi:10.1002/bse.681

Bansal, P. (2005). Evolving sustainability: A longitudinal study of corporate sustainable development. *Strategic Management Journal, 26*(3), 197–218. doi:10.1002/smj.441

Bask, A., Halme, M., Kallio, M., & Kuula, M. (2013). Consumer preferences for sustainability and their impact on supply chain management: The case of mobile phones. *International Journal of Physical Distribution & Logistics Management, 43*(5/6), 380–406. doi:10.1108/IJPDLM-03-2012-0081

Baumgartner, R. J. (2009). Organizational culture and leadership: Preconditions for the development of a sustainable corporation. *Sustainable Development, 17*(2), 102–113. doi:10.1002/sd.405

Berezan, O., Raab, C., Yoo, M., & Love, C. (2013). Sustainable hotel practices and nationality: The impact on guest satisfaction and guest intention to return. *International Journal of Hospitality Management, 34*(2), 227–233. doi:10.1016/j.ijhm.2013.03.010

Beske, P. (2012). Dynamic capabilities and sustainable supply chain management. *International Journal of Physical Distribution & Logistics Management, 42*(4), 372–387. doi:10.1108/09600031211231344

Beske, P., & Seuring, S. (2014). Putting sustainability into supply chain management. *Supply Chain Management: An International Journal, 19*(3), 322–331. doi: 10.1108/SCM-12-2013-0432

Bos-Brouwers, H. E. J. (2010). Corporate sustainability and innovation in SMEs: Evidence of themes and activities in practice. *Business Strategy and the Environment, 19*(7), 417–435. doi: 10.1002/bse.652

Brammer, S., Hoejmose, S., & Marchant, K. (2012). Environmental management in SMEs in the UK: Practices, pressures and perceived benefits. *Business Strategy and the Environment, 21*(7), 423–434. doi:10.1002/bse.717

Brown, K. A. (1996). Workplace safety: A call for research. *Journal of Operations Management, 14*(2), 157–161. doi:10.1016/0272-6963(95)00042-9

Cai, J., Liu, X., Xiao, Z., & Liu, J. (2009). Improving supply chain performance management: A systematic approach to analyzing iterative KPI accomplishment. *Decision Support Systems*, *46*(2), 512–521. doi:10.1016/j.dss.2008.09.004

Caniato, F., Caridi, M., Crippa, L., & Moretto, A. (2011). Environmental sustainability in fashion supply chains: An exploratory case based research. *International Journal of Production Economics*, *135*(2), 659–670. doi:10.1016/j.ijpe.2011.06.001

Cantor, D. E., Corsi, T. M., & Grimm, C. M. (2006). Safety technology adoption patterns in the US motor carrier industry. *Transportation Journal*, *45*(3), 20–45.

Cantor, D. E., Morrow, P. C., & Montabon, F. (2012). Engagement in environmental behaviors among supply chain management employees: An organizational support theoretical perspective. *Journal of Supply Chain Management*, *48*(3), 33–51. doi:10.1111/j.1745-493X.2011.03257.x

Capaldi, N. (2005). Corporate social responsibility and the bottom line. *International Journal of Social Economics*, *32*(5), 408–423. doi:10.1108/03068290510591263

Carter, C. R., Auskalnis, R., & Ketchum, C. (1999). Purchasing from minority business enterprises: A cross-industry comparison of best practices. *Journal of Supply Chain Management*, *35*(1), 28–32. doi:10.1111/j.1745-493X.1999.tb00053.x

Carter, C. R., & Dresner, M. (2001). Environmental purchasing and supply management: Cross-functional development of grounded theory. *Journal of Supply Chain Management*, *37*(3), 12–27. doi:10.1111/j.1745-493X.2001.tb00102.x

Carter, C. R., & Easton, P. L. (2011). Sustainable supply chain management: Evolution and future directions. *International Journal of Physical Distribution & Logistics Management*, *41*(1), 46–62. doi:10.1108/09600031111101420

Carter, C. R., Ellram, L. M., & Tate, W. L. (2007). Structure and influence: A logistics management application of social network analysis. *Journal of Business Logistics*, *28*(1), 137–168. doi:10.1002/j.2158-1592.2007.tb00235.x

Carter, C. R., & Rogers, D. S. (2008). A framework of sustainable supply chain management: Moving toward new theory. *International Journal of Physical Distribution & Logistics Management*, *38*(5), 360–387. doi:10.1108/09600030810882816

Castro, N. R., & Chousa, J. P. (2006). An integrated framework for the financial analysis of sustainability. *Business Strategy and the Environment*, *15*(5), 322–333. doi:10.1002/bse.539

Cetinkaya, B., Cuthbertson, R., Ewer, G., Klaas-Wissing, T., Piotrowicz, W., & Tyssen, C. (2011). *Sustainable supply chain management*. New York, NY: Springer–Verlag.

Chen, I. J., & Paulraj, A. (2004). Towards a theory of supply chain management: The constructs and measurements. *Journal of Operations Management*, *22*(2), 119–150. doi:10.1016/j.jom.2003.12.007

Chiarini, A. (2014). Strategies for developing an environmentally sustainable supply chain: Differences between manufacturing and service sectors. *Business Strategy and the Environment*, *23*(7), 493–504. doi:10.1002/bse.1799

Christmann, P. (2000). Effects of "best practices" of environmental management on cost advantage: The role of complementary assets. *Academy of Management Journal*, *43*(4), 663–680. doi:10.2307/1556360

Comas Martí, J. M., & Seifert, R. W. (2013). Assessing the comprehensiveness of supply chain environmental strategies. *Business Strategy and the Environment*, *22*(2), 339–356. doi:10.1002/bse.1749

Connelly, B. L., Ketchen, D. J., & Hult, G. T. M. (2013). Global supply chain management: Toward a theoretically driven research agenda. *Global Strategy Journal*, *3*(3), 227–243. doi:10.1111/j.2042-5805.2013.01041.x

Cooper, M. C., Lambert, D. M., & Pagh, J. D. (1997). Supply chain management – more than a new name for logistics. *International Journal of Logistics Management*, *8*(1), 1–14. doi:10.1108/09574099710805556

Corsi, T. M., Tuck, J. M., & Gardner, L. L. (1982). Minority motor carriers and the Motor Carrier Act of 1980. *Transportation Journal*, *22*(1), 42–55.

Cousins, P. D., & Lawson, B. (2007). Sourcing strategy, supplier relationships and firm performance: An empirical investigation of UK organizations. *British Journal of Management*, *18*(2), 123–137. doi:10.1111/j.1467-8551.2006.00506.x

Cravens, D. W., Piercy, N. F., & Shipp, S. H. (1996). New organizational forms for competing in highly dynamic environments: The network paradigm. *British Journal of Management*, *7*(3), 203–218. doi:10.1111/j.1467-8551.1996.tb00115.x

Crum, M. R., Dooley, F. J., & Morrow, P. C. (1995). Employee attitudes about railroad injury compensation. *Transportation Journal*, *35*(1), 15–29.

Curkovic, S., & Sroufe, R. (2011). Using ISO 14001 to promote a sustainable supply chain strategy. *Business Strategy and the Environment*, *20*(2), 71–93. doi:10.1002/bse.671

Dadhich, P., Genovese, A., Kumar, N., & Acquaye, A. (2015). Developing sustainable supply chains in the UK construction industry: A case study. *International Journal of Production Economics*, *164*, 271–284. doi:10.1016/j.ijpe.2014.12.012

Daily, B. F., & Huang, S. (2001). Achieving sustainability through attention to human resource factors in environmental management. *International Journal of Operations & Production Management*, *21*(12), 1539–1552. doi:10.1108/01443570110410892

Dangelio, R. M., & Pontrandolfo, P. (2010). From green product definitions and classifications to Green Option Matrix. *Journal of Cleaner Production*, *18*(16/17), 1608–1628. doi:10.1016/j.jclepro.2010.07.007

Das, T. K. (2006). Strategic alliance temporalities and partner opportunism. *British Journal of Management*, *17*(1), 1–21. doi:10.1111/j.1467-8551.2006.00482.x

Diabat, A., Kannan, D., & Mathiyazhagan, K. (2014). Analysis of enablers for implementation of sustainable supply chain management: A textile case. *Journal of Cleaner Production*, *83*(1), 391–403. doi: 10.1016/j.jclepro.2014.06.081

Dobrzykowski, D. D., Tran, O. T., & Hong, P. (2011). Insights into integration for supply chain redesign in service and product-focused firms. *International Journal of Services and Operations Management*, *8*(3), 260–282. doi:10.1504/IJSOM.2011.038972

Dos Santos, M. A. O., Svensson, G., & Padin, C. (2013). Indicators of sustainable business practices: Woolworths in South Africa. *Supply Chain Management: An International Journal*, *18*(1), 104–108. doi:10.1108/13598541311293212

Drumwright, M. E. (1994). Socially responsible organizational buying: Environmental concern as a non-economic buying criterion. *Journal of Marketing*, *58*(2), 1–19. doi:10.2307/1252307

Dyllick, T., & Hockerts, K. (2002). Beyond the business case for corporate sustainability. *Business Strategy and the Environment*, *11*(2), 130–141. doi:10.1002/bse.323

Edgeman, R., & Eskildsen, J. (2014). Modeling and assessing sustainable enterprise excellence. *Business Strategy and the Environment*, *23*(3), 173–187. doi:10.1002/bse.1779

Elkington, J. (1997). *Cannibals with forks: The triple bottom line of the 21st century*. Stony Creek, CT: New Society Publishers.

Elkington, J. (2006). Governance for sustainability. *Corporate Governance*, *14*(6), 522–529. doi:10.1111/j.1467-8683.2006.00527.x

Ellen, P. S., Webb, D. J., & Mohr, L. A. (2006). Building corporate associations: Consumer attributions for corporate social responsibility programs. *Journal of the Academy of Marketing Science*, *34*(2), 147–157. doi:10.1177/0092070305284976

Epstein, M. J., & Widener, S. K. (2011). Facilitating sustainable development decisions: Measuring stakeholder reactions. *Business Strategy and the Environment*, *20*(2), 107–123. doi:10.1002/bse.680

Eskandarpour, M., Dejax, P., Miemczyk, J., & Péton, O. (2015). Sustainable supply chain network design: An optimization-oriented review. *Omega: The International Journal of Management Science*, *54*, 11–32. doi:10.1016/j.omega.2015.01.006

Fabbe-Costes, N., Roussat, C., & Colin, J. (2011). Future sustainable supply chains: What should companies scan? *International Journal of Physical Distribution & Logistics Management*, *41*(3), 228–252. doi:10.1108/09600031111123778

Friedman, T. L. (2005). *The world is flat: A brief history of the twenty-first century*. New York, NY: Farrar, Strauss and Giroux.

Frohlich, M., & Westbrook, R. (2001). Arcs of integration: An international study of supply chain strategies. *Journal of Operations Management*, *19*(2), 185–200. doi:10.1016/S0272-6963(00)00055-3

Fynes, B., de Burca, S., & Mangan, J. (2008). The effect of relationship characteristics on relationship quality and performance. *International Journal of Production Economics*, *111*(1), 56–69. doi:10.1016/j.ijpe.2006.11.019

Gibson, B. J., Mentzer, J. T., & Cook, R. L. (2005). Supply chain management: The pursuit of a consensus definition. *Journal of Business Logistics*, *26*(2), 17–25. doi:10.1002/j.2158-1592.2005.tb00203.x

Gladwin, T. N., Kennelly, J. J., & Krause, T. (1995). Shifting paradigms for sustainable development: Implications for management theory and research. *Academy of Management Review*, *20*(4), 874–907. doi: 10.2307/258959

Gold, S., Seuring, S., & Beske, P. (2010). Sustainable supply chain management and inter-organizational resources: A literature review. *Corporate Social Responsibility and Environmental Management*, *17*(4), 230–245. doi: 10.1002/csr.207

Goldschmidt, K., Harrison, T., Holtry, M., & Reeh, J. (2013). Sustainable procurement: Integrating classroom learning with university sustainability programs. *Decision Sciences Journal of Innovative Education*, *11*(3), 279–294. doi:10.1111/dsji.12007

Golicic, S. L., & Smith, C. D. (2013). A meta-analysis of environmentally sustainable supply chain management practices and firm performance. *Journal of Supply Chain Management*, *49*(2), 78–95. doi:10.1111/jscm.12006

Gouldson, A., & Murphy, J. (2013). *Regulatory realities: The implementation and impact of industrial environmental regulation*. London, UK: Routledge.

Govindan, K., Jafarian, A., & Nourbakhsh, V. (2015). Bi-objective integrating sustainable order allocation and sustainable supply chain network strategic design with stochastic demand using a novel robust hybrid multi-objective metaheuristic. *Computers & Operations Research*, *62*, 112–130. doi:10.1016/j.cor.2014.12.014

Govindan, K., Kaliyan, M., Kannan, D., & Haq, A. N. (2014). Barriers analysis for green supply chain management implementation in Indian industries using analytic hierarchy process. *International Journal of Production Economics*, *147*, 555–568. doi:10.1016/j.ijpe.2013.08.018

Govindan, K., Khodaverdi, R., & Jafarian, A. (2013). A fuzzy multi criteria approach for measuring sustainability performance of a supplier based on triple bottom line approach. *Journal of Cleaner Production*, *47*, 345–354. doi:10.1016/j.jclepro.2012.04.014

Grant, R. M., & Baden-Fuller, C. (2004). A knowledge-accessing theory of strategic alliances. *Journal of Management Studies*, *41*(1), 61–84. doi:10.1111/j.1467-6486.2004.00421.x

Greenberg, A., & Quillian, L. (2012). Managing sustainable production: A framework for integrating sustainability in the manufacturing sector. *Environmental Quality Management*, *21*(4), 25–40. doi:10.1002/tqem.21305

Hahn, T., Figge, F., Pinkse, J., & Preuss, L. (2010). Trade-offs in corporate sustainability: You can't have your cake and eat it. *Business Strategy and the Environment*, *19*(4), 217–229. doi:10.1002/bse.674

Hall, J., & Matos, S. (2010). Incorporating impoverished communities in sustainable supply chains. *International Journal of Physical Distribution & Logistics Management*, *40*(1/2), 124–147. doi:10.1108/09600031011020368

Hall, J., Matos, S., & Silvestre, B. (2012). Understanding why firms should invest in sustainable supply chains: A complexity approach. *International Journal of Production Research, 50*(5), 1332–1348. doi:10.1080/00207543.2011.571930

Halldorsson, A., Kotzab, H., Mikkola, J. H., & Skjøtt-Larsen, T. (2007). Complementary theories to supply chain management. *Supply Chain Management: An International Journal, 12*(4), 284–296. doi:10.1108/13598540710759808

Hampl, N., & Loock, M. (2013). Sustainable development in retailing: What is the impact on store choice? *Business Strategy and the Environment, 22*(3), 202–216. doi:10.1002/bse.1748

Hanson, J. D., Melnyk, S. A., & Calantone, R. J. (2004). Core values and environmental management. *Greener Management International, 2004*(46), 29–40. doi:10.9774/GLEAF.3062.2004.su.00005

Harland, C. (1996). Supply chain management: Relationships, chains and networks. *British Journal of Management, 7*(1), S63–S80. doi:10.1111/j.1467-8551.1996.tb00148.x

Harms, D., Hansen, E. G., & Schaltegger, S. (2013). Strategies in sustainable supply chain management: An empirical investigation of large German companies. *Corporate Social Responsibility and Environmental Management, 20*(4), 205–218. doi:10.1002/csr.1293

Hart, S. L. (1995). A natural-resource-based view of the firm. *Academy of Management Review, 20*(4), 986–1014. doi: 10.2307/258963

Henriques, A., & Richardson, J. (2004). *The triple bottom line: Does it all add up?* London, UK: Earthscan.

Holmes, S. M., Power, M. L., & Walter, C. K. (1996). A motor carrier wellness program: Development and testing. *Transportation Journal, 35*(3), 33–48.

Holt, D., & Ghobadian, A. (2009). An empirical study of green supply chain management practices amongst UK manufacturers. *Journal of Manufacturing Technology Management, 20*(7), 933–956. doi:10.1108/17410380910984212

Hong, P., Kwon, H. B., & Roh, J. J. (2009). Implementation of strategic green orientation in supply chain. *European Journal of Innovation Management, 12*(4), 512–532. doi: 10.1108/14601060910996945

Hong, P., Roh, J. J., & Rawski, G. (2012). Benchmarking sustainability practices: Evidence from manufacturing firms. *Benchmarking: An International Journal*, *19*(4/5), 634–648. doi:10.1108/14635771211258052

Hong, P., Tran, O., & Park, K. (2010). Electronic commerce applications for supply chain integration and competitive capabilities: An empirical study. *Benchmarking: An International Journal*, *17*(4), 539–560. doi:10.1108/14635771011060585

Hsueh, C. F. (2015). A bilevel programming model for corporate social responsibility collaboration in sustainable supply chain management. *Transportation Research Part E, Logistics and Transportation Review*, *73*, 84–95. doi:10.1016/j.tre.2014.11.006

Hubbard, G. (2009). Measuring organizational performance: Beyond the triple bottom line. *Business Strategy and the Environment*, *18*(3), 177–191. doi:10.1002/bse.564

IBM. (2005). *IBM's 2005 innovations in corporate responsibility*. Armonk, NY: IBM Corporation.

Kaenzig, J., Friot, D., Saadé, M., Margni, M., & Jolliet, O. (2011). Using life cycle approaches to enhance the value of corporate environmental disclosures. *Business Strategy and the Environment*, *20*(1), 38–54. doi:10.1002/bse.667

Kasemsap, K. (2015a). Implementing enterprise resource planning. In M. Khosrow-Pour (Ed.), *Encyclopedia of information science and technology* (3rd ed.; pp. 798–807). Hershey, PA: IGI Global. doi:10.4018/978-1-4666-5888-2.ch076

Kasemsap, K. (2015b). The role of cloud computing in global supply chain. In N. Rao (Ed.), *Enterprise management strategies in the era of cloud computing* (pp. 192–219). Hershey, PA: IGI Global. doi:10.4018/978-1-4666-8339-6.ch009

Klassen, R. D. (2000). Exploring the linkage between investment in manufacturing and environmental technologies. *International Journal of Operations & Production Management*, *20*(2), 127–147. doi:10.1108/01443570010304224

Klassen, R. D. (2001). Plant-level environmental management orientation: The influence of management views and plant characteristics. *Production and Operations Management*, *10*(3), 257–275. doi:10.1111/j.1937-5956.2001.tb00374.x

Klassen, R. D., & McLaughlin, C. P. (1996). The impact of environmental management on firm performance. *Management Science*, *42*(8), 1199–1214. doi:10.1287/mnsc.42.8.1199

Koskela, M., & Vehmas, J. (2012). Defining eco-efficiency: A case study on the Finnish forest industry. *Business Strategy and the Environment, 21*(8), 546–566. doi:10.1002/bse.741

Kotzab, H., Teller, C., Grant, D. B., & Sparks, L. (2011). Antecedents for the adoption and execution of supply chain management. *Supply Chain Management: An International Journal, 16*(4), 231–245. doi:10.1108/13598541111139053

Lee, K. H., & Kim, J. W. (2009). Current status of CSR in the realm of supply management: The rise of the Korean electronics industry. *Supply Chain Management: An International Journal, 14*(2), 138–148. doi:10.1108/13598540910942000

Lee, T., & Kashmanian, R. M. (2013). Supply chain sustainability: Compliance- and performance-based tools. *Environmental Quality Management, 22*(4), 1–23. doi:10.1002/tqem.21340

Lehtonen, M. (2004). The environmental – social interface of sustainable development: Capabilities, social capital, institutions. *Ecological Economics, 49*(2), 199–214. doi:10.1016/j.ecolecon.2004.03.019

Li, S., Ragu-Nathan, B., Ragu-Nathan, T. S., & Subba Rao, S. (2006). The impact of supply chain management practices on competitive advantage and organizational performance. *Omega: The International Journal of Management Science, 34*(2), 107–124. doi:10.1016/j.omega.2004.08.002 PMID:17876965

Liao, Y., Hong, P., & Rao, S. (2010). Supply management, supply flexibility and performance outcomes: An empirical investigation of manufacturing firms. *Journal of Supply Chain Management, 46*(3), 6–22. doi:10.1111/j.1745-493X.2010.03195.x

Linton, J. D., Klassen, R., & Jayaraman, V. (2007). Sustainable supply chains: An introduction. *Journal of Operations Management, 25*(6), 1075–1082. doi:10.1016/j.jom.2007.01.012

Luthra, S., Garg, D., & Haleem, A. (2013). Identifying and ranking of strategies to implement green supply chain management in Indian manufacturing industry using Analytical Hierarchy Process. *Journal of Industrial Engineering and Management, 6*(4), 930–962. doi:10.3926/jiem.693

Luthra, S., Garg, D., & Haleem, A. (2014). Green supply chain management: Implementation and performance–a literature review and some issues. *Journal of Advances in Management Research, 11*(1), 20–46. doi:10.1108/JAMR-07-2012-0027

Luthra, S., Luthra, S., & Haleem, A. (2015). Hurdles in implementing sustainable supply chain management: An analysis of Indian automobile sector. *Procedia: Social and Behavioral Sciences*, *189*, 175–183. doi:10.1016/j.sbspro.2015.03.212

Lynagh, P. M., Murphy, P. R., & Poist, R. F. (1996). Career-related perspectives regarding women in logistics: A comparative analysis. *Transportation Journal*, *36*(1), 35–42.

Lysons, K., & Farrington, B. (2012). *Purchasing and supply chain management*. Harlow, UK: Pearson Education.

Markley, M. J., & Davis, L. (2007). Exploring future competitive advantage through sustainable supply chains. *International Journal of Physical Distribution & Logistics Management*, *37*(9), 763–774. doi:10.1108/09600030710840859

McElroy, J. C., Rodriguez, J. M., Griffin, G. C., Morrow, P. C., & Wilson, M. G. (1993). Career stage, time spent on the road, and truckload driver attitudes. *Transportation Journal*, *33*(1), 5–14.

McKinnon, A. C., Stirling, I., & Kirkhope, J. (1993). Improving the fuel efficiency of road freight operations. *International Journal of Physical Distribution & Logistics Management*, *23*(9), 3–11. doi:10.1108/09600039310049736

Meehan, J., & Bryde, D. (2011). Sustainable procurement practice. *Business Strategy and the Environment*, *20*(2), 94–106. doi:10.1002/bse.678

Min, H., & Galle, W. P. (1997). Green purchasing strategies: Trends and implications. *International Journal of Purchasing and Materials Management*, *33*(3), 10–17. doi:10.1111/j.1745-493X.1997.tb00026.x

Mitra, S., & Singhal, V. (2008). Supply chain integration and shareholder value: Evidence from consortium-based industry exchanges. *Journal of Operations Management*, *26*(1), 96–114. doi:10.1016/j.jom.2007.05.002

Mollenkopf, D., Closs, D., Twede, D., Lee, S., & Burgess, G. (2005). Assessing the viability of reusable packaging: A relative cost approach. *Journal of Business Logistics*, *26*(1), 169–197. doi:10.1002/j.2158-1592.2005.tb00198.x

Montabon, F., Melnyk, S. A., Sroufe, R., & Calantone, R. J. (2000). ISO 14000: Assessing its perceived impact on corporate performance. *Journal of Supply Chain Management*, *36*(2), 4–16. doi:10.1111/j.1745-493X.2000.tb00073.x

Muller, C., Vermeulen, W. J. V., & Glasbergen, P. (2012). Pushing or sharing as value-driven strategies for societal change in global supply chains: Two case studies in the British–South African fresh fruit supply chain. *Business Strategy and the Environment, 21*(2), 127–140. doi:10.1002/bse.719

Murphy, P. R., Poist, R. F., & Braunschwieg, C. D. (1996). Green logistics: Comparative views of environmental progressives, moderates, and conservatives. *Journal of Business Logistics, 17*(1), 191–211.

Nidumolu, R., Prahalad, C. K., & Rangaswami, M. R. (2009). Why sustainability is now the key driver of innovation. *Harvard Business Review, 87*(9), 56–64.

Oberhofer, P., & Dieplinger, M. (2014). Sustainability in the transport and logistics sector: Lacking environmental measures. *Business Strategy and the Environment, 23*(4), 236–253. doi:10.1002/bse.1769

Onkila, T. (2011). Multiple forms of stakeholder interaction in environmental management: Business arguments regarding differences in stakeholder relationships. *Business Strategy and the Environment, 20*(6), 379–393. doi: 10.1002/bse.693

Pagell, M., & Shevchenko, A. (2014). Why research in sustainable supply chain management should have no future. *Journal of Supply Chain Management, 50*(1), 44–55. doi:10.1111/jscm.12037

Pagell, M., & Wu, Z. (2009). Building a more complete theory of sustainable supply chain management using case studies of 10 exemplars. *Journal of Supply Chain Management, 45*(2), 37–56. doi:10.1111/j.1745-493X.2009.03162.x

Pagell, M., Wu, Z., & Wasserman, M. E. (2010). Thinking differently about purchasing portfolios: An assessment of sustainable sourcing. *Journal of Supply Chain Management, 46*(1), 57–73. doi:10.1111/j.1745-493X.2009.03186.x

Perrini, F., & Tencati, A. (2006). Sustainability and stakeholder management: The need for new corporate performance evaluation and reporting systems. *Business Strategy and the Environment, 15*(5), 296–308. doi:10.1002/bse.538

Plaza-Úbeda, J. A., Burgos-Jiménez, J., Vazquez, D. A., & Liston-Heyes, C. (2009). The "win–win" paradigm and stakeholder integration. *Business Strategy and the Environment, 18*(8), 487–499. doi:10.1002/bse.593

Ramus, C. A., & Steger, U. (2000). The roles of supervisory support behaviors and environmental policy in employee "eco-initiatives" at leading-edge European companies. *Academy of Management Journal, 43*(4), 605–626. doi:10.2307/1556357

Rivera-Camino, J. (2007). Re-evaluating green marketing strategy: A stakeholder perspective. *European Journal of Marketing*, *41*(11/12), 1328–1358. doi:10.1108/03090560710821206

Roh, J. J., Min, H., & Hong, P. (2011). A co-ordination theory approach to restructuring the supply chain: An empirical study from the focal company perspective. *International Journal of Production Research*, *49*(15), 4517–4541. doi:10.1080/00207543.2010.497506

Rosenau, W. V., Twede, D., Mazzeo, M. A., & Singh, S. P. (1996). Returnable/reusable logistical packaging: A capital budgeting investment decision framework. *Journal of Business Logistics*, *17*(2), 139–165.

Rota, C., Reynolds, N., & Zanasi, C. (2013). Sustainable food supply chains: The role of collaboration and sustainable relationships. *International Journal of Business and Social Science*, *4*(4), 45–53.

Rumelt, R. P. (1974). *Strategy, structure, and economic performance*. Cambridge, MA: Harvard Business School Press.

Sahamie, R., Stindt, D., & Nuss, C. (2013). Transdisciplinary research in sustainable operations: An application to closed-loop supply chains. *Business Strategy and the Environment*, *22*(4), 245–268. doi:10.1002/bse.1771

Sarkis, J. (2001). Manufacturing's role in corporate environmental sustainability. *International Journal of Operations & Production Management*, *21*(5/6), 666–686. doi:10.1108/01443570110390390

Savitz, A. W., & Weber, K. (2006). *The triple bottom line*. San Francisco, CA: Jossey–Bass.

Schaltegger, S., & Wagner, M. (2006). Integrative management of sustainability performance, measurement and reporting. *International Journal of Accounting, Auditing and Performance Evaluation*, *3*(1), 1–19. doi:10.1504/IJAAPE.2006.010098

Schneider, A., & Meins, E. (2012). Two dimensions of corporate sustainability assessment: Towards a comprehensive framework. *Business Strategy and the Environment*, *21*(4), 211–222. doi:10.1002/bse.726

Searcy, C. (2012). Corporate sustainability performance measurement systems: A review and research agenda. *Journal of Business Ethics*, *107*(3), 239–253. doi:10.1007/s10551-011-1038-z

Seuring, S., & Muller, M. (2008). From a literature review to a conceptual framework for sustainable supply chain management. *Journal of Cleaner Production*, *16*(15), 1699–1710. doi:10.1016/j.jclepro.2008.04.020

Sezen, B., & Turkkantos, S. (2013). The effects of relationship quality and lean applications on buyer-seller relationships. *International Journal of Business Performance and Supply Chain Modelling*, *5*(4), 378–400. doi:10.1504/IJBPSCM.2013.058204

Shang, K. C., Lu, C. S., & Li, S. (2010). A taxonomy of green supply chain management capability among electronics-related manufacturing firms in Taiwan. *Journal of Environmental Management*, *91*(5), 1218–1226. doi:10.1016/j.jenvman.2010.01.016 PMID:20181423

Sharp, J. M., Novack, R. A., & Anderson, M. A. (1991). Purchasing hazardous waste transportation service: Federal legal considerations. *Transportation Journal*, *31*(2), 4–14.

Shrivastava, P. (1995). The role of corporations in achieving ecological sustainability. *Academy of Management Review*, *20*(4), 936–960. doi: 10.2307/258961

Shrivastava, P., Ivanaj, S., & Persson, S. (2013). Transdisciplinary study of sustainable enterprise. *Business Strategy and the Environment*, *22*(4), 230–244. doi:10.1002/bse.1773

Sikdar, S. K. (2003). Sustainable development and sustainability metrics. *The American Institute of Chemical Engineers Journal*, *49*(8), 1928–1932. doi:10.1002/aic.690490802

Silvestre, B. S. (2015). Sustainable supply chain management in emerging economies: Environmental turbulence, institutional voids and sustainability trajectories. *International Journal of Production Economics*, *167*, 156–169. doi:10.1016/j.ijpe.2015.05.025

Silvestre, B. S., & Silva Neto, R. (2014). Capability accumulation, innovation, and technology diffusion: Lessons from a base of the pyramid cluster. *Technovation*, *34*(5), 270–283. doi:10.1016/j.technovation.2013.09.007

Slaper, T. F., & Hall, T. (2011). The triple bottom line: What is it and how does it work? *Indiana Business Review*, *86*(1), 4–8.

Soni, G., & Kodali, R. (2010). Internal benchmarking for assessment of supply chain performance. *Benchmarking: An International Journal*, *17*(1), 44–76. doi:10.1108/14635771011022316

Spence, L., & Bourlakis, M. (2009). The evolution from corporate social responsibility to supply chain responsibility: The case of Waitrose. *Supply Chain Management: An International Journal, 14*(4), 291–302. doi:10.1108/13598540910970126

Srivastava, S. K. (2007). Green supply chain management: A state-of-the-art literature review. *International Journal of Management Reviews, 9*(1), 53–80. doi:10.1111/j.1468-2370.2007.00202.x

Srivastava, S. K. (2008). Network design for reverse logistics. *Omega: The International Journal of Management Science, 36*(4), 535–548. doi:10.1016/j.omega.2006.11.012

Teng, C. C., Horng, J. S., Hu, M. L., Chien, L. H., & Shen, Y. C. (2012). Developing energy conservation and carbon reduction indicators for the hotel industry in Taiwan. *International Journal of Hospitality Management, 31*(1), 199–208. doi:10.1016/j.ijhm.2011.06.006

Thabrew, L., Wiek, A., & Ries, R. (2009). Environmental decision making in multi-stakeholder contexts: Applicability of life cycle thinking in development planning and implementation. *Journal of Cleaner Production, 17*(1), 67–76. doi:10.1016/j.jclepro.2008.03.008

Tibor, T., & Feldman, I. (1996). *ISO 14000: A guide to the new environmental management standards.* Burr Ridge, IL: Irwin Professional Publishing.

von Geibler, J., Liedtke, C., Wallbaum, H., & Schaller, S. (2006). Accounting for the social dimension of sustainability: Experiences from the biotechnology industry. *Business Strategy and the Environment, 15*(5), 334–346. doi:10.1002/bse.540

Walker, H., Di Sisto, L., & McBain, D. (2008). Drivers and barriers to environmental supply chain management practices: Lessons from the public and private sectors. *Journal of Purchasing and Supply Management, 14*(1), 69–85. doi:10.1016/j.pursup.2008.01.007

Walker, H., & Jones, N. (2012). Sustainable supply chain management across the UK private sector. *Supply Chain Management: An International Journal, 17*(1), 15–28. doi:10.1108/13598541211212177

Wamalwa, B. P. (2014). Sustainable supply chain management as a strategic tool for competitive advantage in tea industry in Kenya. *Journal of Management and Sustainability, 4*(3), 157–164. doi:10.5539/jms.v4n3p157

Wiedmann, T. O., Lenzen, M., & Barrett, J. R. (2009). Companies on the scale. *Journal of Industrial Ecology*, *13*(3), 361–383. doi:10.1111/j.1530-9290.2009.00125.x

Winter, M., & Knemeyer, A. M. (2013). Exploring the integration of sustainability and supply chain management: Current state and opportunities for future inquiry. *International Journal of Physical Distribution & Logistics Management*, *43*(1), 18–38. doi:10.1108/09600031311293237

Wittstruck, D., & Teuteberg, F. (2012). Understanding the success factors of sustainable supply chain management: Empirical evidence from the electrics and electronics industry. *Corporate Social Responsibility and Environmental Management*, *19*(3), 141–158. doi:10.1002/csr.261

Wong, C. W. Y. (2013). Leveraging environmental information integration to enable environmental management capability and performance. *Journal of Supply Chain Management*, *49*(2), 114–136. doi:10.1111/jscm.12005

Wu, J., & Haasis, H. D. (2013). Converting knowledge into sustainability performance of freight villages. *Logistics Research*, *6*(2/3), 63–88. doi:10.1007/s12159-013-0098-0

Wu, Q., He, Q., Duan, Y., & O'Regan, N. (2012). Implementing dynamic capabilities for corporate strategic change toward sustainability. *Strategic Change*, *21*(5/6), 231–247. doi:10.1002/jsc.1906

Xu, X., & Gursoy, D. (2015). Influence of sustainable hospitality supply chain management on customers' attitudes and behaviors. *International Journal of Hospitality Management*, *49*, 105–116. doi:10.1016/j.ijhm.2015.06.003

Yang, M., Hong, P., & Modi, S. (2011). Impact of lean manufacturing and environmental management on business performance: An international study of manufacturing firms. *International Journal of Production Economics*, *129*(2), 251–261. doi:10.1016/j.ijpe.2010.10.017

Yongvanich, K., & Guthrie, J. (2006). An extended performance reporting framework for social and environmental accounting. *Business Strategy and the Environment*, *15*(5), 309–321. doi:10.1002/bse.541

Zailani, S., Jeyaraman, K., Vengadasan, G., & Premkumar, R. (2012). Sustainable supply chain management (SSCM) in Malaysia: A survey. *International Journal of Production Economics*, *140*(1), 330–340. doi:10.1016/j.ijpe.2012.02.008

Zhu, Q., & Geng, Y. (2013). Drivers and barriers of extended supply chain practices for energy saving and emission reduction among Chinese manufacturers. *Journal of Cleaner Production*, *40*, 6–12. doi:10.1016/j.jclepro.2010.09.017

Zhu, Q., & Sarkis, J. (2004). Relationships between operational practices and performance among early adopters of green supply chain management practices in Chinese manufacturing enterprises. *Journal of Operations Management, 22*(3), 265–289. doi:10.1016/j.jom.2004.01.005

Zhu, Q., Sarkis, J., & Lai, K. H. (2008). Confirmation of a measurement model for green supply chain management practices implementation. *International Journal of Production Economics, 111*(2), 261–273. doi:10.1016/j.ijpe.2006.11.029

ADDITIONAL READING

Brockhaus, S., Kersten, W., & Knemeyer, A. M. (2013). Where do we go from here? Progressing sustainability implementation efforts across supply chains. *Journal of Business Logistics, 34*(2), 167–182. doi:10.1111/jbl.12017

Bruzzone, A. A. G., Anghinolfi, D., Paolucci, M., & Tonelli, F. (2012). Energy-aware scheduling for improving manufacturing process sustainability: A mathematical model for flexible flow shops. *CIRP Annals – Manufacturing Technology, 61*(1), 459–462. doi: 10.1016/j.cirp.2012.03.084

Carbo, J., Langella, I. M., Dao, V. T., & Haase, S. J. (2014). Breaking the ties that bind: From corporate sustainability to socially sustainable systems. *Business and Society Review, 119*(2), 175–206. doi:10.1111/basr.12029

Carbone, V., Moatti, V., & Esposito, V. V. (2012). Mapping corporate responsibility and sustainable supply chains: An exploratory perspective. *Business Strategy and the Environment, 21*(7), 475–494. doi:10.1002/bse.1736

Carter, C. R., & Easton, P. L. (2011). Sustainable supply chain management: Evolution and future directions. *International Journal of Physical Distribution & Logistics Management, 41*(1), 46–62. doi:10.1108/09600031111101420

Colicchia, C., Melacini, M., & Perotti, S. (2011). Benchmarking supply chain sustainability: Insights from a field study. *Benchmarking: An International Journal, 18*(5), 705–732. doi:10.1108/14635771111166839

Dinter, B. (2013). Success factors for information logistics strategy: An empirical investigation. *Decision Support Systems, 54*(3), 1207–1218. doi:10.1016/j.dss.2012.09.001

Dobele, A. R., Westberg, K., Steel, M., & Flowers, K. (2014). An examination of corporate social responsibility implementation and stakeholder engagement: A case study in the Australian mining industry. *Business Strategy and the Environment, 23*(3), 145–159. doi:10.1002/bse.1775

Erol, I., Sencer, S., & Sari, R. (2011). A new fuzzy multi-criteria framework for measuring sustainability performance of a supply chain. *Ecological Economics, 70*(6), 1088–1100. doi:10.1016/j.ecolecon.2011.01.001

Gattiker, T. F., Carter, C. R., Huang, X., & Tate, W. L. (2014). Managerial commitment to sustainable supply chain management projects. *Journal of Business Logistics, 35*(4), 318–337. doi:10.1111/jbl.12073

Hassini, E., Surti, C., & Searcy, C. (2012). A literature review and a case study of sustainable supply chains with a focus on metrics. *International Journal of Production Economics, 140*(1), 69–82. doi:10.1016/j.ijpe.2012.01.042

Hofmann, H., Busse, C., Bode, C., & Henke, M. (2014). Sustainability-related supply chain risks: Conceptualization and management. *Business Strategy and the Environment, 23*(3), 160–172. doi:10.1002/bse.1778

Kashmanian, R. M., & Moore, J. R. (2014). Building greater sustainability in supply chains. *Environmental Quality Management, 23*(4), 13–37. doi:10.1002/tqem.21376

Lee, K. H., & Saen, R. F. (2012). Measuring corporate sustainability management: A data envelopment analysis approach. *International Journal of Production Economics, 140*(1), 219–226. doi:10.1016/j.ijpe.2011.08.024

Mamede, P., & Gomes, C. F. (2014). Corporate sustainability measurement in service organizations: A case study from Portugal. *Environmental Quality Management, 23*(3), 49–73. doi:10.1002/tqem.21370

Manning, L. (2013). Corporate and consumer social responsibility in the food supply chain. *British Food Journal, 115*(1), 9–29. doi:10.1108/00070701311289858

Mefford, R. N. (2011). The economic value of a sustainable supply chain. *Business and Society Review, 116*(1), 109–143. doi:10.1111/j.1467-8594.2011.00379.x

Ngwakwe, C. C. (2012). Rethinking the accounting stance on sustainable development. *Sustainable Development, 20*(1), 28–41. doi:10.1002/sd.462

Rowe, A. L., Nowak, M., Quaddus, M., & Naude, M. (2014). Stakeholder engagement and sustainable corporate community investment. *Business Strategy and the Environment, 23*(7), 461–474. doi:10.1002/bse.1796

Santos, F. C. A., Andrade, E. M., Ferreira, A. C., Leme, P. C. S., & Jabbour, C. J. C. (2013). Practices of environmentally responsible reverse logistics systems in Brazilian companies. *International Journal of Business Performance and Supply Chain Modelling*, *5*(1), 63–85. doi:10.1504/IJBPSCM.2013.051655

Seuring, S. (2013). A review of modeling approaches for sustainable supply chain management. *Decision Support Systems*, *54*(4), 1513–1520. doi:10.1016/j.dss.2012.05.053

Singh, R., & Debnath, M. R. (2012). Modeling sustainable development: India's strategy for the future. *World Journal of Science, Technology and Sustainable Development, 9*(2), 120–135. doi: 10.1108/20425941211244270

Svensson, G., & Wagner, B. (2012). Implementation of a sustainable business cycle: The case of a Swedish dairy producer. *Supply Chain Management: An International Journal*, *17*(1), 93–97. doi:10.1108/13598541211212230

Tonelli, F., Evans, S., & Taticchi, P. (2013). Industrial sustainability: Challenges, perspectives, actions. *International Journal of Business Innovation Research*, *7*(2), 143–163. doi:10.1504/IJBIR.2013.052576

Touboulic, A., Chicksand, D., & Walker, H. (2014). Managing imbalanced supply chain relationships for sustainability: A power perspective. *Decision Sciences*, *45*(4), 577–619. doi:10.1111/deci.12087

Wu, Z., & Pagell, M. (2011). Balancing priorities: Decision-making in sustainable supply chain management. *Journal of Operations Management*, *29*(6), 577–590. doi:10.1016/j.jom.2010.10.001

Yakovleva, N., Sarkis, J., & Sloan, T. (2012). Sustainable benchmarking of supply chains: The case of the food industry. *International Journal of Production Research*, *50*(5), 1297–1317. doi:10.1080/00207543.2011.571926

KEY TERMS AND DEFINITIONS

Competitive Advantage: A superiority gained by an organization when it can provide the same value as its competitors but at a lower price, or can charge higher prices by providing greater value through differentiation.

Logistics: The planning, execution, and control of the procurement, movement, and stationing of personnel, material, and other resources to achieve the objectives of a campaign, plan, project, or strategy.

Stakeholder: A person who has an interest in or investment in something and who is impacted by and cares about how it turns out.

Supply Chain: An entire network of entities, directly or indirectly interlinked and interdependent in serving the same customer.

Supply Chain Management: The management of material and information flow in a supply chain to provide the highest degree of customer satisfaction at the lowest possible cost.

Sustainability: A business strategy that drives long-term corporate growth and profitability by mandating the inclusion of environmental and social issues in the business model.

Triple Bottom Line: The economic, social, and environmental effects of an organization's policies and actions that determine its viability as a sustainable organization.

Compilation of References

Abbasi, M., & Nilsson, F. (2012). Themes and challenges in making supply chains environmentally sustainable. *Supply Chain Management*, *17*(5), 517–530. doi:10.1108/13598541211258582

Abdulmalek, F., & Rajgopal, J. (2007). Analyzing the benefits of lean manufacturing and value stream mapping via simulation: A process sector case study. *International Journal of Production Economics*, *107*(1), 223–236. doi:10.1016/j.ijpe.2006.09.009

Abu Seman, N. A., Zakuan, N., Jusoh, A., & Mohd, S. M. A. (2012). Green Supply Chain Management: A Review and Research Direction. *International Journal of Managing Value and Supply Chains*, *3*(1).

Accorsi, R., Cascini, A., Cholette, S., Manzini, R., & Mora, C. (2014). Economic and environmental assessment of reusable plastic containers: A food catering supply chain case study. *International Journal of Production Economics*, *152*, 88–101. doi:10.1016/j.ijpe.2013.12.014

Agarwal, P., Sahai, M., Mishra, V., Bag, M., & Singh, V. (2011). A review of multi-criteria decision making techniques for supplier evaluation and selection. *International Journal of Industrial Engineering Computations*, *2*(4), 801–810. doi:10.5267/j.ijiec.2011.06.004

Ageron, B., Gunasekaran, A., & Spalanzani, A. (2012). Sustainable supply chain management: An empirical study. *International Journal of Production Economics*, *140*(1), 168–182. doi:10.1016/j.ijpe.2011.04.007

Ahi, P., & Searcy, C. (2013a). A comparative literature analysis of definitions for green and sustainable supply chain management. *Journal of Cleaner Production*, *52*(1), 329–341. doi:10.1016/j.jclepro.2013.02.018

Ahi, P., & Searcy, C. (2013b). A stochastic approach for sustainability analysis under the green economics paradigm. *Stochastic Environmental Research and Risk Assessment*, 1–11.

Aissaoui, N., Haouari, M., & Hassini, E. (2007). Supplier selection and order lot sizing modeling: A review. *Computers & Operations Research*, *34*(12), 3516–3540. doi:10.1016/j.cor.2006.01.016

Al Zaabi, S., Al Dhaheri, N., & Diabat, A. (2013). Analysis of interaction between the barriers for the implementation of sustainable supply chain management. *International Journal of Advanced Manufacturing Technology*, *68*(1/4), 895–905. doi:10.1007/s00170-013-4951-8

Ali, D. (2011). An analysis of the drivers affecting the implementation of green supply chain management. *Resources, Conservation and Recycling*, *55*(6), 659–667. doi:10.1016/j.resconrec.2010.12.002

Ambec, S., & Lanoie, P. (2008). Does it pay to be green? A systematic overview. *The Academy of Management Perspectives*, *22*(4), 45–62. doi:10.5465/AMP.2008.35590353

Amini, M., & Bienstock, C. C. (2014). Corporate sustainability: An integrative definition and framework to evaluate corporate practice and guide academic research. *Journal of Cleaner Production*, *76*, 12–19. doi:10.1016/j.jclepro.2014.02.016

Anand, G., & Kodali, R. (2009). Selection of lean manufacturing systems using the analytic network process – a case study. *Journal of Manufacturing Technology Management*, *20*(2), 258–289. doi:10.1108/17410380910929655

Andersen, M., & Skjoett-Larsen, T. (2009). Corporate social responsibility in global supply chains. *Supply Chain Management: An International Journal*, *14*(2), 75–86. doi:10.1108/13598540910941948

Angeles, R. (2013). Using the Technology-Organization-Environment Framework and Zuboff's Concepts for Understanding Environmental Sustainability and RFID: Two Case Studies. *International Journal of Social. Human Science and Engineering*, *7*(11), 1607–1617.

Angell, L. C., & Klassen, R. D. (1999). Integrating environmental issues into the mainstream: An agenda for research in operations management. *Journal of Operations Management*, *17*(5), 575–598. doi:10.1016/S0272-6963(99)00006-6

Anvari, A., Zulkifli, N., Sorooshian, S., & Boyerhassani, O. (2014). An integrated design methodology based on the use of group AHP-DEA approach for measuring lean tools efficiency with undesirable output. *International Journal of Advanced Manufacturing Technology*, *70*(9-12), 2169–2186. doi:10.1007/s00170-013-5369-z

Araz, C., & Ozkarahan, I. (2007). Supplier evaluation and management system for strategic sourcing based on a new multicriteria sorting procedure. *International Journal of Production Economics*, *106*(2), 585–606. doi:10.1016/j.ijpe.2006.08.008

Arıkan, E., & Jammernegg, W. (2014). The single period inventory model under dual sourcing and product carbon footprint constraint. *International Journal of Production Economics*. DOI: 10.106/j.ijpe.2014.04.027

Arnold, M. G., & Hockerts, K. (2011). The greening Dutchman: Philips' process of green flagging to drive sustainable innovations. *Business Strategy and the Environment*, *20*(6), 394–407. doi: 10.1002/bse.700

Arslan, M., & Turkay, M. (2013). EOQ revisited with sustainability considerations. *Foundations of Computing and Decision Sciences*, *38*(4), 223–249. doi:10.2478/fcds-2013-0011

Ashby, A., Leat, M., & Hudson-Smith, M. (2012). Making connections: A review of supply chain management and sustainability literature. *Supply Chain Management: An International Journal*, *17*(5), 497–516. doi:10.1108/13598541211258573

Attri, R., Dev, N., & Sharma, V. (2013). Interpretive Structural Modeling (ISM) approach: An overview. *Research Journal of Management Sciences*, *2*(2), 3–8.

Azadi, M., Jafarian, M., Saen, R. F., & Mirhedayatian, S. M. (2015). A new fuzzy DEA model for evaluation of efficiency and effectiveness of suppliers in sustainable supply chain management context. *Computers & Operations Research*, *54*, 274–285. doi:10.1016/j.cor.2014.03.002

Azevedo, S., Carvalho, H., & Machado, C. V. (2011). The influence of green practices on supply chain performance: A case study approach. *Transportation Research Part E, Logistics and Transportation Review*, *47*(6), 850–871. doi:10.1016/j.tre.2011.05.017

Bag, S. (2013). Designing the Green Supply Chain Strategies for Indian Manufacturing Firm. *International Journal of Supply Chain Management Systems*, *2*(1), 8–18.

Bag, S. (2013). Designing the green supply chain strategy for Indian manufacturing firm. *Journal of Supply Chain Management Systems*, *2*(1), 8–18.

Bag, S., & Anand, N. (2012). Green Procurement practices and its impact on organization performance: A conceptual framework of Indian manufacturing sector. *UPES Management Review*, *1*(2), 3–11.

Bai, C., & Sarkis, J. (2010). Integrating sustainability into supplier selection with grey system and rough set methodologies. *International Journal of Production Economics*, *124*(1), 252–264. doi:10.1016/j.ijpe.2009.11.023

Bai, C., Sarkis, J., Wei, X., & Koh, L. (2012). Evaluating ecological sustainable performance measures for supply chain management. *Supply Chain Management: An International Journal*, *17*(1), 78–92. doi:10.1108/13598541211212221

Balasubramanian, S. (2012). A Hierarchical Framework of Barriers to Green Supply Chain Management in the Construction Sector. *Journal of Sustainable Development*, *5*(10), 1913-9063.

Balcik, B., Beamon, B. M., & Smilowitz, K. (2008). Last mile distribution in humanitarian relief. *Journal of Intelligent Transportation Systems*, *12*(2), 51–63. doi:10.1080/15472450802023329

Banbury, J. G. (1975). Distribution—the final link in the electricity-supply chain. *Electronics and Power*, *21*(13), 773. doi:10.1049/ep.1975.0796

Banerjee, S. B., & Bonnefous, A. M. (2011). Stakeholder management and sustainability strategies in the French nuclear industry. *Business Strategy and the Environment*, *20*(2), 124–140. doi:10.1002/bse.681

Bansal, P. (2005). Evolving sustainability: A longitudinal study of corporate sustainable development. *Strategic Management Journal*, *26*(3), 197–218. doi:10.1002/smj.441

Barari, S., Gaurav, A., Zhang, C., Mahanty, B., & Tiwari, M.K. (2012). A decision framework for the analysis of green supply chain contracts: An evolutionary game approach. *Experts Systems with Applications, 39*, 2965-2976.

Barrett, H. R., Ilbery, B. W., Brown, A. W., & Binns, T. (1999). Globalization and the changing networks of food supply: The importation of fresh horticultural produce from Kenya into the UK. *Transactions of the Institute of British Geographers, 24*(2), 159–174. doi:10.1111/j.0020-2754.1999.00159.x

Bask, A., Halme, M., Kallio, M., & Kuula, M. (2013). Consumer preferences for sustainability and their impact on supply chain management: The case of mobile phones. *International Journal of Physical Distribution & Logistics Management, 43*(5/6), 380–406. doi:10.1108/IJPDLM-03-2012-0081

Battini, D., Persona, A., & Sgarbossa, F. (2014). A sustainable EOQ model: Theoretical formulation and applications. *International Journal of Production Economics, 149*, 145–153. doi:10.1016/j.ijpe.2013.06.026

Baumgartner, R. J. (2009). Organizational culture and leadership: Preconditions for the development of a sustainable corporation. *Sustainable Development, 17*(2), 102–113. doi:10.1002/sd.405

Bazan, E., Jaber, M. Y., & Zanoni, S. (2016). A review of mathematical inventory models for reverse logistics and the future of its modeling: An environmental perspective. *Applied Mathematical Modelling, 40*(5), 4151–4178.

Beamon, B. M. (1999). Designing the green supply chain. *Logistics Information Management, 12*(4), 332-342.

Beamon, B. M. (1999). Designing the green supply chain. *Logistics Information Management, 12*(4), 332–342. doi:10.1108/09576059910284159

Bei, W., & Linyan, S. (2005). A review of reverse logistics. *Applied Sciences, 7*, 16–29.

Ben Brik, A., Mellahi, K., & Rettab, B. (2013). Drivers of Green Supply Chain in Emerging Economies. *Thunderbird International Business Review, 55*(2), 123–136. doi:10.1002/tie.21531

Benjaafar, S., Li, Y., & Daskin, M. (2010). Carbon footprint and the management of supply chains: Insights from simple models. *IEEE Transactions on Automation Science and Engineering, 10*(1), 99–116. doi:10.1109/TASE.2012.2203304

Berezan, O., Raab, C., Yoo, M., & Love, C. (2013). Sustainable hotel practices and nationality: The impact on guest satisfaction and guest intention to return. *International Journal of Hospitality Management, 34*(2), 227–233. doi:10.1016/j.ijhm.2013.03.010

Bergmiller, G. G. (2006). *Lean manufacturers transcendence to green manufacturing: Correlating the diffusion of lean and green manufacturing systems. (PhD)*. University of South Florida.

Berns, M., Townend, A., Khayat, Z., Balagopal, B., Reeves, M., Hopkins, M. & Kruschwitz, N. (2009). The business of sustainability.What it means to managers now. *MIT Sloan Management Review*, 20–26.

Berry, M. A., & Rondinelli, D. A. (1998). Proactive corporate environment management: A new industrial revolution. *The Academy of Management Executive*, *12*(2), 38–50.

Beske, P. (2012). Dynamic capabilities and sustainable supply chain management. *International Journal of Physical Distribution & Logistics Management*, *42*(4), 372–387. doi:10.1108/09600031211231344

Beske, P., & Seuring, S. (2014). Putting sustainability into supply chain management. *Supply Chain Management: An International Journal*, *19*(3), 322–331. doi: 10.1108/SCM-12-2013-0432

Bhateja, A. K., Babbar, R., Sarbjit, S., & Sachdeva, A. (2012). Study of the critical factor finding's regarding evaluation of green supply chain performance of Indian scenario for manufacturing sector. *International Journal of Computational Engineering& Management*, *15*(1), 74–80.

Bhatja, A. K., Babbar, R., & Zhu, Q. (2011, July). Study of green supply chain management in Indian manufacturing industries: A literature review cum an analytical approach for the measurement of performance. *International Journal of Computational Engineering& Management*, *13*, 84–99.

Bicheno, J. (2004). *The new lean toolbox: towards fast, flexible flow*. Buckingham, UK: PICSIE Books.

Bilgiç, R., & Yilmaz, N. (2013). The correlates of psychological health among the turkish unemployed: Psychological burden of financial help during unemployment. *International Journal of Psychology*, *48*(5), 1000–1008. doi:10.1080/00207594.2012.717702 PMID:23151222

Blackburn, J. D., Guide, V. D. R. Jr, Souza, G. C., & Van Wassenhove, L. N. (2004). Reverse Supply Chains for Commercial Returns. *California Management Review*, *46*(2), 6–22. doi:10.2307/41166207

Bonney, M., & Jaber, M. Y. (2011). Environmentally responsible inventory models: Non-classical models for a non-classical era. *International Journal of Production Economics*, *133*(1), 43–53. doi:10.1016/j.ijpe.2009.10.033

Bonney, M., & Jaber, M. Y. (2013). Developing an input–output activity matrix (IOAM) for environmental and economic analysis of manufacturing systems and logistics chains. *International Journal of Production Economics*, *143*(2), 369–377. doi:10.1016/j.ijpe.2011.12.016

Boon-itt, S., & Wong, C. W. (2011). The moderating effects of technological and demand uncertainties on the relationship between supply chain integration and customer delivery performance. *International Journal of Physical Distribution & Logistics Management*, *41*(3), 253–276.

Bos-Brouwers, H. E. J. (2010). Corporate sustainability and innovation in SMEs: Evidence of themes and activities in practice. *Business Strategy and the Environment*, *19*(7), 417–435. doi: 10.1002/bse.652

Bose, I., & Pal, R. (2012). Do green supply chain management initiatives impact stock prices of firms? *Decision Support Systems*, *52*(3), 624–634. doi:10.1016/j.dss.2011.10.020

Bouchery, Y., Ghaffari, A., Jemai, Z., & Dallery, Y. (2012). Including sustainability criteria in inventory models. *European Journal of Operational Research*, *222*(2), 229–240. doi:10.1016/j.ejor.2012.05.004

Boyer, K. K., Prud'homme, A. M., & Chung, W. (2009). The last mile challenge: Evaluating the effects of customer density and delivery window patterns. *Journal of Business Logistics*, *30*(1), 185–201. doi:10.1002/j.2158-1592.2009.tb00104.x

Bozorgi, A., Pazour, J., & Nazzal, D. (2014). A new inventory model for cold items that considers costs and emissions. *International Journal of Production Economics*. DOI:10.1016/j/ijpe.2014.01.006

Brammer, S., Hoejmose, S., & Marchant, K. (2012). Environmental management in SMEs in the UK: Practices, pressures and perceived benefits. *Business Strategy and the Environment*, *21*(7), 423–434. doi:10.1002/bse.717

Brandenburg, M., Govindan, K., Sarkis, J., & Seuring, S. (2014). Quantitative models for sustainable supply chain management: Developments and directions. *European Journal of Operational Research*, *233*(2), 299–312. doi:10.1016/j.ejor.2013.09.032

Brown, J. R., & Guiffrida, A. L. (2014). Carbon emissions comparison of last mile delivery versus customer pick up. *International Journal of Logistics and Applications*, *17*(6), 503–521. doi:10.1080/13675567.2014.907397

Brown, K. A. (1996). Workplace safety: A call for research. *Journal of Operations Management*, *14*(2), 157–161. doi:10.1016/0272-6963(95)00042-9

Brunt, D., & Butterworth, C. (2001). Waste elimination–a supply chain perspective. In D. Taylor & D. Brunt (Eds.), *Manufacturing Operations and Supply Chain Management, The LEAN Approach* (pp. 79–87). Thomson Learning.

Bruun, P., & Mefford, R. N. (2004). Lean production and the Internet. *International Journal of Production Economics*, *89*(3), 247–260. doi:10.1016/j.ijpe.2003.10.007

Bryceson, K. P. (2006).*E'Issues for Agribusiness: The 'what', 'why', 'how*. CABI.

Buckholz, R. (1991). Corporate responsibility and the good society: From economics to ecology. *Business Horizons*, *34*(4), 19–31. doi:10.1016/0007-6813(91)90003-E

Burgess, K., Singh, P. J., & Koroglu, R. (2006). Supply chain management: A structured literature review and implications for future research. *International Journal of Operations & Production Management*, *26*(7), 703–729. doi:10.1108/01443570610672202

Burns, J. F., & Sivazlian, B. D. (1978). Dynamic analysis of multi-echelon supply systems. *Computers & Industrial Engineering*, 2(4), 181–193. doi:10.1016/0360-8352(78)90010-4

Byrne, P. M., & Deeb, A. (1993). Logistics must meet the 'green' challenge. *Transportation & Distribution, 34*(2), 33–37.

Cai, J., Liu, X., Xiao, Z., & Liu, J. (2009). Improving supply chain performance management: A systematic approach to analyzing iterative KPI accomplishment. *Decision Support Systems, 46*(2), 512–521. doi:10.1016/j.dss.2008.09.004

Caniato, F., Caridi, M., Crippa, L., & Moretto, A. (2011). Environmental sustainability in fashion supply chains: An exploratory case based research. *International Journal of Production Economics, 135*(2), 659–670. doi:10.1016/j.ijpe.2011.06.001

Caniels, M. C. J., Gehrsitz, M. H., & Semeijn, J. (2013). Participation of suppliers in greening supply chains: An empirical analysis of German automotive suppliers. *Journal of Purchasing and Supply Management, 19*(3), 134–143. doi:10.1016/j.pursup.2013.02.005

Cantor, D. E., Corsi, T. M., & Grimm, C. M. (2006). Safety technology adoption patterns in the US motor carrier industry. *Transportation Journal, 45*(3), 20–45.

Cantor, D. E., Morrow, P. C., & Montabon, F. (2012). Engagement in environmental behaviors among supply chain management employees: An organizational support theoretical perspective. *Journal of Supply Chain Management, 48*(3), 33–51. doi:10.1111/j.1745-493X.2011.03257.x

Capaldi, N. (2005). Corporate social responsibility and the bottom line. *International Journal of Social Economics, 32*(5), 408–423. doi:10.1108/03068290510591263

Carbon Trust. (2006). *Carbon Footprint in the supply chain: The next step for business*. Report No. CTC616. Carbon Trust London. Retrieved from: https://www.carbontrust.com/media/84932/ctc616-carbon-footprints-in-the-supply-chain.pdf

Carter, R. (1995, April). The seven Cs of effective suppler evaluation. *Purchasing and Supply Management*, 44-45.

Carter, C. R., Auskalnis, R., & Ketchum, C. (1999). Purchasing from minority business enterprises: A cross-industry comparison of best practices. *Journal of Supply Chain Management, 35*(1), 28–32. doi:10.1111/j.1745-493X.1999.tb00053.x

Carter, C. R., & Dresner, M. (2001). Purchasing's Role in Environmental Management: Cross-Functional Development of Grounded Theory. *Journal of Supply Chain Management, 37*(2), 12–27. doi:10.1111/j.1745-493X.2001.tb00102.x

Carter, C. R., & Easton, P. L. (2011). Sustainable supply chain management: Evolution and future directions. *International Journal of Physical Distribution & Logistics Management, 41*(1), 46–62.

Carter, C. R., & Ellram, L. M. (1998). Reverse logistics: A review of the literature and framework for future investigation. *Journal of Business Logistics, 19*(1), 85–102.

Carter, C. R., Ellram, L. M., & Tate, W. L. (2007). Structure and influence: A logistics management application of social network analysis. *Journal of Business Logistics, 28*(1), 137–168. doi:10.1002/j.2158-1592.2007.tb00235.x

Carter, C. R., & Rogers, D. S. (2008). A framework of sustainable supply chain management: Moving toward new theory. *International Journal of Physical Distribution & Logistics Management, 38*(5), 360–387. doi:10.1108/09600030810882816

Castro, N. R., & Chousa, J. P. (2006). An integrated framework for the financial analysis of sustainability. *Business Strategy and the Environment, 15*(5), 322–333. doi:10.1002/bse.539

Cetinkaya, B. (2011). Developing a Sustainable Supply Chain Strategy. In *Sustainable supply chain management: Practical ideas for moving towards best practice* (pp. 17–55). Worcester, MA: Springer.

Cetinkaya, B., Cuthbertson, R., Ewer, G., Klaas-Wissing, T., Piotrowicz, W., & Tyssen, C. (2011). *Sustainable supply chain management*. New York, NY: Springer–Verlag.

Chaabane, A., Ramudhin, A., & Paquet, M. (2012). Design of sustainable supply chains under the emission trading scheme. *International Journal of Production Economics, 135*(1), 37–49. doi:10.1016/j.ijpe.2010.10.025

Chai, J., Liu, J. N. K., & Ngai, E. W. T. (2013). Application of decision making-techniques in supplier selection: A systematic review of literature. *Expert Systems with Applications, 40*(10), 3872–3885. doi:10.1016/j.eswa.2012.12.040

Chang, C. H. (2011). The influence of corporate environmental ethics on competitive advantage: The mediation role of green innovation. *Journal of Business Ethics, 104*(3), 361–370. doi:10.1007/s10551-011-0914-x

Chan, R. Y. K., & Lau, L. B. Y. (2001). Explaining green purchasing behaviour: A cross-cultural study on American and Chinese consumers. *Journal of International Consumer Marketing, 14*(2/3), 9–41.

Chao, P. (1998). Impact of country-of-origin dimensions on product quality and design quality perceptions. *Journal of Business Research, 42*(1), 1–6. doi:10.1016/S0148-2963(97)00129-X

Chen, M.K., Wai, T-W. & Hung, T-Y. (2012). Component selection system for green supply chain. *Experts Systems with Applications, 39*, 5687-5701.

Chen, I. J., & Paulraj, A. (2004). Towards a theory of supply chain management: The constructs and measurements. *Journal of Operations Management, 22*(2), 119–150. doi:10.1016/j.jom.2003.12.007

Chen, J. C., Cheng, C.-H., Huang, P. T. B., Wang, K.-J., Huang, C.-J., & Ting, T.-C. (2013). Warehouse management with lean and RFID application: A case study. *International Journal of Advanced Manufacturing Technology, 69*(1-4), 531–542. doi:10.1007/s00170-013-5016-8

Chen, X., Benjaafar, S., & Elomri, A. (2013). The carbon-constrained EOQ. *Operations Research Letters, 41*(2), 172–179. doi:10.1016/j.orl.2012.12.003

Chiang, C. Y., Kocabasoglu-Hillmer, C., & Suresh, N. (2012). An empirical investigation of the impact of strategic sourcing and flexibility on firm's supply chain agility. *International Journal of Operations & Production Management*, *32*(1), 49–78. doi:10.1108/01443571211195736

Chiarini, A. (2014). Strategies for developing an environmentally sustainable supply chain: Differences between manufacturing and service sectors. *Business Strategy and the Environment*, *23*(7), 493–504. doi:10.1002/bse.1799

Chiou, T. Y., Chan, H. K., Lettice, F., & Chung, S. H. (2011). The Influence of Greening the Suppliers and Green Innovation on Environmental Performance and Competitive Advantage in Taiwan. *Transportation Research Part E, Logistics and Transportation Review*, *47*, 822–836.

Cho, D. W., Lee, Y. H., Ahn, S. H., & Hwang, M. K. (2012). A framework for measuring the performance of service supply chain management. *Computers & Industrial Engineering*, *62*(3), 801–818. doi:10.1016/j.cie.2011.11.014

Choi, T.-M., & Chiu, C.-H. (2012). Mean-downside-risk and mean-variance newsvendor models: Implications for sustainable fashion retailing. *International Journal of Production Economics*, *135*(2), 552–560. doi:10.1016/j.ijpe.2010.10.004

Chopra, S. (2003). Designing the distribution network in a supply chain. *Transportation Research*, *39*(2), 123–140. doi:10.1016/S1366-5545(02)00044-3

Chopra, S., & Meindl, P. (2010). *Supply chain management: strategy, planning, and operation, 4th*. Upper Saddle River, NJ: Pearson Education, Inc.

Chopra, S., & Peter, M. (2003). *'Supply chain* (2nd ed.). Upper Saddle River, NJ: Prentice-Hall.

Chow, W. S., Madu, C. N., Kuei, C.-H., Lu, M. H., Lin, C., & Tseng, H. (2008). 'Supply Chain Management in the US and Taiwan: An empirical study', Omega. *International Journal of Management Sciences*, *36*, 665–679.

Christmann, P. (2000). Effects of "best practices" of environmental management on cost advantage: The role of complementary assets. *Academy of Management Journal*, *43*(4), 663–680. doi:10.2307/1556360

Comas Martí, J. M., & Seifert, R. W. (2013). Assessing the comprehensiveness of supply chain environmental strategies. *Business Strategy and the Environment*, *22*(2), 339–356. doi:10.1002/bse.1749

Connelly, B. L., Ketchen, D. J., & Hult, G. T. M. (2013). Global supply chain management: Toward a theoretically driven research agenda. *Global Strategy Journal*, *3*(3), 227–243. doi:10.1111/j.2042-5805.2013.01041.x

Consumer. (n.d.). *Consumers Association of Penang*. Retrieved from: www.consumer.org.my

Cooper, M. C., Lambert, D. M., & Pagh, J. D. (1997). Supply chain management – more than a new name for logistics. *International Journal of Logistics Management*, *8*(1), 1–14. doi:10.1108/09574099710805556

Coronado, J. L. (2007). *An Optimization Model for Strategic Supply Chain Design under Stochastic Capacity Disruptions*. (Dissertation). Department of Industrial and Systems Engineering, Texas A&M University, College Station, TX.

Corsi, T. M., Tuck, J. M., & Gardner, L. L. (1982). Minority motor carriers and the Motor Carrier Act of 1980. *Transportation Journal, 22*(1), 42–55.

Costanza, R., & Daly, H. E. (1992). Natural capital and sustainable development. *Conservation Biology, 6*(1), 37–46. doi:10.1046/j.1523-1739.1992.610037.x

Cousins, P. D., & Lawson, B. (2007). Sourcing strategy, supplier relationships and firm performance: An empirical investigation of UK organizations. *British Journal of Management, 18*(2), 123–137. doi:10.1111/j.1467-8551.2006.00506.x

Cravens, D. W., Piercy, N. F., & Shipp, S. H. (1996). New organizational forms for competing in highly dynamic environments: The network paradigm. *British Journal of Management, 7*(3), 203–218. doi:10.1111/j.1467-8551.1996.tb00115.x

Croom, S., Romano, P., & Ginnakis, M. (2000). Supply chain management: An analytic framework for critical literature review. *European J. of Purchasing & Supply Management, 6*(1), 67–83. doi:10.1016/S0969-7012(99)00030-1

Crum, M. R., Dooley, F. J., & Morrow, P. C. (1995). Employee attitudes about railroad injury compensation. *Transportation Journal, 35*(1), 15–29.

CSCMP – Supply Chain Management. (n.d.). *Council of Supply Chain Management Professionals*. Retrieved from http://cscmp.org/about-us/supply-chain-management-definitions

CSCSC – Supply Chain Definitions. (n.d.). *Canadian Supply Chain Sector Council*. Retrieved from http://www.supplychaincanada.org/en/supply-chain

Cullinane, K., & Toy, N. (2000). Identifying influential attributes in freight route/mode choice decisions: A content analysis. *Transportation Research Part E, Logistics and Transportation Review, 36*, 41–53.

Curkovic, S., & Sroufe, R. (2011). Using ISO 14001 to promote a sustainable supply chain strategy. *Business Strategy and the Environment, 20*(2), 71–93. doi:10.1002/bse.671

Dadhich, P., Genovese, A., Kumar, N., & Acquaye, A. (2015). Developing sustainable supply chains in the UK construction industry: A case study. *International Journal of Production Economics, 164*, 271–284. doi:10.1016/j.ijpe.2014.12.012

Daily, B. F., & Huang, S. (2001). Achieving sustainability through attention to human resource factors in environmental management. *International Journal of Operations & Production Management, 21*(12), 1539–1552. doi:10.1108/01443570110410892

Dailymail. (2008). *Cadbury recall all chocolate made in China over fears it may contain contaminated milk*. Retrieved 29 October 2012 from: https://www.dailymail.co.uk/news/article-1063950/cadbury-recall-chocolate-China-fears-contain-contaminated-milk.html

Daly, H. E. (1990). Toward some operational principles of sustainable development. *Ecological Economics*, *2*(1), 1–6. doi:10.1016/0921-8009(90)90010-R

Dangelio, R. M., & Pontrandolfo, P. (2010). From green product definitions and classifications to Green Option Matrix. *Journal of Cleaner Production*, *18*(16/17), 1608–1628. doi:10.1016/j.jclepro.2010.07.007

Darnall, N., Jolley, G. J., & Handfield, R. (2008). Environmental management systems and green supply chain management: Complements for sustainability? *Business Strategy and the Environment*, *17*(1), 30–45. doi:10.1002/bse.557

Das, T. K. (2006). Strategic alliance temporalities and partner opportunism. *British Journal of Management*, *17*(1), 1–21. doi:10.1111/j.1467-8551.2006.00482.x

De Boer, L., Labro, E., & Morlacchi, P. (2001). A review of methods supporting supplier selection. *European Journal of Purchasing & Supply Management*, *7*(2), 75–89. doi:10.1016/S0969-7012(00)00028-9

De Brito, M. P., & Dekker, R. (2003). A Framework for Reverse Logistics [R]. ERIM Report Series[-LIS]. *Research Management*, ERS-2003–ERS-2045.

Degraeve, Z., Labro, E., & Roodhooft, F. (2000). An evaluation of vendor selection models from a total cost of ownership perspective. *European Journal of Operational Research*, *125*(1), 34–58. doi:10.1016/S0377-2217(99)00199-X

Deif, A. M. (2011). A system model for green manufacturing. *Journal of Cleaner Production*, *19*(14), 1553–1559. doi:10.1016/j.jclepro.2011.05.022

Dekker, R., Bloemhof, J., & Mallidis, I. (2012). Operations research for green logistics –An overview of aspects, issues, contributions and challenges. *European Journal of Operational Research*, *219*(3), 671–679. doi:10.1016/j.ejor.2011.11.010

DeLaet, C. (1985). Developing networks and institutions in support of sustainable development. *Landscape Planning*, *12*(3), 301–309. doi:10.1016/0304-3924(85)90008-5

Delmas, M. (2001). Stakeholders and competitive advantage: The case of ISO 14001. *Production and Operations Management*, *10*(3), 343–358. doi:10.1111/j.1937-5956.2001.tb00379.x

Demir, E., Bektaş, T., & Laporte, G. (2014). A review of recent research on green road freight transportation. *European Journal of Operational Research*, *237*(3), 775–793. doi:10.1016/j.ejor.2013.12.033

Dhanda, K. K., & Hill, R. P. (2005). The role of information technology and systems in reverse logistics: A case study. *International Journal of Technology Management*, *31*(1/2), 140–151. doi:10.1504/IJTM.2005.006628

Diabat, A., Govindan, K., & Panicker, V. V. (2012). Supply chain risk management and its mitigation in a food industry. *International Journal of Production Research*, *50*(11), 3039–3050. do i:10.1080/00207543.2011.588619

Diabat, A., Kannan, D., & Mathiyazhagan, K. (2014). Analysis of enablers for implementation of sustainable supply chain management: A textile case. *Journal of Cleaner Production*, *83*(1), 391–403. doi: 10.1016/j.jclepro.2014.06.081

Dickson, G. W. (1966). An analysis of vendor selection systems and decisions. *Journal of Purchasing*, *2*(1), 5–17.

Digalwar, A. K., & Metri, B. A. (2004). Performance Measurement Framework for World Class Manufacturing. *International Journal of Applied Management and Technology*, *3*(2), 83–101.

Dobrzykowski, D. D., Tran, O. T., & Hong, P. (2011). Insights into integration for supply chain redesign in service and product-focused firms. *International Journal of Services and Operations Management*, *8*(3), 260–282. doi:10.1504/IJSOM.2011.038972

Dolcemascolo, D. (2006). *Improving the extended value stream: lean for the entire supply chain*. Productivity Press.

Dombrowski, U., Mielke, T., & Schulze, S. (2012). Lean Production Systems as a Framework for Sustainable Manufacturing. In G. Seliger (Ed.), *Sustainable Manufacturing* (pp. 17–22). Springer Berlin Heidelberg. doi:10.1007/978-3-642-27290-5_3

Dos Santos, M. A. O., Svensson, G., & Padin, C. (2013). Indicators of sustainable business practices: Woolworths in South Africa. *Supply Chain Management: An International Journal*, *18*(1), 104–108. doi:10.1108/13598541311293212

Dotoli, M., Fanti, M. P., Iacobellis, G., & Rotunno, G. (2012). *A lean manufacturing strategy using Value Stream Mapping, the Unified Modeling Language, and discrete event simulation*. Paper presented at the International Conference on Automation Science and Engineering (CASE), Seoul.

Dowlatshahi, S. (2000). Developing a theory of reverse logistics. *Interfaces*, *30*(3), 143–155. doi:10.1287/inte.30.3.143.11670

Drake, M. J., & Marley, K. A. (2014). A century of the EOQ. In T.-M. Choi (Ed.), *Handbook of EOQ Inventory Problems* (pp. 3–22). Springer. doi:10.1007/978-1-4614-7639-9_1

Drake, P. R., Lee, D. M., & Hussain, M. (2013). The lean and agile purchasing portfolio model. *Supply Chain Management: An International Journal*, *18*(1), 3–20. doi:10.1108/13598541311293140

Drumwright, M. (1994). Socially Responsible Organisational Buying: Environmental Concern as a Non-economic Buying Criterion. *Journal of Marketing*, *58*(3), 1–19. doi:10.2307/1252307

Dubai Electricity and Water Authority. (2011). *Green Buildings Regulations and Specifications*. Retrieved on October 12,2012, from http://www.dewa.gov.ae/images/greenbuilding_eng.pdf

Dubey, R., & Bag, S. (2013). Exploring the dimensions of sustainable practices: An empirical study on Indian manufacturing firms. *International Journal of Operations and Quantitative Management*, *19*(2), 101–124.

Dubey, R., Singh, T., Ali, S. S., & Tiwari, S. (2013). Contextual relationships among antecedents of truck freight using ISM and its validating using MICMAC analysis. *Int. J. Logistics and Systems Management*, *20*(1), 42–58. doi:10.1504/IJLSM.2015.065963

Dubey, R., Sonwaney, V., Aital, P., Venkatesh, V., & Ali, S. S. (2013). Antecedents of innovation and contextual relationship. *Int. J. Innovation and Research*, *9*(1), 1–14. doi:10.1504/IJBIR.2015.065950

Dües, C. M., Tan, K. H., & Lim, M. (2013). Green as the new Lean: How to use Lean practices as a catalyst to greening your supply chain. *Journal of Cleaner Production*, *40*(0), 93–100. doi:10.1016/j.jclepro.2011.12.023

Dyllick, T., & Hockerts, K. (2002). Beyond the business case for corporate sustainability. *Business Strategy and the Environment*, *11*(2), 130–141. doi:10.1002/bse.323

Edgeman, R., & Eskildsen, J. (2014). Modeling and assessing sustainable enterprise excellence. *Business Strategy and the Environment*, *23*(3), 173–187. doi:10.1002/bse.1779

Edwards, J., McKinnon, C., Cherrett, T., McLeod, F., & Song, L. (2010). Carbon dioxide benefits of using collection-delivery points for failed home deliveries in the United Kingdom. *Transportation Research Record: Journal of the Transportation Research Board*, (2191), 136-143.

Edwards, J. B., McKinnon, A. C., & Cullinane, S. L. (2010). Comparative analysis of the carbon footprints of conventional and online retailing: A "last mile" perspective. *International Journal of Physical Distribution & Logistics Management*, *40*(1/2), 103–123. doi:10.1108/09600031011018055

El Saadany, A. M. (2009). *Inventory management in reverse logistics with imperfect production, learning, lost sales, subassemblies, and price/quality considerations*. Toronto: Ryerson University.

El Saadany, A. M. A., & Jaber, M. Y. (2010). A production/remanufacturing inventory model with price and quality dependant return rate. *Computers & Industrial Engineering*, *58*(3), 352–362. doi:10.1016/j.cie.2009.01.017

El Saadany, A. M. A., Jaber, M. Y., & Bonney, M. (2011). Environmental performance measures for supply chains. *Management Research Review*, *34*(11), 1202–1221. doi:10.1108/01409171111178756

El Saadany, A. M. A., Jaber, M. Y., & Bonney, M. (2011). Environmental performance measures for supply chains. *Management Research Review*, *34*(1), 1202–1221.

El-Haik, B., & Al-Aomar, R. (2006). *Simulation-based Lean Six-sigma and Design for Six-sigma*. John Wiley & Sons. doi:10.1002/0470047720

Elkington, J. (1997). *Cannibals with forks: The triple bottom line of the 21st century*. Stony Creek, CT: New Society Publishers.

Elkington, J. (1998). *Cannibals with forks- The triple bottom line of the 21st century*. Stoney Creek, CT: New Society.

Elkington, J. (2006). Governance for sustainability. *Corporate Governance, 14*(6), 522–529. doi:10.1111/j.1467-8683.2006.00527.x

Ellen, P. S., Webb, D. J., & Mohr, L. A. (2006). Building corporate associations: Consumer attributions for corporate social responsibility programs. *Journal of the Academy of Marketing Science, 34*(2), 147–157. doi:10.1177/0092070305284976

Ellinger, A. E., & Ellinger, A. D. (2014). Leveraging human resource development expertise to improve supply chain managers' skills and competencies. *European Journal of Training and Development, 38*(1/2), 118–135.

Eltayeb, T. K., & Zailani, S. (2009). Going green through green supply chain initiatives towards environmental sustainability. *Operations and Supply Chain Management, 2*(2), 93–110.

Eltayeb, T. K., Zailani, S., & Ramayah, T. (2011). Green supply chain initiatives among certified companies in Malaysia and environmental sustainability: Investigating the outcomes. *Resources, Conservation and Recycling, 55*, 495–506.

Emiliani, M. L. (2001). Redefining the focus of investment analysts. *The TQM Magazine, 13*(1), 34–51. doi:10.1108/09544780110359400

EPA. (2003). *Lean Manufacturing and the Environment: Research on Advanced Manufacturing Systems and the Environment and Recommendations for Leveraging Better Environmental Performance*. Washington, DC: United States Environmental Protection Agency.

EPA. (2007). *The Lean and Environment Toolkit (U. S. E. P. Agency, Trans.)*. Washington, DC: U.S. Environmental Protection Agency.

EPA. (2009). *The Environmental Professional's Guide to Lean & Six Sigma*. Washington, DC: U.S. Environmental Protection Agency.

EPA. (n.d.). *About ENERGY STAR*. Retrieved on 24 Mar. 2014, from https://www.energystar.gov/about/

Epstein, M. J., & Widener, S. K. (2011). Facilitating sustainable development decisions: Measuring stakeholder reactions. *Business Strategy and the Environment, 20*(2), 107–123. doi:10.1002/bse.680

Eskandarpour, M., Dejax, P., Miemczyk, J., & Péton, O. (2015). Sustainable supply chain network design: An optimization-oriented review. *Omega: The International Journal of Management Science, 54*, 11–32. doi:10.1016/j.omega.2015.01.006

Esper, T. L., Jensen, T. D., Turnipseed, F. L., & Burton, S. (2003). The last mile: An examination of effects of online retail delivery strategies on consumers. *Journal of Business Logistics, 24*(2), 177–203. doi:10.1002/j.2158-1592.2003.tb00051.x

Fabbe-Costes, N., Roussat, C., & Colin, J. (2011). Future sustainable supply chains: What should companies scan? *International Journal of Physical Distribution & Logistics Management, 41*(3), 228–252. doi:10.1108/09600031111123778

Faruk, A. C., Lamming, R. C., Cousins, P. D., & Bowen, F. E. (2001). Analyzing, mapping, and managing environmental impacts along supply chains. *Journal of Industrial Ecology*, *5*(2), 13–36. doi:10.1162/10881980152830114

Fassoula, E. (2005). Reverse logistics as a means of reducing the cost of quality. *Total Quality Management and Business Excellence*, *16*(5), 631–643. doi:10.1080/14783360500077575

Faust, M. (2013). Cashmere: A lux-story supply chain told by retailers to build a competitive sustainable advantage. *International Journal of Retail & Distribution Management*, *41*(11), 973–985.

FDA, U.S. Food and Drug Administration. (n.d.). *Proposed Changes to the Nutrition Facts Label*. Retrieved on 26 Mar. 2014 from http://www.fda.gov/Food/GuidanceRegulation/GuidanceDocumentsRegulatoryInformation/LabelingNutrition/ucm385663.htm

Feller, A., Shunk, D., & Callarman, T. (2006, March). Value chains versus supply chains. *BPTrends*.

Ferguson, M. E., & Toktay, L. B. (2006). The Effect of Competition on Recovery Strategies. *Production and Operations Management*, *15*(3), 351–368. doi:10.1111/j.1937-5956.2006.tb00250.x

Ferretti, I., Zanoni, S., Zavanella, L., & Diana, A. (2007). Greening the aluminium supply chain. *International Journal of Production Economics*, *108*(1), 236–245. doi:10.1016/j.ijpe.2006.12.037

Fiksel, G. (2003). Designing resilient, sustainable systems. *Environmental Science & Technology*, *37*(23), 5330–5339. doi:10.1021/es0344819 PMID:14700317

Fleischmann, M. (2001). *Reverse logistics network structures and design (No. ERS-2001-52-LIS)*. ERIM Report Series Research in Management.

Fleischmann, M., Bloemhof-Ruwaard, J. M., Dekker, R., Van der Laan, E., Van Nunen, J. A., & Van Wassenhove, L. N. (1997). Quantitative models for reverse logistics: A review. *European Journal of Operational Research*, *103*(1), 1–17. doi:10.1016/S0377-2217(97)00230-0

Fliedner, G. (2008). *Sustainability: a new lean principle.* Paper presented at the 39th annual meeting of the decision sciences institute, Baltimore, MD.

Flynn, B. B., Huo, B., & Zhao, X. (2010). The impact of supply chain integration on performance: A contingency and configuration approach. *Journal of Operations Management*, *28*(1), 58–71.

Fonseca, C. M., & Fleming, P. J. (1995). An overview of evolutionary algorithms in multiobjective optimization. *Evolutionary Computation*, *3*(1), 1–16. doi:10.1162/evco.1995.3.1.1

Francesco, T., & Fabio, I. (2010). Shadows and lights of GSCM (Green Supply Chain Management): Determinants and effects of these practices based on a multi-national study. *Journal of Cleaner Production*, *18*(10), 953–962.

Friedman, T. L. (2005). *The world is flat: A brief history of the twenty-first century*. New York, NY: Farrar, Strauss and Giroux.

Frohlich, M., & Westbrook, R. (2001). Arcs of integration: An international study of supply chain strategies. *Journal of Operations Management, 19*(2), 185–200. doi:10.1016/S0272-6963(00)00055-3

Furterer, S. L. (2009). *Lean Six Sigma in service: Applications and case studies*. Boca Raton, FL: CRC Press. doi:10.1201/9781420079104

Fynes, B., de Burca, S., & Mangan, J. (2008). The effect of relationship characteristics on relationship quality and performance. *International Journal of Production Economics, 111*(1), 56–69. doi:10.1016/j.ijpe.2006.11.019

Gaiardelli, P., Saccani, N., & Songini, L. (2007). Performance measurement of the after-sales service network—Evidence from the automotive industry. *Computers in Industry, 58*(7), 698–708. doi:10.1016/j.compind.2007.05.008

Galbreth, M. R., & Ghosh, B. (2013). Competition and sustainability: The impact of consumer awareness. *Decision Sciences, 44*(1), 127–159. doi:10.1111/j.1540-5915.2012.00395.x

Ganeshan, R., & Harrison, T. (1995). *An introduction to supply chain management*. University Park, PA: Penn State University.

Garetti, M., & Taisch, M. (2012). Sustainable manufacturing: Trends and research challenges. *Production Planning and Control, 2*(2-3), 83–104. doi:10.1080/09537287.2011.591619

Gehin, A., Zwolinski, P., & Brissaud, D. (2008). A tool to implement sustainable end-of-life strategies in the product development phase. *Journal of Cleaner Production, 16*(5), 566–576. doi:10.1016/j.jclepro.2007.02.012

Gevaers, R., Van de Voorde, E., & Vanelslander, T. (2011). *Characteristics and typology of last-mile logistics from an innovative perspective in an urban context*. Paper presented at the Transportation Research Board Annual Meeting, Washington, DC.

Ghadge, A., Dani, S., & Kalawsky, R. (2012). Supply chain risk management: Present and future scope. *International Journal of Logistics Management, 23*(3), 313–339. doi:10.1108/09574091211289200

Gibson, B. J., Mentzer, J. T., & Cook, R. L. (2005). Supply chain management: The pursuit of a consensus definition. *Journal of Business Logistics, 26*(2), 17–25. doi:10.1002/j.2158-1592.2005.tb00203.x

Gilbert, S. (2001). *Greening Supply Chain: Enhancing Competitiveness Through Green Productivity*. Taipei: Academic Press.

Gimenez, C., Sierra, V., & Rodon, J. (2012). Sustainable Operations: Their impact on the triple bottom line. *International Journal of Production Economics, 140*(1), 149–159. doi:10.1016/j.ijpe.2012.01.035

Giunipero, L. C., Hooker, R. E., & Denslow, D. (2012). Purchasing and supply management sustainability: Drivers and barriers. *Journal of Purchasing and Supply Management*, *18*(4), 258–269. doi:10.1016/j.pursup.2012.06.003

Gladwin, T. N., Kennelly, J. J., & Krause, T. (1995). Shifting paradigms for sustainable development: Implications for management theory and research. *Academy of Management Review*, *20*(4), 874–907. doi: 10.2307/258959

Glock, C. H. (2012). The joint economic lot size problem: A review. *International Journal of Production Economics*, *135*(2), 671–686. doi:10.1016/j.ijpe.2011.10.026

Glock, C. H., Jaber, M. Y., & Searcy, C. (2012). Sustainability strategies in an EPQ model with price and quality sensitive demand. *The International Journal of Logistics Management*, *23*(3), 340–359. doi:10.1108/09574091211289219

Gobble, M. M. (2012). Innovation and sustainability. *Research Technology Management*, *55*(5), 64–67. doi:10.5437/08956308X5505005

Godfrey, R. (1998). *Ethical purchasing: Developing the supply chain beyond the environment.* In T. Russel (Ed.), *Greener Purchasing: Opportunities and Innovations* (pp. 244–251). Sheffield, UK: Greenleaf Publishing. doi:10.9774/GLEAF.978-1-909493-03-2_20

Gold, S., Seuring, S., & Beske, P. (2010). Sustainable supply chain management and inter-organizational resources: A literature review. *Corporate Social Responsibility and Environmental Management*, *17*(4), 230–245. doi:10.1002/csr.207

Goldschmidt, K., Harrison, T., Holtry, M., & Reeh, J. (2013). Sustainable procurement: Integrating classroom learning with university sustainability programs. *Decision Sciences Journal of Innovative Education*, *11*(3), 279–294. doi:10.1111/dsji.12007

Golicic, S. L., Boerstler, C. N., & Ellram, L. M. (2010). Greening transportation in the supply chain. *Sloan Management Review*, *51*(2), 47–55.

Golicic, S. L., & Smith, C. D. (2013). A meta-analysis of environmentally sustainable supply chain management practices and firm performance. *Journal of Supply Chain Management*, *49*(2), 78–95. doi:10.1111/jscm.12006

González-Benito, J., & González-Benito, O. (2005). A study of the motivations for the environmental transformation of companies. *Industrial Marketing Management*, *34*(5), 462–475. doi:10.1016/j.indmarman.2004.08.005

Gonzalo, F., & Fernando, R. (2010). A methodology to identify sustainability indicators in construction project management—Application to infrastructure projects in Spain. *Ecological Indicators*, *10*(6), 1193–1201. doi:10.1016/j.ecolind.2010.04.009

Goodman, R. W. (2005). Whatever you call it, just don't think of last-mile logistics, last. *Global Logistics and Supply Chain Strategies*, *9*(12), 46–51.

Gorane, S. J., & Ravi, K. (2013). Modelling the SCM enablers: An integrated ISM-fuzzy MICMAC approach. *Asia Pacific Journal of Marketing and Logistics*, *25*(2), 263–286. doi:10.1108/13555851311314059

Gort, R. E. (2008). *Lean And Sustainability: How Can They Reinforce Each Other? (MBA)*. Tilburg, The Netherlands: Tilburg University and Eindhoven University of Technology.

Gouldson, A., & Murphy, J. (2013). *Regulatory realities: The implementation and impact of industrial environmental regulation*. London, UK: Routledge.

Govindan, K., Jafarian, A., & Nourbakhsh, V. (2015). Bi-objective integrating sustainable order allocation and sustainable supply chain network strategic design with stochastic demand using a novel robust hybrid multi-objective metaheuristic. *Computers & Operations Research*, *62*, 112–130. doi:10.1016/j.cor.2014.12.014

Govindan, K., Kaliyan, M., Kannan, D., & Haq, A. N. (2014). Barriers analysis for green supply chain management implementation in Indian industries using analytic hierarchy process. *International Journal of Production Economics*, *147*, 555–568. doi:10.1016/j.ijpe.2013.08.018

Govindan, K., Khodaverdi, R., & Jafarian, A. (2013). A fuzzy multi criteria approach for measuring sustainability performance of a supplier based on triple bottom line approach. *Journal of Cleaner Production*, *47*, 345–354. doi:10.1016/j.jclepro.2012.04.014

Govindan, K., Rajendran, S., Sarkis, J., & Murugesan, P. (2013). Multi criteria decision making approaches for green supplier evaluation and selection: A literature review. *Journal of Cleaner Production*. doi:10.1016/j.jclepro.2013.06.046

Grant, R. M., & Baden-Fuller, C. (2004). A knowledge-accessing theory of strategic alliances. *Journal of Management Studies*, *41*(1), 61–84. doi:10.1111/j.1467-6486.2004.00421.x

Greenberg, A., & Quillian, L. (2012). Managing sustainable production: A framework for integrating sustainability in the manufacturing sector. *Environmental Quality Management*, *21*(4), 25–40. doi:10.1002/tqem.21305

Green, K. W. Jr, Zelbst, P. J., Meacham, J., & Bhadauria, V. S. (2012). 'Green supply chain management practices: Impact on performance', Supply Chain Management. *International Journal (Toronto, Ont.)*, *17*(30), 290–305.

Green, K. W. Jr, Zelbst, P. J., Meacham, J., & Bhadauria, V. S. (2012). Green supply chain management practices: Impact on performance. *Supply Chain Management: An International Journal*, *17*(3), 290–305. doi:10.1108/13598541211227126

Guide, V. D. R., Harrison, T. P., & Wassenhove, L. N. V. (2003). The Challenge of Closed-Loop Supply Chains. *Interfaces*, *33*(6), 3–6. doi:10.1287/inte.33.6.3.25182

Guide, V. D. R., & Wassenhove, L. N. V. (2002). The reverse supply chain. *Harvard Business Review*, *80*(2), 25–26. PMID:12964466

Guiffrida, A. L. (1999). *A Cost-Based Model for Evaluating Vendor Delivery Performance*. (Thesis). Department of Industrial Engineering, State University of New York at Buffalo, Buffalo, NY.

Gülsün, B., Bıyık, G., & Özgen, D. (2006). *Reverse logistics: A survey*. 10th International Research/Expert Conference TMT 2006, Barcelona-Lloret de Mar, Spain.

Gunady, M. G. A., Biswas, W., Solah, V. A., & James, A. P. (2012). Evaluating the global warming potential of the fresh produce supply chain for strawberries, romaine/cos lettuces (lactuca sativa), and button mushrooms (agaricus bisporus) in Western Australia using life cycle assessment (LCA). *Journal of Cleaner Production*, *28*, 81–87. doi:10.1016/j.jclepro.2011.12.031

Gundlach, G. T., Bolumole, Y. A., Eltantawy, R. A., & Frankel, R. (2006). The changing landscape of supply chain management, marketing channels of distribution, logistics and purchasing. *Journal of Business and Industrial Marketing*, *21*(7), 428–438. doi:10.1108/08858620610708911

Gunesekaran, A., & Spalanzani, A. (2012). Sustainability of manufacturing and services: Investigations for research and applications. *International Journal of Production Economics*, *140*(1), 35–47. doi:10.1016/j.ijpe.2011.05.011

Gungor, A., & Gupta, S. M. (1999). Issues in environmentally conscious manufacturing and product recovery: A survey. *Computers & Industrial Engineering*, *36*(4), 811–853. doi:10.1016/S0360-8352(99)00167-9

Gupta, S., & Desai, P. O. D. (2011). Sustainable supply chain management: Review and research opportunities. *IIMB Management Review*, *23*(4), 234–245. doi:10.1016/j.iimb.2011.09.002

Gurtu, A., Searcy, C., & Jaber, M. Y. (2015). An analysis of keywords used in the literature on green supply chains. *Management Research Review*, *38*(2), 166–194. doi:10.1108/MRR-06-2013-0157

Gurumurthy, A., & Kodali, R. (2011). Design of lean manufacturing systems using value stream mapping with simulation: A case study. *Journal of Manufacturing Technology Management*, *22*(4), 444–473. doi:10.1108/17410381111126409

Gutés, C. M. (1996). The concept of weak sustainability. *Ecological Economics*, *17*(3), 147–156. doi:10.1016/S0921-8009(96)80003-6

H'Mida, S., & Lakhal, S. Y. (2007). A model for assessing the greenness effort in a product supply chain. *International Journal of Global Environmental Issues*, *7*(1), 4–24. doi:10.1504/IJGENVI.2007.012074

Hagelaar, G. J. L. F., & van der Vorst, J. G. A. J. (2002). Environmental supply chain management: Using life cycle assessment to structure supply chains. *International Food and Agribusiness Management Review*, *4*(4), 399–412. doi:10.1016/S1096-7508(02)00068-X

Hahn, T., Figge, F., Pinkse, J., & Preuss, L. (2010). Trade-offs in corporate sustainability: You can't have your cake and eat it. *Business Strategy and the Environment*, *19*(4), 217–229. doi:10.1002/bse.674

Hajmohammad, S., Vachon, S., Klassen, R. D., & Gavronski, I. (2013). Reprint of Lean management and supply management: Their role in green practices and performance. *Journal of Cleaner Production*, *56*, 86–93. doi:10.1016/j.jclepro.2013.06.038

Halldorsson, A., Kotzab, H., Mikkola, J. H., & Skjøtt-Larsen, T. (2007). Complementary theories to supply chain management. *Supply Chain Management: An International Journal*, *12*(4), 284–296. doi:10.1108/13598540710759808

Hall, J., & Matos, S. (2010). Incorporating impoverished communities in sustainable supply chains. *International Journal of Physical Distribution & Logistics Management*, *40*(1/2), 124–147. doi:10.1108/09600031011020368

Hall, J., Matos, S., & Silvestre, B. (2012). Understanding why firms should invest in sustainable supply chains: A complexity approach. *International Journal of Production Research*, *50*(5), 1332–1348. doi:10.1080/00207543.2011.571930

Hampl, N., & Loock, M. (2013). Sustainable development in retailing: What is the impact on store choice? *Business Strategy and the Environment*, *22*(3), 202–216. doi:10.1002/bse.1748

Hanson, J. D., Melnyk, S. A., & Calantone, R. J. (2004). Core values and environmental management. *Greener Management International*, *2004*(46), 29–40. doi:10.9774/GLEAF.3062.2004.su.00005

Hanss, D., & Böhm, G. (2012). Sustainability seen from the perspective of consumers. *International Journal of Consumer Studies*, *36*(6), 678–687. doi:10.1111/j.1470-6431.2011.01045.x

Haque, B., & James-moore, M. (2004). Applying lean thinking to new product introduction. *Journal of Engineering Design*, *15*(1), 1–31. doi:10.1080/0954482031000150125

Hardin, G. (2009). The Tragedy of the Commons. *Journal of Natural Resources Policy Research*, *1*(3), 243–253. doi:10.1080/19390450903037302

Harland, C. (1996). Supply chain management: Relationships, chains and networks. *British Journal of Management*, *7*(1), S63–S80. doi:10.1111/j.1467-8551.1996.tb00148.x

Harms, D., Hansen, E. G., & Schaltegger, S. (2013). Strategies in sustainable supply chain management: An empirical investigation of large German companies. *Corporate Social Responsibility and Environmental Management*, *20*(4), 205–218. doi:10.1002/csr.1293

Harris, F. W. (1913). How many parts to make at once? *Factory, the Magazine of Management*, *135-136*, 152.

Harris, J. (2006). New Paths to Business Value: linking environment, health performance to strategic sourcing. In J. Sarkis (Ed.), *Greening the supply chain* (pp. 39–65). Worcester, MA: Springer. doi:10.1007/1-84628-299-3_3

Hart, S. L. (1995). A natural-resource-based view of the firm. *Academy of Management Review*, *20*(4), 986–1014. doi: 10.2307/258963

Hassini, E., Surti, C., & Searcy, C. (2012). A literature review and a case study of sustainable supply chains with a focus on metrics. *International Journal of Production Economics*, *140*(1), 69–82. doi:10.1016/j.ijpe.2012.01.042

Hauschild, M. Z., Jeswiet, J., & Alting, L. (2004). Design for environment—do we get the focus right? *CIRP Annals-Manufacturing Technology*, *53*(1), 1–4. doi:10.1016/S0007-8506(07)60631-3

Henriques, A., & Richardson, J. (2004). *The triple bottom line: Does it all add up?* London, UK: Earthscan.

Herrmann, C., Thiede, S., Stehr, J., & Bergmann, L. (2008). An environmental perspective on Lean Production. In M. Mitsuishi, K. Ueda, & F. Kimura (Eds.), *Manufacturing Systems and Technologies for the New Frontier* (pp. 83–88). Springer London. doi:10.1007/978-1-84800-267-8_16

Hervani, A.A., Helms, M.M., & Sarkis. J. (2005). Performance measurement for green supply chain management. *Benchmarking: An International Journal, 12*(4), 330–353.

Hervani, A. A., Helms, M. M., & Sarkis, J. (2005). 'Performance measurement for green supply chain management', Benchmarking. *International Journal (Toronto, Ont.)*, *12*(4), 330–353.

Hervani, A. A., Helms, M. M., & Sarkis, J. (2005). "Performance measurement for green supply chain management" Benchmarking. *International Journal (Toronto, Ont.)*, *12*(4), 330–353.

Hervani, A. A., Helms, M. M., & Sarkis, J. (2005). Performance measurement for green supply chain management. *Benchmarking*, *12*(4), 330–353. doi:10.1108/14635770510609015

Hicks, B. J. (2007). Lean information management: Understanding and eliminating waste. *International Journal of Information Management*, *27*(4), 233–249. doi:10.1016/j.ijinfomgt.2006.12.001

Hoffman, A. J. (2008). From Heresy to Dogma: An Institutional History of Corporate Environmentalism. Stanford Business Books.

Ho, J.C., & Shalishali, M.K., Tseng, T.L. & Ang, D.S. (2009). Opportunities in green supply chain management. *The Coastal Business Journal*, *8*(1), 18–31.

Hojmose, S., & Brammer, S. (2012). Green supply chain management: The role of trust and top management in B2B and B2C markets. *Industrial Marketing Management*, *41*(4), 609–620. doi:10.1016/j.indmarman.2012.04.008

Holmes, S. M., Power, M. L., & Walter, C. K. (1996). A motor carrier wellness program: Development and testing. *Transportation Journal*, *35*(3), 33–48.

Holt, D., & Ghobadian, A. (2009). An empirical study of green supply chain management practices amongst UK manufacturers. *Journal of Manufacturing Technology Management*, *20*(7), 933–956. doi:10.1108/17410380910984212

Holweg, M. (2007). The genealogy of lean production. *Journal of Operations Management*, *25*(2), 420–437. doi:10.1016/j.jom.2006.04.001

Hong, P., Kwon, H. B., & Roh, J. J. (2009). Implementation of strategic green orientation in supply chain. *European Journal of Innovation Management*, *12*(4), 512–532. doi: 10.1108/14601060910996945

Hong, P., Roh, J. J., & Rawski, G. (2012). Benchmarking sustainability practices: Evidence from manufacturing firms. *Benchmarking: An International Journal*, *19*(4/5), 634–648. doi:10.1108/14635771211258052

Hong, P., Tran, O., & Park, K. (2010). Electronic commerce applications for supply chain integration and competitive capabilities: An empirical study. *Benchmarking: An International Journal*, *17*(4), 539–560. doi:10.1108/14635771011060585

Hopp, W. J., & Spearman, M. L. (2004). Commissioned Paper To Pull or Not to Pull: What Is the Question? *Manufacturing & Service Operations Management*, *6*(2), 133–148. doi:10.1287/msom.1030.0028

Hopp, W., & Ebrahim-Khanjari, N. (2012). Trust and information sharing in supply chains. *Production and Operations Management*, *21*(3), 444–464. doi:10.1111/j.1937-5956.2011.01284.x

Hosseini, A. (2007). Identification of Green Management of system's factors: - A Conceptualized Model. *International Journal of Management Science and Engineering Management*, *2*(3), 221–228.

Hsu, C. W., & Hu, A. H. (2008). Green Supply Chain Management in the Electronic Industry. *International Journal of Science and Technology*, *5*(2), 205–216.

Hsueh, C. F. (2015). A bilevel programming model for corporate social responsibility collaboration in sustainable supply chain management. *Transportation Research Part E, Logistics and Transportation Review*, *73*, 84–95. doi:10.1016/j.tre.2014.11.006

Hua, G., Cheng, T. C. E., & Wang, S. (2011). Managing carbon footprints in inventory management. *International Journal of Production Economics*, *132*(2), 178–185. doi:10.1016/j.ijpe.2011.03.024

Huang, M., Smilowitz, K., & Balcik, B. (2011). Models for relief routing: Equity, efficiency and efficacy. *Transportation Research Part E, Logistics and Transportation Review*, *48*(1), 2–18. doi:10.1016/j.tre.2011.05.004

Hubbard, G. (2009). Measuring organizational performance: Beyond the triple bottom line. *Business Strategy and the Environment*, *18*(3), 177–191. doi:10.1002/bse.564

Hugos, M. H. (2006). *Essentials of supply chain management* (2nd ed.). John Wiley and Sons.

Hussain, M., Tsironis, L., & Ajmal, M. M. (2011). A QFD strategy for improving customer satisfaction: Case study of telecom companies of Pakistan. *Asian Journal on Quality*, *12*(3), 282–295. doi:10.1108/15982681111187119

Hutchins, M. J., & Sutherland, J. W. (2008). An exploration of measures of social sustainability and their application to supply chain decisions. *Journal of Cleaner Production*, *16*(15), 1688–1698. doi:10.1016/j.jclepro.2008.06.001

Hwa T.J. (2001). *Green Productivity and Supply Chain Management, Greening Supply Chain: Enhancing Competitiveness Through Green Productivity*. Taipei: Academic Press.

IBM. (2005). *IBM's 2005 innovations in corporate responsibility*. Armonk, NY: IBM Corporation.

IEA. (n.d.). *CO2 Emissions from Fuel Combustion2013*. IEA. Retrieved from http://www.iea.org/statistics

Ilgin, M. A., & Gupta, S. M. (2010). Environmentally conscious manufacturing and product recovery (ECMPRO): A review of the state of the art. *Journal of Environmental Management*, *91*(3), 563–591. doi:10.1016/j.jenvman.2009.09.037 PMID:19853369

Ilin, V., Simić, D., & Simić, S. (2012). Lean RFID approach enhancing the information and material flows in emergency department. *Journal of Medical Informatics & Technologies*, *21*, 49–55.

Imai, M. (1986). *Kaizen: the key to Japan's competitive success* (1st ed.). New York: McGraw-Hill.

Inman, R. A. (1999). Environmental management: New challenges for production and inventory managers. *Production and Inventory Management Journal*, *40*(3), 46–49.

IPCC. (2007). *Climate change 2007: The physical science basis*. Retrieved 1 December 2013 from: <http://www.ipcc.ch/pdf/assessment report/ar4/wg1/ar4-wg1-spm.pdf>

Jabbour, A. B. L., & Jabbour, C. J. (2009). Are supplier selection criteria going green? Case studies of companies in Brazil. *Industrial Management & Data Systems*, *109*(4), 477–495. doi:10.1108/02635570910948623

Jaber, M.Y., & Zolfaghari, S. (2008). Quantitative models for centralised supply chain coordination. In *Supply Chain Theory and Applications*. Vienna, Austria: I-Tech Education and Publishing.

Jaber, M. Y. (Ed.). (2009). *Inventory management: non-classical views*. CRC Press. doi:10.1201/9781420079982

Jaber, M. Y., Glock, C. H., & El Saadany, A. M. A. (2013). Supply chain coordination with emissions reduction incentives. *International Journal of Production Research*, *51*(1), 69–82. doi:10.1080/00207543.2011.651656

Jacobs, P. (1985). A sustainable society through sustainable development: Towards a regional development strategy for northern Quebec. *Landscape Planning*, *12*(3), 267–283. doi:10.1016/0304-3924(85)90006-1

Jain, D. (2014). RFID and its impact on supply chain. *International Journal of Logistics & Supply Chain Management Perspectives*, *2*(4), 741–756.

Jain, V., Wadhwa, S., & Deshmukh, S. G. (2009). Select supplier-related issues in modelling a dynamic supply chain: Potential, challenges and direction for future research. *International Journal of Production Research*, *47*(11), 3013–3039. doi:10.1080/00207540701769958

Jin, Y., Olhofer, M., & Sendhoff, B. (2001). *Dynamic weighted aggregation for evolutionary multi-objective optimization: Why does it work and how?*. Academic Press.

Jonatan, P., & Marcel, D. (2009). Overcoming Barriers to Sustainability: An Explanation of Residential Builders Reluctance to Adopt Clean Technologies. *Business Strategy and the Environment, 18*(8), 515–527. doi:10.1002/bse.615

Joshi, G. N., & Dave, M. B. (2012). Green Manufacturing for Greener Tomorrow. In *International Conference on Technology and Business Management.*

Kaenzig, J., Friot, D., Saadé, M., Margni, M., & Jolliet, O. (2011). Using life cycle approaches to enhance the value of corporate environmental disclosures. *Business Strategy and the Environment, 20*(1), 38–54. doi:10.1002/bse.667

Kainuma, Y., & Tawara, N. (2006). A multiple attribute utility theory approach to lean and green supply chain management. *International Journal of Production Economics, 101*(1), 99–108. doi:10.1016/j.ijpe.2005.05.010

Kam, B., Smyrnios, G.C. & Walker, R.H. (2003). Strategic business operations, freight transport and eco-efficiency: a conceptual model. *Greening the Supply Chain, Greenleaf, Sheffield*

Kang, S.-H., Kang, B., Shin, K., Kim, D., & Han, J. (2012). A theoretical framework for strategy development to introduce sustainable supply chain management. *Procedia: Social and Behavioral Sciences, 40*, 631–635. doi:10.1016/j.sbspro.2012.03.241

Kannan, G., & Haq, A. N. (2007). Analysis of interactions of criteria and sub-criteria for the selection of supplier in the built-in-order supply chain environment. *International Journal of Production Research, 45*(17), 3831–3852. doi:10.1080/00207540600676676

Kaplan, R., & Norton, D. (1992). The balanced scorecard: Measures that drive performance. *Harvard Business Review, 70*(1), 71–79. PMID:10119714

Kasemsap, K. (2015a). Implementing enterprise resource planning. In M. Khosrow-Pour (Ed.), *Encyclopedia of information science and technology* (3rd ed.; pp. 798–807). Hershey, PA: IGI Global. doi:10.4018/978-1-4666-5888-2.ch076

Kasemsap, K. (2015b). The role of cloud computing in global supply chain. In N. Rao (Ed.), *Enterprise management strategies in the era of cloud computing* (pp. 192–219). Hershey, PA: IGI Global. doi:10.4018/978-1-4666-8339-6.ch009

Kaur, M., Sandhu, M., Mohan, N., & Sandhu, P. S. (2011). RFID Technology Principles, Advantages, Limitations & Its Applications. *International Journal of Computer and Electrical Engineering, 3*(1), 151–157. doi:10.7763/IJCEE.2011.V3.306

Kaya, M., & Özer, Ö. (2009). Quality risk in outsourcing: Noncontractible product quality and private quality cost information. *Naval Research Logistics, 56*(7), 669–685. doi:10.1002/nav.20372

Keiner, M. (2006). *The future of sustainability*. Springer Netherlands. doi:10.1007/1-4020-4908-0

Keoleian, G.A. (1993). The application of life cycle assessment to design. *Environmental Life Cycle Assessment and its Applications, 1*(3-4), 143 – 149.

Khalid, N. K. M., Hashim, A. Y. B., & Salleh, M. R. (2014). On Value Stream Mapping and Its Industrial Significance. *Journal of Industrial and Intelligent Information*, *2*(2), 88–90. doi:10.12720/jiii.2.2.88-90

Khan, M., Jaber, M. Y., & Glock, C. H. (2012). *Impact of learning on the environmental performance of a two level supply chain.* Paper presented at the 17th International Symposium on Inventories, Budapest, Hungary.

Khan, J., & Haleem, A. (2013). An integrated ISM and Fuzzy MICMAC approach for modeling of the enablers of Technology Management. *Indian Journal of Applied Research*, *3*(7), 236–242. doi:10.15373/2249555X/JULY2013/74

Khosla, A., Prakash, S., & Revi, A. (1986). A transcultural view of sustainable development. The landscape of design. *Landscape and Urban Planning*, *13*(C), 401–410. doi:10.1016/0169-2046(86)90057-5

King, A. A., & Lenox, M. J. (2001). Lean and Green? An empirical examination of the relationship between lean production and environmental performance. *Production and Operations Management*, *10*(3), 244–256. doi:10.1111/j.1937-5956.2001.tb00373.x

Klassen, R. D. (2000). Exploring the linkage between investment in manufacturing and environmental technologies. *International Journal of Operations & Production Management*, *20*(2), 127–147. doi:10.1108/01443570010304224

Klassen, R. D. (2001). Plant-level environmental management orientation: The influence of management views and plant characteristics. *Production and Operations Management*, *10*(3), 257–275. doi:10.1111/j.1937-5956.2001.tb00374.x

Klassen, R. D., & McLaughlin, C. P. (1993). TQM and environmental excellence in manufacturing. *Industrial Management & Data Systems*, *93*(6), 14–22. doi:10.1108/02635579310040924

Klassen, R. D., & McLaughlin, C. P. (1996). The impact of environmental management on firm performance. *Management Science*, *42*(8), 1199–1214. doi:10.1287/mnsc.42.8.1199

Kobayashi, I. (1990). *20 keys to workplace improvement* (W. S. Warren, Trans.). Cambridge, MA: Productivity Press, Inc.

Kogg, B. (2003). Greening a cotton-textile supply chain. *Greener Management International*, *43*(September), 53–64.

Kolk, A., & Pinkse, J. (2005). Business responses to climate change: Identifying emergent strategies. *California Management Review*, *47*(3), 6–20. doi:10.2307/41166304

Koskela, M., & Vehmas, J. (2012). Defining eco-efficiency: A case study on the Finnish forest industry. *Business Strategy and the Environment*, *21*(8), 546–566. doi:10.1002/bse.741

Kotzab, H., Teller, C., Grant, D. B., & Sparks, L. (2011). Antecedents for the adoption and execution of supply chain management. *Supply Chain Management: An International Journal*, *16*(4), 231–245. doi:10.1108/13598541111139053

Kull, T. J., Boyer, K., & Calantone, R. (2007). Last-mile supply chain efficiency: An analysis of learning curves in online ordering. *International Journal of Operations & Production Management, 27*(4), 409–434. doi:10.1108/01443570710736985

Kumar, V. (2007). Analyzing and optimizing the environmental performance of supply chains. In *Proceedings of ACEEE Summer Study on Energy Efficiency in Industry T.* Surya Technologies, Inc. Retrieved from www.suryatech.com

Kumar, A., & Rahman, S. (2014). RFID-enabled process reengineering of closed-loop supply chains in the healthcare industry of Singapore. *Journal of Cleaner Production*, (0). doi:10.1016/j.jclepro.2014.04.037

Kuo, T. C., Huang, S. H., & Zhang, H. C. (2001). Design for manufacture and design for 'X': Concepts, applications, and perspectives. *Computers & Industrial Engineering, 41*(3), 241–260. doi:10.1016/S0360-8352(01)00045-6

Lambert, D. M., Cooper, M. C., & Pagh, J. D. (1998). Supply chain management: Implementation issues and research opportunities. *International Journal of Logistics Management, 9*(2), 1–19. doi:10.1108/09574099810805807

Lambert, D. M., Stock, J. R., & Ellram, L. M. (1998). *Fundamentals of logistics management.* Boston, MA: Irwin/McGraw-Hill.

Lamming, R., & Hampson, J. (1996). The environment as a supply chain management issue. *British Journal of Management, 7*(s1), s45–s62. doi:10.1111/j.1467-8551.1996.tb00147.x

Langenwalter, G. (2007). *Life is Our Ultimate Customer: From Lean to Sustainability.* Retrieved 20 May 2014, from AME http://www.ame.org/search/node/Life%20is%20Our%20Ultimate%20Customer%3A%20From%20Lean%20to%20Sustainability

Large, R. O., & Thomsen, C. G. (2011). Drivers of Green Supply Chain Management Performance: Evidence from Germany. *Journal of Purchasing and Supply Management, 17*, 176–184.

Larson, T., & Greenwood, R. (2004). Perfect complements: Synergies between lean production and eco-sustainability initiatives. *Environmental Quality Management, 13*(4), 27–36. doi:10.1002/tqem.20013

Lau, K. H. (2011). Benchmarking green logistics performance with a composite index. *Benchmarking, 18*(6), 873–896. doi:10.1108/14635771111180743

Lee, K. H., & Kim, J. W. (2009). Current status of CSR in the realm of supply management: The rise of the Korean electronics industry. *Supply Chain Management: An International Journal, 14*(2), 138–148. doi:10.1108/13598540910942000

Lee, K.-H. (2011). Integrating carbon footprint into supply chain management: The case of Hyundai Motor Company (HMC) in the automobile industry. *Journal of Cleaner Production, 19*(11), 1216–1223. doi:10.1016/j.jclepro.2011.03.010

Lee, S. (2008). Drivers for the participation of small and medium-sized suppliers in green supply chain initiatives. *Supply Chain Management: An International Journal*, *13*(3), 185–198. doi:10.1108/13598540810871235

Lee, S., & Klassen, R. D. (2008). Drivers and enablers that foster environmental management capabilities in small- and medium-sized suppliers in supply chains. *Production and Operations Management*, *17*(6), 573–586. doi:10.3401/poms.1080.0063

Lee, T., & Kashmanian, R. M. (2013). Supply chain sustainability: Compliance- and performance-based tools. *Environmental Quality Management*, *22*(4), 1–23. doi:10.1002/tqem.21340

Lehtonen, M. (2004). The environmental – social interface of sustainable development: Capabilities, social capital, institutions. *Ecological Economics*, *49*(2), 199–214. doi:10.1016/j.ecolecon.2004.03.019

Lepoutre, J., & Heene, A. (2006). Investigating the impact of firm size on small business social responsibility: A critical review. *Journal of Business Ethics*, *67*(3), 257–273. doi:10.1007/s10551-006-9183-5

Leung, J., Cheung, W., & Chu, S.-C. (2014). Aligning RFID applications with supply chain strategies. *Information & Management*, *51*(2), 260–269. doi:10.1016/j.im.2013.11.010

Lian, Y. H., & Van Landeghem, H. (2002). *An application of simulation and value stream mapping in lean manufacturing*. Paper presented at the 14th European Simulation Symposium: Simulation in Industry (ESS), Dresden, Germany.

Liao, Y., Hong, P., & Rao, S. (2010). Supply management, supply flexibility and performance outcomes: An empirical investigation of manufacturing firms. *Journal of Supply Chain Management*, *46*(3), 6–22. doi:10.1111/j.1745-493X.2010.03195.x

Liker, J. K. (2004). *The Toyota way: 14 management principles from the world's greatest manufacturer*. New York: McGraw-Hill.

Lin, C. Y., & Ho, Y. H. (2008). An Empirical Study On Logistics Service Providers' Intention To Adopt Green Innovations. *Journal of Technology Management & Innovation*, *3*(1), 17–26.

Linn, M. W., Sandifer, R., & Stein, S. (1985). Effects of unemployment on mental and physical health. *American Journal of Public Health*, *75*(5), 502–506. doi:10.2105/AJPH.75.5.502 PMID:3985238

Lin, R. (2012). An integrated model for supplier selection under a fuzzy situation. *International Journal of Production Economics*, *138*(1), 55–61. doi:10.1016/j.ijpe.2012.02.024

Linton, J. D., Klassen, R., & Jayaraman, V. (2007). Sustainable supply chains: An introduction. *Journal of Operations Management*, *25*(6), 1075–1082. doi:10.1016/j.jom.2007.01.012

Lippman, S. (2001). Supply Chain Environmental Management. *Environmental Quality Management*, *11*(2), 11–14. doi:10.1002/tqem.1301

Li, S., Ragu-Nathan, B., Ragu-Nathan, T. S., & Subba Rao, S. (2006). The impact of supply chain management practices on competitive advantage and organizational performance. *Omega: The International Journal of Management Science*, *34*(2), 107–124. doi:10.1016/j.omega.2004.08.002 PMID:17876965

Liu, S., Kasturiratne, D., & Moizer, J. (2012). A hub and spoke model for multi dimensional integration of green marketing and sustainable supply chain management. *Industrial Marketing Management*, *41*(4), 581–588. doi:10.1016/j.indmarman.2012.04.005

Liu, X., Yang, J., Qu, S., Wang, L., Shishime, T., & Bao, C. (2011). Sustainable Production:Practices and Determinant Factors of Green Supply Chain Management of Chinese Companies. *Business Strategy and the Environment.*

Lo, C. W., & Leung, S. W. (2000). Environmental agency and public opinion in Guangzhou: The limits of a popular approach to environmental governance. *The China Quarterly*, *163*, 677–704. doi:10.1017/S0305741000014612

Locher, D. A. (2011). *Value stream mapping for lean development: a how-to guide for streamlining time to market*. CRC Press.

Longo, F., & Mirabelli, G. (2008). An advance supply chain management tool based on modeling and simulation. *Computers & Industrial Engineering*, *54*(3), 570–588. doi:10.1016/j.cie.2007.09.008

Lowry, K., & Carpenter, R. A. (1985). Institutionalizing sustainable development: Experiences in five countries. *Environmental Impact Assessment Review*, *5*(3), 239–254. doi:10.1016/0195-9255(85)90004-6

Lu, J.-W., Chang, N.-B., & Liao, L. (2013). Environmental Informatics for Solid and Hazardous Waste Management: Advances, Challenges, and Perspectives. *Critical Reviews in Environmental Science and Technology*, *43*(15), 1557–1656. doi:10.1080/10643389.2012.671097

Luthra, S., Garg, D., & Haleem, A. (2013). Identifying and ranking of strategies to implement green supply chain management in Indian manufacturing industry using Analytical Hierarchy Process. *Journal of Industrial Engineering and Management*, *6*(4), 930–962. doi:10.3926/jiem.693

Luthra, S., Garg, D., & Haleem, A. (2014). Green supply chain management: Implementation and performance–a literature review and some issues. *Journal of Advances in Management Research*, *11*(1), 20–46. doi:10.1108/JAMR-07-2012-0027

Luthra, S., Kumar, V., Kumar, S., & Haleem, A. (2009). 'Green supply chain management issues: A literature review approach', Journal of Information. *Knowledge and Research in Mechanical Engineering*, *1*(1), 12–20.

Luthra, S., Kumar, V., Kumar, S., & Haleem, A. (2011). Barriers to implement green supply chain management in automobile industry using interpretative structural modeling technique-An Indian perspective. *Journal of Industrial Engineering and Management*, *4*(2), 231–257. doi:10.3926/jiem.2011.v4n2.p231-257

Luthra, S., Luthra, S., & Haleem, A. (2015). Hurdles in implementing sustainable supply chain management: An analysis of Indian automobile sector. *Procedia: Social and Behavioral Sciences*, *189*, 175–183. doi:10.1016/j.sbspro.2015.03.212

Lynagh, P. M., Murphy, P. R., & Poist, R. F. (1996). Career-related perspectives regarding women in logistics: A comparative analysis. *Transportation Journal*, *36*(1), 35–42.

Lysons, K., & Farrington, B. (2012). *Purchasing and supply chain management*. Harlow, UK: Pearson Education.

Manikas, A., & Godfrey, M. (2010). Inducing green behavior in a manufacturer. *Global Journal of Business Research*, *4*(2), 27–38.

Manuj, I., & Mentzer, J. T. (2008). Global supply chain risk management strategies. *Int. J. Physical Distribution & Logistics Management*, *38*(3), 192–223. doi:10.1108/09600030810866986

Markley, M. J., & Davis, L. (2007). Exploring future competitive advantage through sustainable supply chains. *International Journal of Physical Distribution & Logistics Management*, *37*(9), 763–774. doi:10.1108/09600030710840859

Marsillac, E. L. (2008). Environmental impacts on reverse logistics and green supply chains: Similarities and integration. *International Journal of Logistics Systems and Management*, *4*(2), 422.

Martínez-Jurado, P. J., & Moyano-Fuentes, J. (2013). Lean Management, Supply Chain Management and Sustainability: A Literature Review. *Journal of Cleaner Production*, (0). doi:10.1016/j.jclepro.2013.09.042

Maxwell, D., & Van der Vorst, R. (2003). Developing sustainable products and services'. *Journal of Cleaner Production*, *11*(8), 883–895. doi:10.1016/S0959-6526(02)00164-6

McDonald, T., Van Aken, E. M., & Rentes, A. F. (2002). Utilising Simulation to Enhance Value Stream Mapping: A Manufacturing Case Application. *International Journal of Logistics Research and Applications*, *5*(2), 213–232. doi:10.1080/13675560210148696

McElroy, J. C., Rodriguez, J. M., Griffin, G. C., Morrow, P. C., & Wilson, M. G. (1993). Career stage, time spent on the road, and truckload driver attitudes. *Transportation Journal*, *33*(1), 5–14.

McFadden, D. (1978). Cost, Revenue and Profit Functions. In M. Fuss & D. McFadden (Eds.), *Production Economics: A Dual Approach to Theory and Applications*. Elsevier.

McIntyre, K., Smith, H. A., Henham, A., & Pretlove, J. (1998). Logistical performance measurement and greening supply chains: Diverging mindsets. *The International Journal of Logistics Management*, *9*(1), 57–68. doi:10.1108/09574099810805744

McIntyre, K., Smith, H., Henham, A., & Pretlove, J. (1998). 'Environmental performance indicators for integrated supply chains: The case of Xerox Ltd', Supply Chain Management. *International Journal (Toronto, Ont.)*, *3*(3), 149–156.

McKinnon, A. C., Stirling, I., & Kirkhope, J. (1993). Improving the fuel efficiency of road freight operations. *International Journal of Physical Distribution & Logistics Management*, *23*(9), 3–11. doi:10.1108/09600039310049736

McKinnon, A. C., & Tallam, D. (2003). Unattended delivery to the home: An assessment of the security implications. *International Journal of Retail and Distribution Applications*, *31*(1), 30–41. doi:10.1108/09590550310457827

McLeod, F., Cherrett, T., & Song, L. (2006). Transport impacts of local collection/delivery points. *International Journal of Logistics Research and Applications*, *9*(3), 307–317. doi:10.1080/13675560600859565

Meehan, J., & Bryde, D. (2011). Sustainable procurement practice. *Business Strategy and the Environment*, *20*(2), 94–106. doi:10.1002/bse.678

Mehta, V., & Shah, H. (2005). Characteristics of a work organization from a lean perspective. *Engineering Management Journal*, *17*(2), 14–20. doi:10.1080/10429247.2005.11415282

Melton, T. (2005). The Benefits of Lean Manufacturing: What Lean Thinking has to Offer the Process Industries. *Chemical Engineering Research & Design*, *83*(6), 662–673. doi:10.1205/cherd.04351

Mentzer, J. T., Dewitt, W., & Keebler, J. S. (2001). Defining supply chain management. *Journal of Business Logistics*, *22*(2), 1–25.

Mentzer, J. T., DeWitt, W., Keebler, J. S., Min, S., Nix, N. W., Smith, C. D., & Zacharia, Z. G. (2001). Defining supply chain management. *Journal of Business Logistics*, *22*(2), 1–25. doi:10.1002/j.2158-1592.2001.tb00001.x

Mentzer, J. T., Stank, T. P., & Esper, T. L. (2008). Supply chain management and its relationship to logistics, marketing, production, and operations management. *Journal of Business Logistics*, *29*(1), 31–46. doi:10.1002/j.2158-1592.2008.tb00067.x

Min, H., & Galle, W. (1997). Green purchasing strategies: Trends and implications. *International Journal of Purchasing and Materials*, *4*, 410–417.

Min, H., & Galle, W. P. (1997). Green purchasing strategies: Trends and implications. *International Journal of Purchasing and Materials Management*, *33*(3), 10–17. doi:10.1111/j.1745-493X.1997.tb00026.x

Min, H., & Galle, W. P. (2001). Green purchasing practices of US firms'. *International Journal of Operations & Production Management*, *21*(9), 1222–1238. doi:10.1108/EUM0000000005923

Mirhedayatian, S. M., Azadi, M., & Sean, R. F. (2014). A novel network data envelopment analysis model for evaluating green supply chain management. *International Journal of Production Economics*, *147*, 544–554. doi:10.1016/j.ijpe.2013.02.009

Mitra, S. (2004). Managing Environmental Issues in Supply Chains. In B. S. Sahay (Ed.), *Emerging Issues in Supply Chain Management, IIM Calcutta working paper series* (pp. 60–68). Macmillan.

Mitra, S., & Singhal, V. (2008). Supply chain integration and shareholder value: Evidence from consortium-based industry exchanges. *Journal of Operations Management*, *26*(1), 96–114. doi:10.1016/j.jom.2007.05.002

Mohammed, J., & Sadique, A. (2011). Lean Value Stream Manufacturing for Sustainability. In G. Seliger, M. M. K. Khraisheh, & I. S. Jawahir (Eds.), *Advances in Sustainable Manufacturing* (pp. 365–370). Springer Berlin Heidelberg. doi:10.1007/978-3-642-20183-7_52

Mohanraj, R., Sakthivel, M., & Vinodh, S. (2011). QFD integrated value stream mapping: An enabler of lean manufacturing. *International Journal of Productivity and Quality Management*, *7*(4), 501–522. doi:10.1504/IJPQM.2011.040546

Molamohamadi, Z., & Ismail, N. (2013). Developing a New Scheme for Sustainable Manufacturing. *International Journal of Materials. Mechanics and Manufacturing*, *1*(1), 1–5. doi:10.7763/IJMMM.2013.V1.1

Mollenkopf, D., Closs, D., Twede, D., Lee, S., & Burgess, G. (2005). Assessing the viability of reusable packaging: A relative cost approach. *Journal of Business Logistics*, *26*(1), 169–197. doi:10.1002/j.2158-1592.2005.tb00198.x

Mollenkopf, D., Stolze, H., Tate, W. L., & Ueltschy, M. (2010). Green, lean, and global supply chains. *International Journal of Physical Distribution & Logistics Management*, *40*(1), 14–41.

Monden, Y. (1998). *Toyota production system: an integrated approach to just-in-time*. Norcross, GA: Engineering & Management Press.

Montabon, F., Melnyk, S. A., Sroufe, R., & Calantone, R. J. (2000). ISO 14000: Assessing its perceived impact on corporate performance. *Journal of Supply Chain Management*, *36*(2), 4–16. doi:10.1111/j.1745-493X.2000.tb00073.x

Moreira, F., Alves, A. C., & Sousa, R. M. (2010). Towards Eco-efficient Lean Production Systems. In Á. Ortiz, R. Franco, & P. Gasquet (Eds.), *Balanced Automation Systems for Future Manufacturing Networks* (Vol. 322, pp. 100–108). Springer Berlin Heidelberg. doi:10.1007/978-3-642-14341-0_12

Mossakowski, K. N. (2009). The influence of past unemployment duration on symptoms of depression among young women and men in the United States. *American Journal of Public Health*, *99*(10), 1826–1832. doi:10.2105/AJPH.2008.152561 PMID:19696382

Mostafa, S., Dumrak, J., & Soltan, H. (2013). A framework for lean manufacturing implementation. *Production & Manufacturing Research*, *1*(1), 44–64. doi:10.1080/21693277.2013.862159

Mudgal, R. K., Shankar, R., Talib, P., & Raj, T. (2009). Greening the supply chain practices: An Indian perspective of enablers' relationships. *International Journal of Advanced Operations Management*, *1*(2/3), 151–176. doi:10.1504/IJAOM.2009.030671

Mukhopadhyay, A., & Goswami, A. (2014). Economic production quantity models for imperfect items with pollution costs. *Systems Science & Control Engineering: An Open Access Journal*, *2*(1), 368–378. doi:10.1080/21642583.2014.912571

Muller, C., Vermeulen, W. J. V., & Glasbergen, P. (2012). Pushing or sharing as value-driven strategies for societal change in global supply chains: Two case studies in the British–South African fresh fruit supply chain. *Business Strategy and the Environment, 21*(2), 127–140. doi:10.1002/bse.719

Mumford, E. (1985). Defining system requirements to meet business needs: A case study example. *The Computer Journal, 28*(2), 97–104. doi:10.1093/comjnl/28.2.97

Murphy, P. R., Poist, R. F., & Braunschwieg, C. D. (1996). Green logistics: Comparative views of environmental progressives, moderates, and conservatives. *Journal of Business Logistics, 17*(1), 191–211.

Nagel, M. H., & Tomiyama, T. (2004). *Intelligent sustainable manufacturing systems, management of the linkage between sustainability and intelligence - an overview.* Paper presented at the IEEE International Conference on Systems, Man & Cybernetics.

Narasimhan, R., & Carter, J. R. (1998). *Environmental supply chain management.* Arizona State University.

Narasimhan, R., Swink, M., & Kim, S. W. (2006). Disentangling leanness and agility: An empirical investigation. *Journal of Operations Management, 24*(5), 440–457. doi:10.1016/j.jom.2005.11.011

National Research Council (US). (1999). Committee on Industrial Environmental Performance Metrics. Industrial Environmental Performance Metrics: Challenges and Opportunities. National Academy Press.

Naylor, B. J., Naim, M. M., & Berry, D. (1999). Leagility: Integrating the lean and agile manufacturing paradigms in the total supply chain. *International Journal of Production Economics, 62*(1–2), 107–118. doi:10.1016/S0925-5273(98)00223-0

NBC News. (2008). *Cadbury recalls Chinese-made chocolates.* Retrieved 29 October 2012 from: http://www.msnbc.msn.com/id/26938647/ns/health-food-safety/t/cadbury-recalls-chinese-made-chocolates/

Nelson, D. M., Marsillac, E., & Rao, S. S. (2012). Antecedents and Evolution of the Green Supply Chain. *Journal of Operations and Supply Chain Management*, 29 – 43.

Neto, J. Q. F., Bloemhof-Ruwaard, J. M., van Nunen, J. A. E. E., & Heck, E. V. (2008). Designing and evaluating sustainable logistics networks. *International Journal of Production Economics, 111*(2), 195–208. doi:10.1016/j.ijpe.2006.10.014

Nidumolu, R., Prahalad, C. K., & Rangaswami, M. R. (2009). Why sustainability is now the key driver of innovation. *Harvard Business Review, 87*(9), 56–64.

Nidumolu, R., Prahalad, C. K., & Rangaswami, M. R. (2009). Why Sustainability is now the key driver of innovation. *Harvard Business Review, 87*(9), 56–64.

Ninlawan, C., Seksan, P., Tossapol, K., & Pilada, W. (2010). The implementation of Green Supply Chain Management Practices in Electronic Industry.*Proceedings of the International MultiConference of Engineers and Computer Scientists.*

Oberhofer, P., & Dieplinger, M. (2014). Sustainability in the transport and logistics sector: Lacking environmental measures. *Business Strategy and the Environment, 23*(4), 236–253. doi:10.1002/bse.1769

Ohno, T. (1988). *Toyota production system: beyond large-scale production.* Portland, OR: Productivity Press.

Onkila, T. (2011). Multiple forms of stakeholder interaction in environmental management: Business arguments regarding differences in stakeholder relationships. *Business Strategy and the Environment, 20*(6), 379–393. doi: 10.1002/bse.693

Orsato, R. (2006). Competitive environmental strategies: When does it pay to be green? *California Management Review, 48*(2), 127–143. doi:10.2307/41166341

Overcash, M. (2012). A comparison of reusable and disposable perioperative textiles: Sustainability state-of-the-art 2012. *Anesthesia and Analgesia, 114*(5), 1055–1066. doi:10.1213/ANE.0b013e31824d9cc3 PMID:22492184

Pagell, M., & Shevchenko, A. (2014). Why research in sustainable supply chain management should have no future. *Journal of Supply Chain Management, 50*(1), 44–55. doi:10.1111/jscm.12037

Pagell, M., & Wu, Z. (2009). Building a more complete theory of sustainable supply chain management using case studies of 10 exemplars. *Journal of Supply Chain Management, 45*(2), 37–56. doi:10.1111/j.1745-493X.2009.03162.x

Pagell, M., Wu, Z., & Wasserman, M. E. (2010). Thinking differently about purchasing portfolios: An assessment of sustainable sourcing. *Journal of Supply Chain Management, 46*(1), 57–73. doi:10.1111/j.1745-493X.2009.03186.x

Pal, O., Gupta, A. K., & Garg, R. K. (2013). Supplier selection criteria and methods in supply chains: A review. *International Journal of Management Sciences, 7*(10), 137–143.

Papista, E., & Krystallis, A. (2013). Investigating the types of value and cost of green brands: Proposition of a conceptual framework. *Journal of Business Ethics, 115*(1), 75–92. doi:10.1007/s10551-012-1367-6

Parnell, J. A. (2008). Sustainable strategic management: Construct, parameters, research directions. *International Journal of Sustainable Strategic Management, 1*(1), 35–45. doi:10.1504/IJSSM.2008.018125

Paulraj, A. (2009). Environmental motivations: A classification scheme and its impact on environmental strategies and practices. *Business Strategy and the Environment, 18*(7), 453–468. doi:10.1002/bse.612

Pavnaskar, S. J., Gershenson, J. K., & Jambekar, A. B. (2003). Classification scheme for lean manufacturing tools. *International Journal of Production Research, 41*(13), 3075–3090. doi:10.1080/0020754021000049817

Peattie, K., & Ratnayaka, M. (1992). Responding to the green movement. *Industrial Marketing Management, 21*(2), 103–110. doi:10.1016/0019-8501(92)90004-D

Pereira, J.V. (2009). The new supply chain's frontier: Information management. *Int. J. of Information Management, 29*, 372-379.

Perrini, F., & Tencati, A. (2006). Sustainability and stakeholder management: The need for new corporate performance evaluation and reporting systems. *Business Strategy and the Environment, 15*(5), 296–308. doi:10.1002/bse.538

Petroleum Times. (1951). New oil jetty is largest in Thames Estuary. *Petroleum Times, 55*(1399), 216.

Plaza-Úbeda, J. A., Burgos-Jiménez, J., Vazquez, D. A., & Liston-Heyes, C. (2009). The "win–win" paradigm and stakeholder integration. *Business Strategy and the Environment, 18*(8), 487–499. doi:10.1002/bse.593

Porter, M. E., & Kramer, M. R. (2006). The link between competitive advantage and corporate social responsibility. *Harvard Business Review, 84*(12), 78–92. PMID:17183795

Powell, D., & Skjelstad, L. (2012). RFID for the extended lean enterprise. *International Journal of Lean Six Sigma, 3*(3), 172–186. doi:10.1108/20401461211282691

Prajogo, D., & Olhager, J. (2012). Supply chain integration and performance: The effects of long-term relationships, information technology and sharing, and logistics integration. *International Journal of Production Economics, 135*(1), 514–522. doi:10.1016/j.ijpe.2011.09.001

Prusak, L. (2010). What Can't Be Measured. *Harvard Business Review Blog Network.* Retrieved 16 January 2013 from: http://blogs.hbr.org/davenport/2010/10/what_cant_be_measured.html

Punakivi, P., Yrjölä, H., & Holmström, J. (2001). Solving the last mile issue: Reception box or delivery box? *International Journal of Physical Distribution & Logistics Management, 31*(6), 427–439. doi:10.1108/09600030110399423

Ramadan, M., Zelong, Wang, & Noche, B. (2012). *RFID- Enabled dynamic Value Stream Mapping.* Paper presented at the International Conference on Service Operations and Logistics, and Informatics (SOLI), Suzhou.

Ramanathan, U., & Gunasekaran, A. (2014). Supply chain collaboration: Impact of success in long-term partnerships. *International Journal of Production Economics, 147*, 252–259. doi:10.1016/j.ijpe.2012.06.002

Ramesh, V., Prasad, K. V. S., & Srinivas, T. R. (2008). Implementation of a Lean Model for Carrying out Value Stream Mapping in a Manufacturing Industry. *Journal of Industrial and Systems Engineering, 2*(3), 180–196.

Ramus, C. A., & Steger, U. (2000). The roles of supervisory support behaviors and environmental policy in employee "eco-initiatives" at leading-edge European companies. *Academy of Management Journal*, *43*(4), 605–626. doi:10.2307/1556357

Rao, S., & Riahi, K. (2006). The role of non-CO 2 greenhouse gases in climate change mitigation: Long-term scenarios for the 21st century. *The Energy Journal*, 177-200.

Rao, P. (2002). Greening the supply chain: A new initiative in South East Asia. *International Journal of Operations & Production Management*, *22*(6), 632–655. doi:10.1108/01443570210427668

Rao, P., & Holt, D. (2005). Do green supply chains lead to competitiveness and economic performance? *International Journal of Operations & Production Management*, *25*(9), 898–916. doi:10.1108/01443570510613956

Ravi, V. & Shankar, R. (2005). Analysis of interactions among the barriers of reverse logistics. *International Journal of Technological Forecasting and Social Change, 72*(8), 1011-1029.

Rawabdeh, I. A. (2005). A model for the assessment of waste in job shop environments. *International Journal of Operations & Production Management*, *25*(8), 800–822. doi:10.1108/01443570510608619

Rawal, A. (2009). RFID: The Next Generation Auto-ID Technology. *Microwave Journal*, *52*(3), 58–62.

Rebitzer, G., Ekvall, T., Frischknecht, R., Hunkeler, D., Norris, G., Rydberg, T., & Pennington, D. W. et al. (2004). Life cycle assessment: Part 1: Framework, goal and scope definition, inventory analysis, and applications. *Environment International*, *30*(5), 701–720. doi:10.1016/j.envint.2003.11.005 PMID:15051246

Reefke, H., & Trochhi, M. (2013). Balanced scorecard for sustainable supply chains: Design and development guidelines. *International Journal of Productivity and Performance Management*, *62*(8), 805–826. doi:10.1108/IJPPM-02-2013-0029

Rettab, B., & Ben Brik, A. (2008). *Green Supply Chain in Dubai. Dubai Chamber Centre for Responsible Business, Dubai.* Retrieved October 14, 2012, from http://www.dubaichamber.com/wp.../GREEN-SUPPLY-CHAIN-IN-DUBAI.pdf

Reverse Logistics Trends. (n.d.). Retrieved from: www.reverselogsiticstrends.com

Richards, D. J. (1997). *The industrial green game: Implications for environmental design and management*. National Academy of Engineering.

Richey, R. G. Jr, Chen, H., Upreti, R., Fawcett, S. E., & Adams, F. G. (2009). The moderating role of Barriers on the relationship between drivers to supply chain integration and firm performance. *International Journal of Physical Distribution & Logistics Management*, *39*(10), 826–840.

Richey, R. G., Stefan, E. G., & Patricia, J. D. (2005). The role of resource commitment and innovation in reverse logistics performance. *International Journal of Physical Distribution & Logistics Management*, *35*(4), 233–257. doi:10.1108/09600030510599913

Rivera-Camino, J. (2007). Re-evaluating green marketing strategy: A stakeholder perspective. *European Journal of Marketing*, *41*(11/12), 1328–1358. doi:10.1108/03090560710821206

Roberts, J. (1996). Green consumers in the 1990s: Profile and implications for advertising. *Journal of Business Research*, *36*(3), 217–231. doi:10.1016/0148-2963(95)00150-6

Rogers, D.S., & Tibben-Lembke, R. S. (1998). *Going backwards: Reverse logistics trends and practices*. Academic Press.

Rogers, D. S., & Tibben-Lembke, R. S. (1999). *Going backwards: reverse logistics trends and practices* (Vol. 2). Pittsburgh, PA: Reverse Logistics Executive Council.

Rohana, A., Nooririnah, O., Isa, H., Kamat, S., & Mehad, M. (2013). Lean Waste Analysis and Improvement Using Dynamic Value Stream Mapping. *Global Engineers & Technologists Review*, *3*(2), 1–8.

Roh, J. J., Min, H., & Hong, P. (2011). A co-ordination theory approach to restructuring the supply chain: An empirical study from the focal company perspective. *International Journal of Production Research*, *49*(15), 4517–4541. doi:10.1080/00207543.2010.497506

Rohrmus, D., Döricht, V., Weinert, N., Holst, J.-C., & Kiener, C. (2013). Green Cycle Factory. In A. Azevedo (Ed.), *Advances in Sustainable and Competitive Manufacturing Systems* (pp. 1653–1663). Springer International Publishing. doi:10.1007/978-3-319-00557-7_133

Romer, J. R. (2012). Single-use plastic carryout bags: An icon of waste. *Sustainability*, *5*(6), 341–343. doi:10.1089/SUS.2012.9921

Rosenau, W. V., Twede, D., Mazzeo, M. A., & Singh, S. P. (1996). Returnable/reusable logistical packaging: A capital budgeting investment decision framework. *Journal of Business Logistics*, *17*(2), 139–165.

Rosenblatt, M. J. (1986). The dynamics of plant layout. *Management Science*, *32*(1), 76–86. doi:10.1287/mnsc.32.1.76

Rosič, H., & Jammernegg, W. (2013). The economic and environmental performance of dual sourcing: A newsvendor approach. *International Journal of Production Economics*, *143*(1), 109–119. doi:10.1016/j.ijpe.2012.12.007

Rota, C., Reynolds, N., & Zanasi, C. (2013). Sustainable food supply chains: The role of collaboration and sustainable relationships. *International Journal of Business and Social Science*, *4*(4), 45–53.

Rothenberg, S., Pil, F. K., & Maxwell, J. (2001). Lean, green, and the quest for superior environmental performance. *Production and Operations Management*, *10*(3), 228–243. doi:10.1111/j.1937-5956.2001.tb00372.x

Rother, M., & Shook, J. (1999). *Learning to see: value stream mapping to create value and eliminate muda*. Brookline, MA: Lean Enterprise Institute, Inc.

Rother, M., & Shook, J. (2003). *Learning to see: value stream mapping to add value and eliminate muda*. Cambridge, MA: Lean Enterprise Institute.

Routroy, S. (2009). Antecedents and drivers for green supply chain management implementation in manufacturing environment. *ICFAI Journal of Supply Chain Management, 6*(1), 20–35.

Ru-Jen, L., Rong-Huei, C., & Thi-Hang, N. (2011). Green supply chain management performance in automobile manufacturing industry under uncertainty. *Procedia: Social and Behavioral Sciences, 25*, 233–245. doi:10.1016/j.sbspro.2011.10.544

Rumelt, R. P. (1974). *Strategy, structure, and economic performance*. Cambridge, MA: Harvard Business School Press.

Ryan, P. (2010). *Sustainable Logistics: Towards the development of environmentally conscious supply chains*. (Bachelor of Science Thesis). University of Limerick.

Sahamie, R., Stindt, D., & Nuss, C. (2013). Transdisciplinary research in sustainable operations: An application to closed-loop supply chains. *Business Strategy and the Environment, 22*(4), 245–268. doi:10.1002/bse.1771

Sang, M. L., Sung, T. K., & Donghyun, C. (2012). Green supply chain management and organizational performance. *Industrial Management & Data Systems, 112*(8), 1148–1180. doi:10.1108/02635571211264609

Saremi, S., & Taghizadeh, M. (2013). RFID Adoption by Supply Chain Organizations in Malaysia. *International Proceedings of Economics Development & Research, 59*.

Sarkis, J. (2001). Manufacturing's role in corporate environmental sustainability. *International Journal of Operations & Production Management, 21*(5/6), 666–686. doi:10.1108/01443570110390390

Sarkis, J. (2003). A strategic decision framework for green supply chain management. *Journal of Cleaner Production, 11*(4), 397–409. doi:10.1016/S0959-6526(02)00062-8

Sarkis, J. (2012). A boundaries and flows perspective of green supply chain management. *Supply Chain Management: An International Journal, 17*(2), 202–216. doi:10.1108/13598541211212924

Sarkis, J., Zhu, Q., & Lai, K. H. (2011). An organizational theoretic review of green supply chain management literature. *International Journal of Production Economics, 130*(1), 1–15. doi:10.1016/j.ijpe.2010.11.010

Savitz, A. W., & Weber, K. (2006). *The triple bottom line*. San Francisco, CA: Jossey–Bass.

Sawhney, R., Teparakul, P., Bagchi, A., & Li, X. (2007). En-Lean: A framework to align lean and green manufacturing in the metal cutting supply chain. *International Journal of Enterprise Network Management, 1*(3), 238–260. doi:10.1504/IJENM.2007.012757

Schaltegger, S., & Burritt, R. L. (2014). Measuring and managing sustainability performance of supply chains. Review and sustainability supply chain management framework. *Supply Chain Management: An International Journal, 19*(3), 2–2. doi:10.1108/SCM-02-2014-0083

Schaltegger, S., & Wagner, M. (2006). Integrative management of sustainability performance, measurement and reporting. *International Journal of Accounting, Auditing and Performance Evaluation*, *3*(1), 1–19. doi:10.1504/IJAAPE.2006.010098

Schieberl, J., & Nickles, M. (2014). Outsourcing U.S. jobs abroad: Why? *The International Business & Economics Research Journal*, *13*(2), 253.

Schmidtke, D., Heiser, U., & Hinrichsen, O. (2014). A simulation-enhanced value stream mapping approach for optimisation of complex production environments. *International Journal of Production Research*, 1–15. doi:10.1080/00207543.2014.917770

Schneider, A., & Meins, E. (2012). Two dimensions of corporate sustainability assessment: Towards a comprehensive framework. *Business Strategy and the Environment*, *21*(4), 211–222. doi:10.1002/bse.726

Schrettle, S., Hinz, A., MaikeScherrer-Rathje, M. & Friedli, T. (2014). Turning sustainability into action: Explaining firms' sustainability efforts and their impact on firm performance. *International Journal of Production Economics, 147*(A), 73–84

Searcy, C. (2012). Corporate sustainability performance measurement systems: A review and research agenda. *Journal of Business Ethics*, *107*(3), 239–253. doi:10.1007/s10551-011-1038-z

Seksan, P., Ninlawan, C., Tossapol, K., & Pilada, W. (2010). The implementation of green supply chain management practices in electronics industry. *Proceedings of the International Multi Conference of Engineers and Computer Scientists.*

Seth, D., & Gupta, V. (2005). Application of value stream mapping for lean operations and cycle time reduction: An Indian case study. *Production Planning and Control*, *16*(1), 44–59. doi:10.1080/09537280512331325281

Seth, D., Seth, N., & Goel, D. (2008). Application of value stream mapping (VSM) for minimization of wastes in the processing side of supply chain of cottonseed oil industry in Indian context. *Journal of Manufacturing Technology Management*, *19*(4), 529–550. doi:10.1108/17410380810869950

Seuring, S., & Müller, M. (2008). From a literature review to a conceptual framework for sustainable supply chain management. *Journal of Cleaner Production*, *16*(15), 1699–1710. doi:10.1016/j.jclepro.2008.04.020

Sezen, B., & Turkkantos, S. (2013). The effects of relationship quality and lean applications on buyer-seller relationships. *International Journal of Business Performance and Supply Chain Modelling*, *5*(4), 378–400. doi:10.1504/IJBPSCM.2013.058204

Shah, R., & Ward, P. T. (2007). Defining and developing measures of lean production. *Journal of Operations Management*, *25*(4), 785–805. doi:10.1016/j.jom.2007.01.019

Shang, K. C., Lu, C. S., & Li, S. (2010). A taxonomy of green supply chain management capability among electronics-related manufacturing firms in Taiwan. *Journal of Environmental Management*, *91*, 1218–1226. PMID:20181423

Shang, K. C., Lu, C. S., & Li, S. (2010). A taxonomy of green supply chain management capability among electronics-related manufacturing firms in Taiwan. *Journal of Environmental Management, 91*(5), 1218–1226. doi:10.1016/j.jenvman.2010.01.016 PMID:20181423

Sharma, S., & Vredenburg, H. (1998). Proactive corporate environmental strategy and the development of competitively valuable organizational capabilities. *Strategic Management Journal, 19*(8), 729–753. doi:10.1002/(SICI)1097-0266(199808)19:8<729::AID-SMJ967>3.0.CO;2-4

Sharp, J. M., Novack, R. A., & Anderson, M. A. (1991). Purchasing hazardous waste transportation service: Federal legal considerations. *Transportation Journal, 31*(2), 4–14.

Sheng, E. (2012). Evaluating the cost of sustainability. *WWD, 204*(86), 8.

Sheu, J. B., Chou, Y. H., & Hu, C. C. (2005). An integrated logistics operational model for green-supply chain management. *Transportation Research Part E, Logistics and Transportation Review, 41*(4), 287–313. doi:10.1016/j.tre.2004.07.001

Shi, M., & Yu, W. (2013). Supply chain management and financial performance: Literature review and future directions. *International Journal of Operations & Production Management, 33*(10), 123–1317. doi:10.1108/IJOPM-03-2012-0112

Shi, V. G., Koh, S. C. L., Baldwin, J., & Cucchiella, F. (2012). Natural resource based green supply chain management. *Supply Chain Management: An International Journal, 17*(1), 54–67.

Shrivastava, P. (1995). The role of corporations in achieving ecological sustainability. *Academy of Management Review, 20*(4), 936–960. doi: 10.2307/258961

Shrivastava, P., Ivanaj, S., & Persson, S. (2013). Transdisciplinary study of sustainable enterprise. *Business Strategy and the Environment, 22*(4), 230–244. doi:10.1002/bse.1773

Shukla, A. C., Deshmukh, S. G., & Kanda, A. (2009). Environmentally responsive supply chains: Learning from the Indian auto sector. *Journal of Advances in Management Research, 6*(2), 154–171. doi:10.1108/09727980911007181

Siikavirta, H., Punakivi, M., Karkkainen, M., & Linnanen, L. (2003). Effects of e-commerce on greenhouse gas emissions. *Journal of Industrial Ecology, 6*(2), 83–97. doi:10.1162/108819802763471807

Sikdar, S. K. (2003). Sustainable development and sustainability metrics. *The American Institute of Chemical Engineers Journal, 49*(8), 1928–1932. doi:10.1002/aic.690490802

Silvestre, B. S. (2014). (in press). A Hard Nut to Crack! Implementing Supply Chain Sustainability in an Emerging Economy. *Journal of Cleaner Production.*

Silvestre, B. S. (2015). Sustainable supply chain management in emerging economies: Environmental turbulence, institutional voids and sustainability trajectories. *International Journal of Production Economics, 167*, 156–169. doi:10.1016/j.ijpe.2015.05.025

Silvestre, B. S., & Silva Neto, R. (2014). Capability accumulation, innovation, and technology diffusion: Lessons from a base of the pyramid cluster. *Technovation*, *34*(5), 270–283. doi:10.1016/j.technovation.2013.09.007

Simpson, B. J. K., & Radford, S. K. (2012). Consumer perceptions of sustainability: A free elicitation study. *Journal of Nonprofit & Public Sector Marketing*, *24*(4), 272–291. doi:10.1080/10495142.2012.733654

Simpson, D., Power, D., & Samson, D. (2007). Greening the automotive supply chain: A relationship perspective. *International Journal of Operations & Production Management*, *27*(1), 28–48. doi:10.1108/01443570710714529

Singh, B., Garg, S. K., & Sharma, S. K. (2011). Value stream mapping: Literature review and implications for Indian industry. *International Journal of Advanced Manufacturing Technology*, *53*(5-8), 799–809. doi:10.1007/s00170-010-2860-7

Singh, R. K., Murty, H. R., Gupta, S. K., & Dikshit, A. K. (2009). An overview of sustainability assessment methodologies. *Ecological Indicators*, *9*(2), 189–212. doi:10.1016/j.ecolind.2008.05.011

Slaper, T. F., & Hall, T. (2011). The triple bottom line: What is it and how does it work? *Indiana Business Review*, *86*(1), 4–8.

Smith, A. D. (2005). Reverse logistics programs: Gauging their effects on CRM and online behavior. *Vine*, *33*(3), 166–181. doi:10.1108/03055720510634216

Solding, P., & Gullander, P. (2009). *Concepts for simulation based Value Stream Mapping.* Paper presented at the 2009 Winter Simulation Conference (WSC), Austin, TX. doi:10.1109/WSC.2009.5429185

Song, L., Guan, W., Cherrett, T., & Baowen, L. (2013). Quantifying the greenhouse gas emissions of local collection-and-delivery points for last-mile deliveries. *Transportation Research Record: Journal of the Transportation Research Board*, (2340), 66-73.

Song, J., & Leng, M. (2012). Analysis of single-period problem under carbon emissions policies. In T.-M. Choi (Ed.), *Handbook of Newsvendor Problems Models, Extensions and Applications* (pp. 297–313). New York: Springer. doi:10.1007/978-1-4614-3600-3_13

Soni, G., & Kodali, R. (2010). Internal benchmarking for assessment of supply chain performance. *Benchmarking: An International Journal*, *17*(1), 44–76. doi:10.1108/14635771011022316

Sourani, A., & Sohail, M. (2011). Barriers to addressing sustainable construction in public procurement strategies. *Engineering Sustainability*, *164*(4), 229–237. doi:10.1680/ensu.2011.164.4.229

Spence, L., & Bourlakis, M. (2009). The evolution from corporate social responsibility to supply chain responsibility: The case of Waitrose. *Supply Chain Management: An International Journal*, *14*(4), 291–302. doi:10.1108/13598540910970126

Srisorn, W. (2013). The benefit of green logistics to organization. *International Journal of Social. Human Science and Engineering*, *7*(8), 1182–1185.

Srivastava, S. K. (2007). 'Green supply-chain management: A state-of-the-art literature review'. *International Journal of Management Reviews*, *9*(1), 53–80.

Srivastava, S. K. (2007). Green supply-chain management: A state-of-the-art literature review. *International Journal of Management Reviews*, *9*(1), 53–80. doi:10.1111/j.1468-2370.2007.00202.x

Srivastava, S. K. (2008). Network design for reverse logistics. *Omega: The International Journal of Management Science*, *36*(4), 535–548. doi:10.1016/j.omega.2006.11.012

Stank, T. P., Goldsby, T. J., Vickery, S. K., & Savitskie, K. (2003). Logistical service performance: Estimating its influence on market share. *Journal of Business Logistics*, *24*(1), 27–55. doi:10.1002/j.2158-1592.2003.tb00031.x

Stapleton, O., Pedraza Martinez, A., & Van Wassenhove, L. N. (2009). *Last mile vehicle supply chain in the international federation of Red Cross and Red Crescent societies*. Working Paper. INSEAD, Fontainebleau, France.

Steven, M. (2004). Networks in reverse logistics. In H. Dyckhoff, R. Lackes, & J. Reese (Eds.), Supply Chain Management and Reverse Logistics (pp. 163–180). Berlin: Springer-Verlag. doi:10.1007/978-3-540-24815-6_8

Stevens, G. C. (1989). Integrating the supply chain. *International Journal of Physical Distribution & Logistics Management*, *19*(8), 3–8. doi:10.1108/EUM0000000000329

Stock, J. R. (1998). *Development and Implementation of Reverse logistics programs*. Council of Logistics Management.

Stock, J. R., & Lambert, D. M. (2001). *Strategic logistics management* (4th ed.). McGraw-Hill.

Storey, J., Emberson, C., Godsell, J., & Harrison, A. (2006). Supply chain management: Theory, Practice and future challenges. *International Journal of Operations & Production Management*, *26*(7), 754–774.

Storey, J., Emberson, C., Godsell, J., & Harrison, A. (2006). Supply chain management: Theory, practice and future challenges. *International Journal of Operations & Production Management*, *26*(7), 754–774. doi:10.1108/01443570610672220

Suering, S. (2013). A review of modeling approaches for sustainable supply chain management. *Decision Support Systems*, *54*(4), 1513–1520. doi:10.1016/j.dss.2012.05.053

Su, J. (2013). Strategic sourcing in the textile and apparel industry. *Industrial Management & Data Systems*, *113*(1), 23–38. doi:10.1108/02635571311289647

Sullivan, W. G., McDonald, T. N., & Van Aken, E. M. (2002). Equipment replacement decisions and lean manufacturing. *Robotics and Computer-integrated Manufacturing*, *18*(3–4), 255–265. doi:10.1016/S0736-5845(02)00016-9

Supply Chain and Logistics Association Canada, Retail Council of Canada. (2009). *Green Supply Chain Management: Retail Chains and Consumer Product Goods – A Canadian Perspective*. Industry Canada.

Supply Chain and Logistics. (2009). *"Green" supply chain movement hits a speed bump*. Retrieved from: www.industryweek.com

Sushil. (2012). Interpreting the Interpretive Structural Model. *Global Journal of Flexible Systems Management, 13*(2), 87-106.

Svensson, G. (2002). The theoretical foundation of supply chain management: A functionalist theory of marketing. *International Journal of Physical Distribution & Logistics Management, 32*(9), 734–754. doi:10.1108/09600030210452422

Talluri, S., DeCampos, H. A., & Hult, G. T. M. (2013). Supplier rationalization: A sourcing decision model. *Decision Sciences, 44*(1), 57–86. doi:10.1111/j.1540-5915.2012.00390.x

Tao, Z., Guiffrida, A. L., & Troutt, M. D. (2010). A green cost based economic production/order quantity model.*Proceedings of the First Annual Kent State International Symposium on Green Supply Chains.*

Tapiero, C. S. (2007). Consumers risk and quality control in a collaborative supply chain. *European Journal of Operational Research, 182*(2), 683–694. doi:10.1016/j.ejor.2006.07.039

Taticchi, P., Tonelli, F., & Pasqualino, R. (2013). Performance measurement of sustainable supply chains: A literature review and a research agenda. *International Journal of Productivity and Performance Management, 62*(8), 782–804. doi:10.1108/IJPPM-03-2013-0037

Teng, C. C., Horng, J. S., Hu, M. L., Chien, L. H., & Shen, Y. C. (2012). Developing energy conservation and carbon reduction indicators for the hotel industry in Taiwan. *International Journal of Hospitality Management, 31*(1), 199–208. doi:10.1016/j.ijhm.2011.06.006

Thabrew, L., Wiek, A., & Ries, R. (2009). Environmental decision making in multi-stakeholder contexts: Applicability of life cycle thinking in development planning and implementation. *Journal of Cleaner Production, 17*(1), 67–76. doi:10.1016/j.jclepro.2008.03.008

The Guardian. (2006). *Product recall costs Cadbury Euro 20 million*. Retrieved 29 October 2012 from: http://www.guardian.co.uk/business/2006/aug/02/food.foodanddrink

Thun, J. H., & Müller, A. (2010). An empirical analysis of green supply chain management in the German automotive industry. *Business Strategy and the Environment, 19*(2), 119–132.

Tibben-Lembke, R. S. (2002). Life after death: Reverse logistics and the product life cycle. *International Journal of Physical Distribution & Logistics Management, 32*(3), 223–244. doi:10.1108/09600030210426548

Tibor, T., & Feldman, I. (1996). *ISO 14000: A guide to the new environmental management standards*. Burr Ridge, IL: Irwin Professional Publishing.

Tice, J., Ahouse, L., & Larson, T. (2005). Lean production and EMSs: Aligning environmental management with business priorities. *Environmental Quality Management, 15*(2), 1–12. doi:10.1002/tqem.20075

Toptal, A., Özlü, H., & Konur, D. (2014). Joint decisions on inventory replenishment and emission reduction investment under different emission regulations. *International Journal of Production Research, 52*(1), 243–269. doi:10.1080/00207543.2013.836615

Torielli, R. M., Abrahams, R. A., Smillie, R. W., & Voigt, R. C. (2011). Using lean methodologies for economically and environmentally sustainable foundries. *China Foundry, 8*(1), 74–88.

Tranfield, D., Denyer, D., & Smart, P. (2003). Towards a methodology for developing evidence-Informed management knowledge by means of systematic review. *British Journal of Management, 14*(3), 207–222. doi:10.1111/1467-8551.00375

Tseng, M. (2013). Modeling sustainable production indicators with linguistic preferences. *Journal of Cleaner Production, 40*, 46–56. doi:10.1016/j.jclepro.2010.11.019

Tsoulfas & Pappis. (2006). Environmental principles applicable to supply chain design. *Journal of Cleaner Production, 14*(18), 1602.

Ubeda, S., Arcelus, F. J., & Faulin, J. (2011). Green logistics at Eroski: A case study. *International Journal of Production Economics, 131*(1), 44–51. doi:10.1016/j.ijpe.2010.04.041

Ülkü, M. A. (2012). Dare to care: Shipment consolidation reduces not only cost, but also environmental damage. *International Journal of Production Economics, 139*(2), 438–446. doi:10.1016/j.ijpe.2011.09.015

UN Global Compact. (2010). *Report on UN Global Compact Management Model: Framework for implementation of human rights, labour, environment and anti-corruption.* Retrieved 20 May 2014 from: http://www.unglobalcompact.org/docs/news_events/9.1_news_archives/2010_06_17/UN_Global_Compact_Management_Model.pdf

UNEP Report. (2011). Retrieved from http://www.unep.org/greeneconomy/greeneconomyreport

UNFCCC. (n.d.). *Issues - bunker fuels.* UNFCCC. Retrieved on 7 March 2014, from http://unfccc.int/methods/emissions_from_intl_transport/items/1057.php

United Nations Environmental Program. (2007). *Building Standards and Climate Change: Status, Challenges and Opportunities.* Retrieved October 30, 2012, from http://unep.fr/shared/publications/.../DTIx0916xPA-BuildingsClimate.pdf

Uygun, Y., & Straub, N. (2012). *Human-centred Model for Application of Lean Production Systems in Networks. In Enabling Manufacturing Competitiveness and Economic Sustainability* (pp. 660–665). Springer. doi:10.1007/978-3-642-23860-4_108

Vachon, S. (2007). Green supply chain practices and the selection of environmental technologies. *International Journal of Production Research, 45*(18/19), 4357–4379. doi:10.1080/00207540701440303

Vachon, S., & Klassen, R. D. (2006). Extending green practices across the supply chain: The impact of upstream and downstream integration. *International Journal of Operations & Production Management*, *26*(7), 795–821. doi:10.1108/01443570610672248

Vachon, S., & Klassen, R. D. (2008). Environmental management and manufacturing performance: The role of collaboration in the supply chain. *International Journal of Production Economics*, *111*(2), 229–315.

van der Veen, J. A., & Venugopal, V. (2014). Economic and environmental performance of the firm: Synergy or trade-off? Insights from the EOQ model. In T.-M. Choi (Ed.), *Handbook of EOQ Inventory Problems* (pp. 121–137). Springer. doi:10.1007/978-1-4614-7639-9_6

Van Hoek, R. I. (1999). From reversed logistics to green supply chains. *Supply Chain Management International Journal (Toronto, Ont.)*, *4*(3), 129–135.

Van Veldhuizen, D. A., & Lamont, G. B. (2000). Multiobjective evolutionary algorithms: Analyzing the state-of-the-art. *Evolutionary Computation*, *8*(2), 125–147. doi:10.1162/106365600568158 PMID:10843518

Van Woensel, T., Creten, R., & Vandaele, N. (2001). Managing the environmental externalities of traffic logistics: The issue of emissions. *Production and Operations Management*, *10*(2), 207–223. doi:10.1111/j.1937-5956.2001.tb00079.x

Varki, S., & Colgate, M. (2001). The role of price perception in an integrated model of behavioral intentions. *Journal of Service Research*, *3*(2), 232–240. doi:10.1177/109467050133004

Vinodh, S., Arvind, K. R., & Somanaathan, M. (2011). Tools and techniques for enabling sustainability through lean initiatives. *Clean Technologies and Environmental Policy*, *13*(3), 469–479. doi:10.1007/s10098-010-0329-x

von Geibler, J., Liedtke, C., Wallbaum, H., & Schaller, S. (2006). Accounting for the social dimension of sustainability: Experiences from the biotechnology industry. *Business Strategy and the Environment*, *15*(5), 334–346. doi:10.1002/bse.540

Wahab, M. I. M., Mamun, S. M. H., & Ongkunaruk, P. (2011). EOQ models for a coordinated two-level international supply chain considering imperfect items and environmental impact. *International Journal of Production Economics*, *134*(1), 151–158. doi:10.1016/j.ijpe.2011.06.008

Walden, J. (2005). *Reverse logistics: Important or irritant?* Retrieved 18 September 2012 from: http://www.honeycombconnect.com/Supply_Chain_Management/document

Walker, H., Di Sisto, L., & McBain, D. (2008). Drivers and barriers to environmental supply chain management practices: Lessons from the public and private sectors. *Journal of Purchasing and Supply Management*, *14*(1), 69–85. doi:10.1016/j.pursup.2008.01.007

Walker, H., & Jones, N. (2012). Sustainable supply chain management across the UK private sector. *Supply Chain Management: An International Journal*, *17*(1), 15–28. doi:10.1108/13598541211212177

Walton, S. V., Handfield, R. B., & Melnyk, S. A. (1998). The green supply chain: Integrating suppliers into environmental management processes. *International Journal of Purchasing and Materials Management*, *34*(2), 2–11. doi:10.1111/j.1745-493X.1998.tb00042.x

Wamalwa, B. P. (2014). Sustainable supply chain management as a strategic tool for competitive advantage in tea industry in Kenya. *Journal of Management and Sustainability*, *4*(3), 157–164. doi:10.5539/jms.v4n3p157

Wang, F., Lai, X., & Shi, N. (2011). A multi-objective optimization for green supply chain network design. *Decision Support Systems*, *51*(2), 262–269. doi:10.1016/j.dss.2010.11.020

Wang, G., Wang, Y., & Zhao, T. (2008). Analysis of interactions among the barriers to energy saving in China. *Energy Policy*, *36*(6), 1879–1889. doi:10.1016/j.enpol.2008.02.006

Wang, Z., & Sarkis, J. (2013). Investigating the relationship of sustainable supply chain management with corporate financial performance. *International Journal of Productivity and Performance Management*, *62*(8), 871–888. doi:10.1108/IJPPM-03-2013-0033

Ware, N., Sing, S., & Banwet, D. (2012). Supplier selection problem: A state-of-the-art review. *Management Science Letters*, *2*(5), 1465–1490. doi:10.5267/j.msl.2012.05.007

Warfield, J. N. (1974). Structuring complex systems. Columbus, OH: Battele Memorial Ins.

Warfield, J. N. (1994). *A science of generic design: Managing complexity through systems design*. Iowa State University press.

Warfield, J. N. (1999). Twenty laws of complexity: Science applicability in organizations. *Systems Research and Behavioral Science*, *16*(1), 3–40. doi:10.1002/(SICI)1099-1743(199901/02)16:1<3::AID-SRES241>3.0.CO;2-F

WBR. (2009). *World Bank Report of United Arab Emirates*. Retrieved November 12, 2012, from http://data.worldbank.org/country/united-arab-emirates

WCED. (1987). *Report of the World Commission on Environment and Development: Our Common Future*. United Nations.

Weber, C. A., Current, J. R., & Benton, W. C. (1991). Vendor selection criteria and methods. *European Journal of Operational Research*, *50*(1), 2–18. doi:10.1016/0377-2217(91)90033-R

Wee, H., Lee, M., Yu, J., & Wang, C. (2011). 'Optimal replenishment policy for a deteriorating green product: Life cycle costing analysis'. *International Journal of Production Economics*, *133*(2), 608–611.

Werbach, A. (2009). *Strategy for sustainability*. Harvard Business Press.

Westing, A. H. (1986). International cooperation for sustainable development and peace. *Environmental Conservation*, *13*(2), 97–99. doi:10.1017/S0376892900036663

Whetten, D. A. (1989). What constitutes a theoretical contribution? *Academy of Management Review*, *14*(4), 490–495. doi:10.2307/258554

Wiedmann, T. O., Lenzen, M., & Barrett, J. R. (2009). Companies on the scale. *Journal of Industrial Ecology*, *13*(3), 361–383. doi:10.1111/j.1530-9290.2009.00125.x

Wiedmann, T., & Minx, J. (2008). A definition of 'carbon footprint. In C. C. Pertsova (Ed.), *Ecological Economics Research Trends* (pp. 1–11). Hauppauge, NY: Nova Science Publishers.

Wills, B. (2011). *Green intentions: creating a green value stream to compete and win.* Boca Raton, FL: CRC Press.

Winter, M., & Knemeyer, A. M. (2013). Exploring the integration of sustainability and supply chain management: Current state and opportunities for future inquiry. *International Journal of Physical Distribution & Logistics Management*, *43*(1), 18–38. doi:10.1108/09600031311293237

Wittstruck, D., & Teuteberg, F. (2011). Understanding the success factors of sustainable supply chain management: Empirical evidence from the electrics and electronics industry. *Corporate Social Responsibility and Environmental Management*, *19*(3), 141–158. doi:10.1002/csr.261

Wolf, C., & Seuring, S. (2010). Environmental impacts as buying criteria for third party logistical services. *International Journal of Physical Distribution & Logistics Management*, *40*(1), 84–102. doi:10.1108/09600031011020377

Womack, J. P., & Jones, D. T. (2003). Lean thinking: Banish waste and create wealth in your corporation (2nd ed.). New York: Free Press, Simon & Schuster, Inc.

Wong, C. W. Y. (2013). Leveraging environmental information integration to enable environmental management capability and performance. *Journal of Supply Chain Management*, *49*(2), 114–136. doi:10.1111/jscm.12005

Woods, L., & Bakshi, B. R. (2014). Reusable vs. disposable cups revisited: Guidance in life cycle comparisons addressing scenario, model, and parameter uncertainties for the US consumer. *International Journal of Life Cycle Assessment*, *19*(4), 931–940. doi:10.1007/s11367-013-0697-7

Wu, C., & Barnes, D. (2011). A literature review of decision-making models and approaches for partner selection in agile supply chains. *Journal of Purchasing and Supply Management*, *17*(4), 256–274. doi:10.1016/j.pursup.2011.09.002

Wu, G. C., Ding, J. H., & Chen, P. S. (2012). The effects of GSCM drivers and institutional pressures on GSCM practices in Taiwan's textile and apparel industry. *International Journal of Production Economics*, *135*(2), 618–636.

Wu, J., & Haasis, H. D. (2013). Converting knowledge into sustainability performance of freight villages. *Logistics Research*, *6*(2/3), 63–88. doi:10.1007/s12159-013-0098-0

Wu, Q., He, Q., Duan, Y., & O'Regan, N. (2012). Implementing dynamic capabilities for corporate strategic change toward sustainability. *Strategic Change*, *21*(5/6), 231–247. doi:10.1002/jsc.1906

Wu, Z., & Pagell, M. (2011). Balancing priorities: Decision-making in sustainable supply chain management. *Journal of Operations Management*, *29*(6), 577–590. doi:10.1016/j.jom.2010.10.001

Wycherley, I. (1999). Greening supply chains: The case of the Body Shop International. *Business Strategy and the Environment*, *8*(2), 120–127. doi:10.1002/(SICI)1099-0836(199903/04)8:2<120::AID-BSE188>3.0.CO;2-X

Xu, J., & Wang, H. (2008). Build competitive advantage through the integration of sustainable supply chains. In *Proceedings of 2008 IEEE International Conference on Service Operations and Logistics, and Informatics, IEEE/SOLI 2008.*

Xu, X., & Gursoy, D. (2015). Influence of sustainable hospitality supply chain management on customers' attitudes and behaviors. *International Journal of Hospitality Management*, *49*, 105–116. doi:10.1016/j.ijhm.2015.06.003

Yakovleva, N., Sarkis, J., & Sloan, T. (2012). Sustainable benchmarking of supply chains: The case of the food industry. *International Journal of Production Research*, *50*(5), 1297–1317. doi:10.1080/00207543.2011.571926

Yang, C. L., Lin, P. S., Chan, Y., & Sheu, C. (2009). Mediated effect of environmental management on manufacturing competitiveness: An empirical study. *International Journal of Production Economics*, *123*(1), 210–220. doi:10.1016/j.ijpe.2009.08.017

Yang, M., Hong, P., & Modi, S. (2011). Impact of lean manufacturing and environmental management on business performance: An international study of manufacturing firms. *International Journal of Production Economics*, *129*(2), 251–261. doi:10.1016/j.ijpe.2010.10.017

Yates, L. (2008). Sustainable consumption: The consumer perspective. *Consumer Policy Review*, *18*(4), 96–99.

Yongvanich, K., & Guthrie, J. (2006). An extended performance reporting framework for social and environmental accounting. *Business Strategy and the Environment*, *15*(5), 309–321. doi:10.1002/bse.541

Yu Lin, C., & Hui Ho, Y. (2008). An Empirical Study on Logistics services provider, intention to adopt Green Innovations. *Journal of Technology. Management and Innovation*, *3*(1), 17–26.

Zailani, S., Jeyaraman, K., Vengadasan, G., & Premkumar, R. (2012). Sustainable supply chain management (SSCM) in Malaysia: A survey. *International Journal of Production Economics*, *140*(1), 330–340. doi:10.1016/j.ijpe.2012.02.008

Zanoni, S., Bettoni, L., & Glock, C. H. (2013). Energy implications in the single-vendor single-buyer integrated production inventory model. In C. Emmanoulisdis, M. Taisch, & D. Kiritsis (Eds.), *Advances in Production Management Systems. Competitive Manufacturing for Innovative Products and Services* (pp. 57–64). Berlin: Springer. doi:10.1007/978-3-642-40352-1_8

Zanoni, S., Mazzoldi, L., & Jaber, M. Y. (2014). Vendor-managed inventory with consignment stock agreement for single vendor–single buyer under the emission-trading scheme. *International Journal of Production Research*, *52*(1), 20–31. doi:10.1080/00207543.2013.812812

Zhang, B., & Xu, L. (2013). Multi-item production planning with carbon cap and trade mechanism. *International Journal of Production Economics*, *144*(1), 118–127. doi:10.1016/j.ijpe.2013.01.024

Zhiyong, T., Lingyu, H., & Guicheng, S. (2014). Carbon footprint and order quantity in logistics. *Journal of Industrial Engineering and Management*, *7*(2), 475–490. doi:10.3926/jiem.1031

Zhong, H. (2006). *Model and algorithms for supply chain Network with Bi-directional floks*. The State University of New Jersey.

Zhu, Q., & Geng, Y. (2013). Drivers and barriers of extended supply chain practices for energy saving and emission reduction among Chinese manufacturers. *Journal of Cleaner Production*, *40*, 6–12. doi:10.1016/j.jclepro.2010.09.017

Zhu, Q., Geng, Y., Fujita, T., & Hashimoto, S. (2010). Green Supply Chain Management in Leading Manufacturers: Case Studies in Japanese Large Companies. *Management Research Review*, *33*(4), 380–392.

Zhu, Q., & Sarkis, J. (2004). 'Relationships between operational practices and performance among early adopters of green supply chain management practices in Chinese manufacturing enterprises'. *Journal of Operations Management*, *22*(3), 265–289.

Zhu, Q., & Sarkis, J. (2004). Relationships between operational practices and performance among early adopters of green supply chain management practices in Chinese manufacturing enterprises. *Journal of Operations Management*, *22*(3), 265–289. doi:10.1016/j.jom.2004.01.005

Zhu, Q., & Sarkis, J. (2006). An inter-sectoral comparison of green supply chain management in China: Drivers and practices. *Journal of Cleaner Production*, *14*(5), 472–486. doi:10.1016/j.jclepro.2005.01.003

Zhu, Q., & Sarkis, J. (2007). The moderating effects of institutional pressures on emergent green supply chain practices and performance. *International Journal of Production Research*, *45*(18-19), 4333–4355. doi:10.1080/00207540701440345

Zhu, Q., Sarkis, J., & Geng, Y. (2005). Green supply chain management in China: Pressure, practices and performance. *International Journal of Operations & Production Management*, *25*(5), 449–468. doi:10.1108/01443570510593148

Zhu, Q., Sarkis, J., & Kee-hung, L. (2007). Green supply chain management: Pressures, practices and performance within the Chinese automobile industry. *Journal of Cleaner Production*, *15*(11-12), 1041–1052. doi:10.1016/j.jclepro.2006.05.021

Zhu, Q., Sarkis, J., & Lai, K. (2013). Institutional based antecedents and performance outcomes of internal and external of green supply chain management practices. *Journal of Purchasing and Supply Management*, *19*(2), 106–117. doi:10.1016/j.pursup.2012.12.001

Zhu, Q., Sarkis, J., & Lai, K. H. (2008). Confirmation of a measurement model for green supply chain management practices implementation. *International Journal of Production Economics*, *111*(2), 261–273. doi:10.1016/j.ijpe.2006.11.029

Zhu, Q., Sarkis, J., Lai, K., & Geng, Y. (2008). The role of organizational size in the adoption of green supply chain management practices in China. *Corporate Social Responsibility and Environmental Management*, *15*(6), 322–337. doi:10.1002/csr.173

Zsidisin, G. A., & Siferd, S. P. (2001). Environmental purchasing: A framework for theory development. *European Journal of Purchasing & Supply Management*, *7*(1), 61–73. doi:10.1016/S0969-7012(00)00007-1

About the Contributors

Neeraj Anand is working as a Professor and HOD – QT/RM/Operations since July, 2012 with UPES, Dehradun. His area of specialization is Quantitative Techniques, Research Methodology and Operations. Other areas of interest are Supply chain management and Project Management. He has 20 years experience in academics and 3 years in industry. He is a certified Internal Auditor from DNV and has performed the role of Management Representative for ISO 9000 certification for UPES from its inception in 2006 to 2010 and again took over the charge in June 2013. He has successfully faced surveillance audits by DNV. He has been actively involved in Extension/Training and Coordinated an Executive Program for Centre for Aviation Studies for executives of University of Applied Science, Frankfurt a Resource person for "Supply Chain Management" for ISPe and delivered sessions on "Project Management" for engineering graduates of THDC, Tehri (U.K.) in 2011. He has conducted 4 MDPs on and Project Management and SCM. He carried out a research project with UHHDC (Uttrakhand Handloom & Handicraft Development Corporation) and another with a UREDA (Uttaranchal Renewable Energy Development Agency). Conducted ESDP (Entrepreneurship and Skill Development Program) sponsored by Ministry of MSME in partnership with Indian Institute of Entrepreneurship, Guwahati in 2010. Organized ICMI 2013 (International Conference on Management of Infrastructure) in Feb 2013. Convener of the national conference on "Innovation and Entrepreneurship" in 2010. He is a life member of Society of Operations Management and an affiliate Member- Energy Institute, London (U.K). Member – BOS (Board of Studies) and Core Member – FRC (Faculty Research Committee) at UPES.

Surajit Bag has completed his PhD in 2014 from College of Management and Economic Studies, University of Petroleum & Energy Studies, U.K. He has trained in Case Study Teaching from Indian Institute of Management Calcutta. He has trained in Total Qual¬ity Management from Indian Institute of Technology Kanpur. He has also mastered Multivariate data analysis under Indian Institute of Technology, Kharagpur. Presently he is working in the position of Procurement Head with

Tega Industries South Africa Pty Ltd, South Africa. He is having more than seven years of industry and research experience. His areas of research interest are Sustainable Supply Chain Management, Supply Chain Risk Management, Humanitarian logistics strategy and Business excellence. He has attended several National and International conferences and has published 40 papers in reputed journals. He is a member of professional bodies such as Indian Rubber Institute (IRI), Operational Research Society of India (ORSI), Society of Operations Management (SOM), AIMS International and Asian Council of Logistics Management (ACLM).

Sreejith Balasubramanian is the Institutional Research Officer at UOWD. He has more than 10 years of experience that includes teaching, research, consultancy and project management. Sreejith received his Bachelor's degree in Electronics and Communication Engineering from University of Calicut. He went on to obtain a Master's degree in International Business from UOW. He is also pursuing a Doctor of Philosophy (PhD) in business from Middlesex University, UK. His research interest is in the area of sustainability and carbon footprint modelling. His work has been published in leading international journals and conferences.

Ehab Bazan holds a Doctor of Philosophy (Ph.D.) degree from Ryerson University in Toronto, Canada, in Mechanical and Industrial Engineering. He has completed his Masters of Applied Science at Ryerson University in Mechanical and Industrial Engineering and his Bachelors of Science in Mechanical Engineering (Design and Production) from Ain Shams University in Cairo, Egypt. He is currently the Operations Manager at a large manufacturing and remanufacturing firm in the automotive industry located in the Greater Toronto Area in Ontario, Canada managing over 150 employees. His previous industrial experience stems from a wide variety of consulting projects both in Egypt and Canada. Teaching experience includes both student academic coaching and teaching at the high school and university levels as well specialized trainings for engineering and management professionals. His current research interests include: production economics, supply chain management, inventory systems, green supply chain management, and environmentally responsible supply chains. His work has been published in internationally renowned journals such as the International Journal of Production Economics, Computers and Industrial Engineering and Applied Mathematical Modelling. He is a member of the Ontario Society of Professional Engineers.

Jay Brown is an Assistant Professor of Operations Management in the Information Systems and Operations Management department at Loyola University Maryland and teaches in the area of Operations Management at the undergraduate and graduate level. He holds a Ph.D. in Operations Management from Kent State University. His research interests include stochastic modeling, the last mile problem, supply chain management, and mathematical programming with publications in such journals as the International Journal of Logistics Research and Applications and the Journal of Manufacturing Systems. His prior work experience includes production control, supply chain management, and managing an operations department. He is a member of DSI, INFORMS, ISM, and POMS.

Maxim A. Bushuev is currently an assistant professor of Management at Kent State University – Geauga. Dr. Bushuev received a B.S. degree in Automation Engineering from Voronezh State Technological Academy in 2001, and a PhD degree in Technical Sciences from the same university in 2006. He received MBA degree from Chadron State College in 2009 and a Ph.D. degree in Operations Management from Kent State University in 2013.

Jantanee Dumrak is a lecturer for Global Management Program at Torrens University Australia. She has Doctorate in Project Management from University of South Australia. She has worked as a thesis supervisor and examiner for Master of Project Management Program at the School of Natural and Built Environments, University of South Australia. Before joining academia, Jantanee worked as a project coordinator for collaborative reproductive health projects funded by the United Nations and the Government of Thailand. She is interested in project and program management of development projects.

Yudi Fernando is a Senior Lecturer in Operations Management at the Graduate School of Business, Universiti Sains Malaysia (USM). He received his PhD from the School of Management at USM. Dr. Yudi also worked for several years in the electronics industry. His research focus is in the area of sustainability supply chain, green logistics, green production and service management. Dr. Yudi currently teaches operations strategy: sustainability and supply chain management, business research methodology, advance business statistics subjects at MBA/DBA levels. He has also been supervising several PhD/DBA students with research topics on operations and supply chain management. He has been a member of editorial boards for International Journal of Operations Research and Information Systems (IJORIS). Dr. Yudi has actively participated as an expert reviewer for international conferences. Thus far, he has published in several journals such International Journal of Services and Operations Management, International Journal of Applied Logistics, Food Control, International Journal of Information Management, International Journal Logistics Systems and Management, International Journal of Operations and Supply Chain Management, International Journal of Productivity and Quality Management, International Journal of Value Chain Management, and so on.

Alfred L. Guiffrida is an Associate Professor of Management in the Department of Management and Information Systems at Kent State University. He holds his Ph.D., M.S. and B.S. degrees all in Industrial Engineering from the State University of New York at Buffalo and an MBA in Management Science from Virginia Tech. His papers have been published in Applied Mathematical Modelling, The Engineering Economist, European Journal of Operational Research, Industrial Marketing Management, Information Systems and Operational Research, International Journal of Integrated Supply Management, International Journal of Production Economics, International Journal of Production Research, Journal of Business and Economics, Journal of Intelligent Manufacturing, and International Transactions in Operations Research. Dr. Guiffrida's research interests are in the areas of operations and supply chain management.

Amulya Gurtu is an Assistant Professor of Supply Chain Management in the Austin E. Cofrin School of Business at the University of Wisconsin, Green Bay, where he started the supply chain management program. He obtained his PhD from Ryerson University, Canada, MBA from Western University, Canada and MS (Engineering) from Bhopal University, India. His current research is on supply chains, global manufacturing, and on environmental and social sustainability. His research has been published in the *Journal of Environment & Development, Management Research Review, Applied Mathematical Modelling*, and *Sustainable Production and Consumption.* He is a manuscript reviewer for many leading journals including the *International Journal of Production Economics,* the *Journal of Cleaner Production,* andthe *Journal of Business Ethics.*

Mohamad Y. Jaber is a Full Professor of Industrial Engineering at Ryerson University. He obtained his PhD in Manufacturing and Operations Management from The University of Nottingham, UK. His research interests are: modelling learning and forgetting curves and their applications and logistics and inventory systems. He has published many papers in internationally renowned journals. He is an Editor for Applied Mathematical Modelling, Computers and Industrial Engineering, and Ain Shams Engineering Journal, and also the editor of the books Inventory Management: Non-Classical Views and Learning Curves: Theory, Models, and Applications. He is a member of several professional societies.

Nitty Hirawaty Kamarulzaman is an Associate Professor in the Department of Agribusiness and Bioresource Economics, Faculty of Agriculture, Universiti Putra Malaysia. Her research interests include supply chain management, sustainable logistics, reverse logistics, agribusiness marketing, and consumer purchasing behavior.

Kijpokin Kasemsap received his BEng degree in Mechanical Engineering from King Mongkut's University of Technology Thonburi, his MBA degree from Ramkhamhaeng University, and his DBA degree in Human Resource Management from Suan Sunandha Rajabhat University. He is a Special Lecturer at Faculty of Management Sciences, Suan Sunandha Rajabhat University based in Bangkok, Thailand. He is a Member of International Association of Engineers (IAENG), International Association of Engineers and Scientists (IAEST), International Economics Development and Research Center (IEDRC), International Association of Computer Science and Information Technology (IACSIT), International Foundation for Research and Development (IFRD), and International Innovative Scientific and Research Organization (IISRO). He also serves on the International Advisory Committee (IAC) for International Association of Academicians and Researchers (INAAR). He has numerous original research articles in top international journals, conference proceedings, and book chapters on business management, human resource management, and knowledge management published internationally.

Andrey Kretinin is a Ph.D. candidate in the Department of Marketing and Entrepreneurship at Kent State University. His primary research envelops the domains of strategic management, entrepreneurship, family business, and international business, with focus on franchising, its structure, governance, and internationalization.

Sherif Mostafa is a lecturer and a PhD candidate in construction and project management at School of Natural and Built Environments, University of South Australia. He obtained MSc in Industrial Engineering and B.Sc. in Production Engineering and Mechanical Design from Mansoura University in 2010 and 2003. Sherif is a member of Barbra Hardy Institute, Project Management Institute (South Australia Chapter), Institute of Industrial Engineers, the Institution of Engineers Australia, and Lean Construction Institute Australia. Sherif is interested in lean and agile manufacturing/construction, offsite manufacturing, modelling and simulation of leagile supply chain, lean six sigma, sustainability and green manufacturing, Multi Criteria Decision Making (MCDM), and Ant Colony Optimisation (ACO). For further details, refer to his staff homepage at http://people.unisa.edu.au/Sherif.Mostafa.

Krishan Kumar Pandey is Professor & Vice Dean of Jindal Global Business School. He did his post-graduation and Doctorate in Statistics and a second post-graduation in Management. 14 years ago, he started his career as a teacher & consultant and worked for reputed firms and prestigious universities in India. Prior to Joining JGBS, Dr. Pandey was Sr. Associate Professor & Head of Decision Sciences at University of Petroleum and Energy Studies, Dehradun. Dr. Pandey has also worked as Assistant Dean (Research) at COMES at UPES. His research interest areas are Energy & Environment, Renewable Energy, Data Mining, Business Analytics, Small Area Estimation, Sampling etc. Dr. Pandey has published 30 research papers in international and national journals of repute. As per Elsevier research dashboard portal till June 2016, eight of his top research papers are downloaded /viewed around 19,275 times with 240 citations and his h- index is 6. His work has been published in top reputed journals in the area of Renewable Energy & Energy Policy like "Applied Energy", "Energy Policy", "Renewable and Sustainable Energy Reviews", "Renewable Energy", "Energy Systems" etc., and has participated in several international & national conferences. He has authored two books published by VDM Verlag, Germany. Dr. Pandey has guided fourteen Ph.D. Scholars from across the industry and academia in different areas of his specialization. Dr. Pandey was the founding Editor in Chief, for UPES Management Review- A biannual International Journal of Core Sector (ISSN 2250 – 0723). He is also acting as an editorial board member & reviewer for various reputed journals published by Elsevier, Emerald, T& F etc. Dr. Pandey has held sessions at the European Union's project for capability building of the Indian aviation sector, jointly executed by UPES and Fachhochschule Frankfurt am Main University of Applied Sciences, Frankfurt, Germany. He has conducted training sessions for executives from aviation and energy sectors- Oil India, IOCL, HPCL, BPCL, ONGC, GAIL, IBM etc. He has organized many national and international conferences as a convener and Co convener. Dr. Pandey has received many awards and cash prizes.

Cory Searcy is an Associate Professor and Director of the Environmental Applied Science and Management Program at Ryerson University. He is also an Associate at the International Institute for Sustainable Development. His current research focuses on sustainability indicators, supply chain management, and integrated management systems. His research has been published in several leading academic journals, including the Journal of Cleaner Production, the Journal of Business Ethics, and the International Journal of Production Economics.

Balan Sundarakani is an Associate Professor and Program Director for Master of Science in Logistics in the Faculty of Business and Management at the University of Wollongong (Australia) in Dubai. Previously he was working as an Assistant Professor with UOWD from May 2009 till Aug 2011. He has 12 years of teaching and research experience in the area of Supply Chain Management across the various universities including the National Institute of Technology, Tiruchirappalli, IIT, Roorkee, Hindustan University, Chennai and the National University of Singapore, Singapore. He also served as an Adjunct Research Fellow with the Chair of Logistics Management with Swiss Federal Institute of Technology (ETH), Zurich, Switzerland.

Gowri Vijayan holds a PhD degree in Agribusiness from the Department of Agribusiness and Bioresource Economics, Faculty of Agriculture, Universiti Putra Malaysia. Her research interests include agribusiness, sustainable supply chain management, green logistics, reverse logistics, strategic management, marketing, agricultural marketing, and Netchain management.

Index

V

W

www.ingramcontent.com/pod-product-compliance
Ingram Content Group UK Ltd.
Pitfield, Milton Keynes, MK11 3LW, UK
UKHW050947290926
15470UKWH00005B/1689

9 781522 506355